Sensors and Actuators

Sensors and Actuators

Francisco André Corrêa Alegria

Instituto Superior Técnico, University of Lisbon, Portugal
& Instituto de Telecomunicações, Portugal

NEW JERSEY • LONDON • SINGAPORE • BEIJING • SHANGHAI • HONG KONG • TAIPEI • CHENNAI • TOKYO

Published by

World Scientific Publishing Co. Pte. Ltd.
5 Toh Tuck Link, Singapore 596224
USA office: 27 Warren Street, Suite 401-402, Hackensack, NJ 07601
UK office: 57 Shelton Street, Covent Garden, London WC2H 9HE

British Library Cataloguing-in-Publication Data
A catalogue record for this book is available from the British Library.

This English edition is based on a translation from the Portuguese language edition:
Sensores e Atuadores
by Francisco Alegria

SENSORS AND ACTUATORS

ISBN 978-981-124-249-6 (hardcover)
ISBN 978-981-124-250-2 (ebook for institutions)
ISBN 978-981-124-251-9 (ebook for individuals)

For any available supplementary material, please visit
https://www.worldscientific.com/worldscibooks/10.1142/12426#t=suppl

To my father and mother

The author is with Instituto Superior Técnico of the Universidade de Lisboa and with Instituto de Telecomunicações.

Preface

Today, man lives surrounded by technology. The use of electronic systems in our daily lives is ubiquitous, whether in our homes and cars, in the workplace, or moments of fun. The interface between these systems and the world we live in consists of sensors and actuators. The sensors collect information about the magnitude of phenomena existing in Nature, such as temperature, speed, and force, and encode it in an electrical signal. For their part, the actuators receive an electrical signal containing specific information and act in the environment producing the most varied changes such as movement, light, and heat. Being the age in which we live in an age of information, all aspects of our life benefit from collecting and using data.

This book is for the university student who wants to obtain training in electrical engineering. Most systems created and used contain sensors and actuators. It is, therefore, essential for an engineer in this area to know how sensors and actuators work, how to use them, and which types are suitable for each application. Furthermore, an engineer should know the advantages and disadvantages of each of the thousands of different kinds on the market and be able to build a sensor or actuator from scratch, when necessary.

This book distinguishes itself from other books that cover this subject in that the information is not presented in a comprehensive form. It is, instead, organized around the physical and chemical principles on which the different sensors and actuators are based. This approach reflects the desire to provide the student with more in-depth knowledge about each device. The purpose is for him/her to understand why a given sensor or actuator has the characteristics that it has and if it is suitable for a specific application. It is not enough to know what a given sensor measures or what changes an actuator causes, but also how it is used.

From the physical or chemical principle used, all other important aspects are addressed, such as the conditions under which a given sensor/actuator works or what type of electrical signal is produced. Other important aspects include which electronic circuit should be used to create an electrical signal containing the information about the desired quantity and the technical terms in the datasheets of these devices. Another essential aspect that is addressed in this book is the comparison between different sensors and actuators that measure or act on the same type of quantity. It is necessary to know how to weigh in other aspects such as accuracy, response time, size, price, reliability, etc., when choosing a sensor or actuator for a given application.

This book is organized into twelve chapters. The first two are more introductory, covering topics such as sensors and actuators and the technical terms used to describe them (Chapter 1). Chapter 2 presents a brief overview of the construction of sensors and actuators using micro-fabrication techniques.

The following seven chapters constitute the main body of this book, with each chapter dedicated to a different class of physical or chemical phenomena: phenomena based on the electric field (Chapter 3), based on electrical resistance (Chapter 4), based on the magnetic field (Chapter 5), based on mechanical phenomena (Chapter 6), based on thermal phenomena (Chapter 7), based on electromagnetic radiation (Chapter 8) and based on chemical phenomena (Chapter 9).

The last three chapters are again different. Chapter 10 deals superficially with the theme of sensor and/or actuator networks, which is so important today. Chapter 11, called "Summary," aims to overcome some shortcomings of the approach followed throughout the book by focusing on comparing different types of sensors that measure the same physical quantity. For this purpose, attention is focused on the three most common types of sensors: displacement sensors, temperature sensors, and force sensors. At the end of this chapter, a summary of the electronic signal conditioning circuits covered throughout the book is made. Each type of sensor can generally be used with more than one signal conditioning circuit. A chose was made to introduce these circuits throughout the body of the book as the different sensors and actuators were studied to bring theory to practice and make the presentation more captivating for the reader. However, systematization of the different types of signal conditioning circuits is justified and a comparison, between other circuits of the same kind, made.

The last chapter (Chapter 12) suggests six practical assignments that can be proposed to students in laboratory classes. This aspect of training is fundamental in an area where, in addition to knowledge, "knowing how to do" is essential. Only with direct contact with the sensors and actuators is it possible to develop the fundamental skills of an engineer. Prime among them is the ability to diagnose problems. It is not possible to write a book presenting all the issues that are encountered in practice and how to solve them. Only with firsthand experience will future engineers be able to develop these skills.

The variety of different quantities that can be measured or acted upon is enormous. The content of this book was chosen so that the subjects covered can be taught in an academic semester (14 weeks with 3 hours of theoretical classes per week) of a university course. We tried to include a great diversity of topics, as expected from an introductory text, with more or less detail depending on the subject. It is intended that the reader who finishes reading this book has acquired a solid foundation on the theme of sensors and actuators and can increase his/her future knowledge independently.

Francisco Alegria
June 2021

Contents

Chapter 1

Introduction

1.1 Motivation

Since the beginning of the industrial revolution, in the late XVIII century, Man has built machines to help him in his tasks, leading to a substantial increase in productivity. With the rise of computers and the Internet, in the XX century, these machines have become increasingly sophisticated and capable of carrying out even more complex tasks. That, in turn, has led to much greater control of its environment. With the aim of increasing even further that control, new ways of gathering information about the physical phenomena that condition our existence have been developed together with new ways to actuate upon our world. A closed cycle is thus formed that flows from Nature to the machines/computers and back to Nature (Figure 1.1).

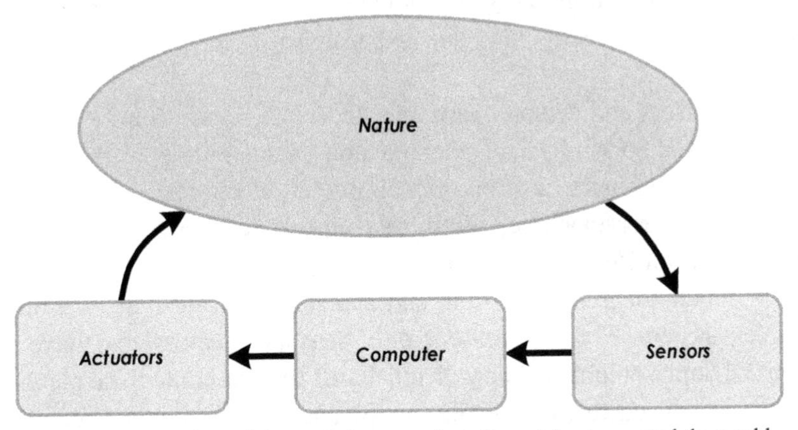

Figure1.1 – Overview of the typical system that allows Man to control the world.

The existence of this cycle is possible due to the availability of sensors capable of associating physical phenomena with electrical signals containing information that can be stored and processed by computers. Likewise, the availability of actuators that are able to transform this data in the computer into a real change in the world around us has allowed increasingly the creation of closed loops that are basically control systems aimed at manipulating the Universe according to the desires of Man.

In the context of a degree in Electrotechnical or Electronics Engineering, this course complements the training given in electronics course that focus on the creation of machines and computers with the courses in the area of mathematics and physics that are dedicated to the development of theories that explain the workings of Nature and that allow the creation of information processing algorithms running on computers, with the courses in computers and software engineering where it is taught how to implement these algorithms on computers and finally with the courses in "systems" that deal with the integration of all these components into a single hardware and software system capable of performing a particular function.

1.2 Definition of Sensor and Actuator

A transducer is a device that converts one form of energy into another. A typical example is a mercury thermometer: thermal energy is converted into mechanical energy in the expansion of mercury within the thermometer.

Both sensors and actuators are transducers. A sensor is a device that, when exposed to a physical phenomenon (temperature, displacement, force, etc.), produces a proportional output electrical signal [1]. A thermocouple, for example, reacts to temperature changes and has an output that is an electrical voltage.

This book focuses on sensors and actuators that are part of a more complex system. For this reason, only sensors/actuators that have an electrical input/output are considered here. In this context, a mercury thermometer used to measure the temperature of the human body and whose indication consists of the level of mercury inside the glass thermometer containing a graduated scale is not considered a sensor. However, in other wider contexts, it could be.

Nature itself contains sensors that generally have an electrochemical nature; that is, its output is an electric signal formed by ion transport. The eye, for instance, converts light intensity into a signal present in the nerve fibers that connect it to the brain. The sensors created by man usually have an output that is based on electrons. Note that both the ions and electrons have an electric charge and thus can be used to carry an electrical signal containing information. This signal is then processed by electronic circuits (amplified, filtered, quantified, etc.) to ensure the integrity of information relayed and to turn it into a form that allows for easy processing, storage, and communication, which today is almost always done digitally using computers, both personal and embedded in electronic circuits. Therefore, the ultimate goal of a sensor is to provide information about the amount of a certain phenomenon/physical parameter in digital form. It is in this perspective that devices arise that have not only the sensor element but also the electronic signal conditioning (amplification, filtering, etc.) and conversion from analog to digital, as well as capabilities for transmitting information (Figure 1.2) – the integrated sensor.

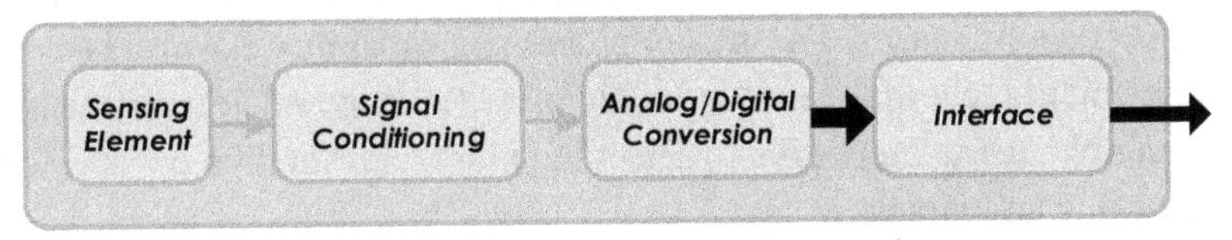

Figure 1.2 – Typical architecture of an integrated sensor.

The signal conditioning circuit can have different purposes. It can adjust the signal level (amplification or attenuation), remove unwanted frequency components and compensate for the undesired influence of certain quantities (typically temperature). It can also be used to make the relationship between the output voltage and the quantity sensed more linear.

The integrated sensor evolved into a *smart* sensor (Figure 1.3) with the incorporation of an element capable of processing information. This element is typically a microcontroller capable of converting the signal from analog to digital, processing the signal digitally (filtering, for example), make calculations based on the information contained in the signal, and interface (digitally) with the outside world. The "intelligence" of the sensor is, therefore, due to the presence of this microcontroller and the possibility that it brings in terms of self-diagnosis, self-identification, and self-adaptation. A smart sensor, therefore, has numerous advantages, allowing, for example, optimization of its operation in terms of accuracy, speed, and energy consumption. At the same time, it is more reliable than its "non-intelligent" counterparts. It also has the ability to change its functioning if the system to which they are attached so determines.

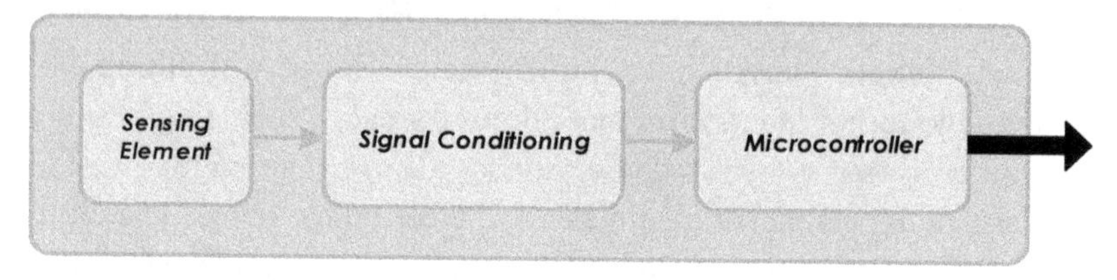

Figure 1.3 – Typical composition of a smart sensor.

The ability to process information can be used, for example, to detect events, compare the measured value with a preset decision level or make judgements based on the accumulated measurements, such as whether it is worthwhile to store the measured values and whether it is worth sending them outside the sensor. The greater processing capability can also be used to implement techniques for automatic classification and learning using, for example, neural networks to select the most relevant data or represent it in a compact manner.

One of the biggest advantages of having a smart sensor is the ability to integrate them into a network of sensors. As will be seen later, various features are required on the part of a sensor so that it can be part of a network. Many of these features have to do with the intelligence of the sensor, in particular the ability to be addressed, to store information, to compress, and to encrypt the data to be transmitted, among others.

Looking at the integration of wireless sensor networks, we can see in Figure 1.4 a block diagram of a smart sensor with more detail than the diagram of Figure 1.3. There is a radio communication module and a battery for powering the sensor without requiring a physical connection to the network. The radio module is what typically consumes more energy and thus should be used sparingly. One way to do this is to store the acquired data in memory and only transmit it when the memory is full or when requested from the outside. This possibility is part of a different capacity of a smart sensor: to be able to receive data from the outside, either in the form of commands that change its configuration, either in the form of data to synchronize, for example, its internal clock reference. Communication with the outside requires the implementation of communication protocols that may, in some cases, be complex, depending on the size and topology of the network. Nowadays, one tries to use standardized protocols designed specifically for sensors in order to make the exchange of information as flexible as possible.

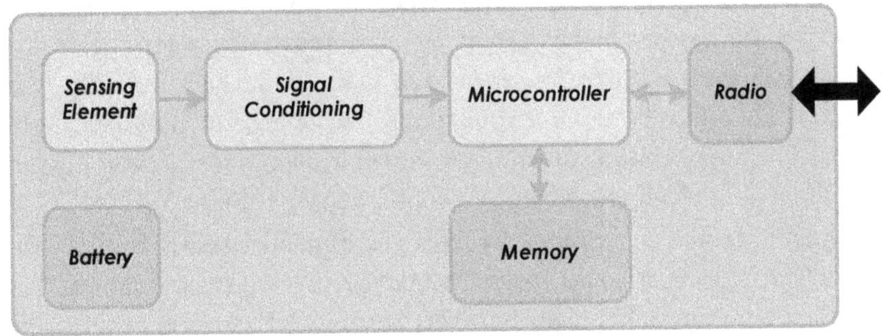

Figure 1.4 – Typical composition of a smart sensor capable of being integrated into a wireless network.

An important feature of intelligent sensors is their capability to perform validation of the acquired data. This can be done analytically, using a mathematical model to which the data should adjust and compare this model with the actual data acquired in order to detect anomalies. An example would be to impose the condition that the value measured by a temperature sensor cannot vary by more than 5°C per second. If a sensor were to measure a temperature of 23°C and, in the next second, it breaks down and measures a value in the lower end of the scale, like −50°C, the sensor would be able to identify the occurrence of a failure.

This same procedure can be done with data from different sensors that are able to communicate with each other. The rule may be, for example, a given sensor in a network may not have a value that more than 50% different from the average of the values of all the other sensors. This set of sensors can be scattered in an area in which the variation of the measured quantity is known not to vary very significantly, or it may even be that a set of identical sensors is placed at the same location to have redundancy.

A smart sensor can, if it is designed for that, modify itself to eliminate eventual errors. For example, in the case of a temperature sensor, a calibration procedure in which the user must place the sensor at a preset temperature and then send a command to the sensor can be used to indicate that the sensor should be measuring a given value of temperature. The sensor is then responsible for calculating one or more correction factors that store in nonvolatile memory and begin to use in future measurements. Another possibility is self-calibration, where the smart sensor has included an actuator that causes the quantity to be measured to take a certain known value. It can thus compare the measured value with the value that was expected. This may be done, for example, in the case of magnetic field sensors. They may contain a coil through which a known current is passed that will create a magnetic field used for the sensor calibration.

An actuator is a device that receives a command, usually in the form of an electrical signal, and makes a change in Nature, producing, for example, a force, a sound, heat, etc. [1] (Figure 1.5). Examples of actuators are motors, relays, valves, heaters, lamps, etc. An actuator is, therefore, also a transducer for transforming one form of energy into another form of energy. A loudspeaker, for example, is a transducer that converts electrical energy into magnetic energy and subsequently into mechanical energy, which is used to generate acoustic waves. This example shows that a sensor or actuator can be just a simple device containing one sensor/actuator or a complex system containing various sensors and actuators working together.

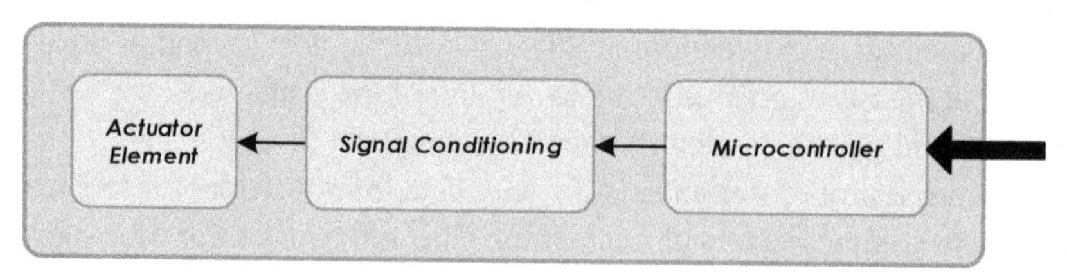

Figure 1.5 – Typical composition of a smart actuator.

Note the distinction made here between sensor, actuator, and transducer. A device that produces a continuous voltage proportional to the temperature of the air around it, for example, is a sensor (of temperature). A device that rotates a shaft by an amount proportional to the electrical signal applied to it, for example, is an actuator. Both cases are examples of transducers.

Another example of a transducer that cannot be considered by itself a sensor or actuator is for, example, an iron rod. The length of the rod depends on the temperature. It can therefore be said that transduction occurs between temperature and length. Since electrical quantities are not involved, it is not a sensor nor an actuator in the context of this book. The iron rod, however, could be used in a more complex device also comprising a displacement transducer that converts an electrical voltage to make a temperature sensor.

Please note that some authors call actuators only devices that produce movement, such as engines, for example. Here, however, as noted, an actuator is any transducer that produces a change in a physical quantity, even if not mechanical. Light-emitting diodes (LEDs) and Peltier cells are therefore considered here actuators (for light and heat in these cases).

Hence, from now on, only the terms "sensor" and "actuator" are going to be used with the following meaning:

• "Sensor" is a device that produces an electrical signal proportional to the value of a physical quantity.

• "Actuator" is a device that changes the value of a physical quantity according to an electric signal.

Electrical signals may be in the form of voltage or current. The information contained in these signals can be coded in its amplitude, frequency, phase, pulse time, or a digital code (in the case of quantized signals).

One of the goals of this book is to teach how to identify, in sensors and actuators, the transducers present, the physical quantities involved, and the electrical signals used as input (sensor) or output (actuators).

1.3 The Domain of Physical Phenomena

Sensors can be classified according to the domain in which the quantity being measured belongs, as shown in Figure 1.6.

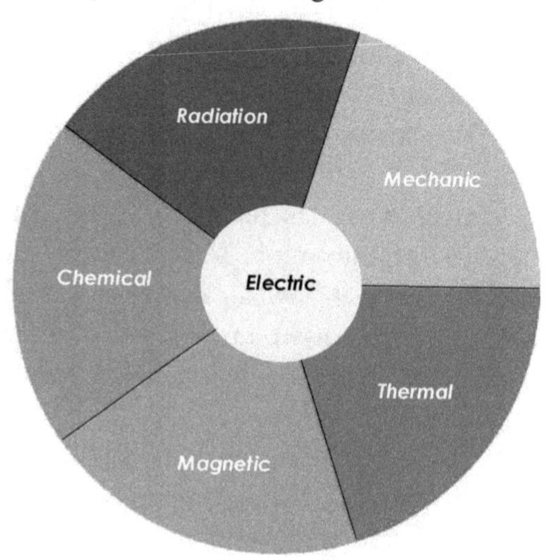

Figure 1.6 – The domains into which the quantities sensed or affected by sensors and actuators can be classified.

In Table 1.1, you can see the different physical phenomena where quantities of different domains are involved. In the shaded cells, you can see the phenomena that involve electrical quantities and thus can be used to make sensors and actuators.

Table 1.1 – Physical Phenomena involving quantities from different domains. The phenomena typically used in sensors and actuators are the ones highlighted.

INPUT	RADIATION	MECHANICAL	THERMAL	ELECTRIC	MAGNETIC	CHEMICAL
RADIATION	Photo luminance	Radiative Pressure	Heating by Radiation	Photo conductivity	Photo-magnetism	Photo-chemical
MECHANIC	Photo-elastic Effect	Moment Conservation	Heat by Friction	Piezo electricity	Magnetostriction	Pressure-induced explosion
THERMAL	Incandescence	Thermal Expansion	Heat Conduction	Seebeck Effect	Curie-Weiss Law	Endothermic reaction
ELECTRIC	Larmor Radiation	Piezo electricity	Peltier Effect	PN Junction	Ampere Law	Electrolysis
MAGNETIC	Faraday Effect	Magnetostriction	Ettinghouse Effect	Hall Effect	Magnetic Induction	
CHEMICAL	Chemical Luminance	Explosive Reaction	Exothermal Reaction	Volta Effect		Chemical Reaction

Note that a sensor can use more than one physical phenomenon, and hence, in principle, any phenomena listed in Table 1.1 can be used in a sensor or actuator, provided they are used in conjunction with a phenomenon involving the quantities in the electric domain.

1.4 Classification of Sensors and Actuators

1.4.1 *Regarding the Energy Source*

One way to classify sensors is regarding the power source they use to operate. Two types can be distinguished: **modulating** sensors and **self-generating** sensors [6], [9] (Figure 1.7).

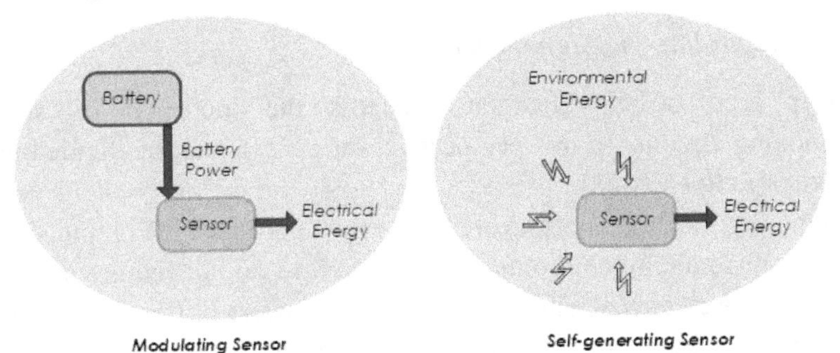

Figure 1.7 – Different types of sensors according to their power source (modulating/self-generating).

Modulating sensors, also called parametric sensors, work by changing a given parameter of an electrical circuit, such as resistance, capacitance, or inductance. This circuit receives power from the outside, from a battery or signal generator, and produces an electrical signal that depends on the value of the voltage or current applied.

One example of a modulating sensor is the strain gauge — the electrical resistance of a strain gauge changes with deformation. To produce a signal, you must place the gauge into an electric circuit such as a Wheatstone bridge, which is powered by a constant voltage source.

On the other hand, self-generating sensors produce an electrical signal without the need to be connected to a power supply. The energy they use to create the output signal comes from the physical phenomenon itself. A temperature sensor such as a thermocouple extracts thermal energy from the surroundings to produce a voltage. The sensor itself produces electricity which is the output signal. There is no need to connect it to a battery or signal generator. Another example is the piezoelectric sensor which produces a voltage between its electrodes that depends on the strain to which it is subjected to.

One should try, when using a sensor, to cause as a negligible disturbance as possible in the environment so as not to alter the value of the quantity that is being measured. In this aspect, the modulating sensors are more appropriate, having an external power supply, since they do not use as much energy from the environment as the auto-generating sensors do.

1.4.2 *Regarding the Signal Conditioning*

Another type of classification concerns the inclusion of signal conditioning circuits. In this perspective, the sensors can be divided into **passive** or **active** [7], [8].

In the first case, the sensors obtain energy to produce the output signal from the surrounding environment and therefore do not require a power supply for the electronic circuit where they are inserted. Being passive is, therefore, synonymous with being a self-generator.

Examples of passive sensors are electrodynamic microphones, thermocouples, photodiodes, and piezoelectric sensors in general.

Passive sensors have the advantage of not requiring a power supply and the disadvantage that the output signals are typically weak and often require amplification using an active electronic circuit.

The active sensors, in turn, get their power from the electronic circuit where they are placed. Being active is thus synonymous with being modulating.

Examples of active sensors are capacitive sensors, image sensors, displacement sensors, like the linear variable differential transducer (LVDT), strain gauges, and temperature sensors using temperature-dependent resistors.

Active sensors have the advantage of allowing a signal gain from input to output and the disadvantage of requiring an external source of energy and the installation of the corresponding electric cables.

Note that sometimes you will incorrectly find in the literature active sensors being called passive and vice-versa as alerted in [8]. This mistake, made, for example, in [9], is explained by the fact that they consider that if a sensor generates a signal, then it is "active." It is logical reasoning but not adopted by the sensors industry in general.

Another view, also found in the literature [1], [10], is to consider the active/passive classification as applying to the measurement system and not to the sensor itself. An example is a radar that uses an electromagnetic wave and the time of flight between two points to measure the distance to an object. This system introduces energy into the environment through the electromagnetic wave produced, and therefore it is called active. In contrast, where the energy is drawn from the environment, such as in thermocouples, they are considered passive.

This view is the opposite of the classification adopted here into modulating and self-generating sensors, and that considers where the energy for the sensor comes from. The classification of active/passive used in [1] and [10] is based on the reverse flow of energy — from the sensor to the environment. On top of that, those classifications concern measurement systems and not individual sensors. Although it is a legitimate point of view, it is not the one followed here nor in the industry in general.

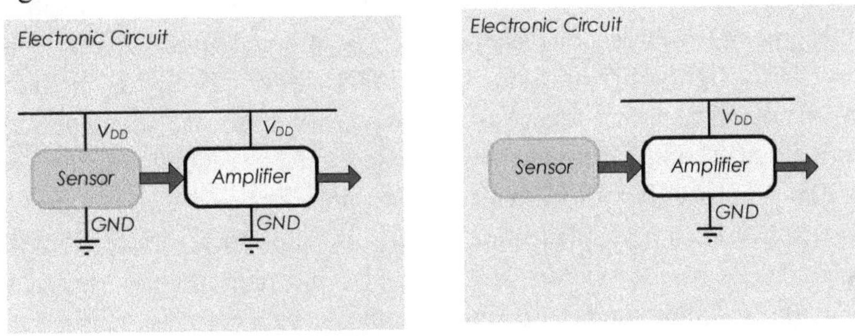

Figure 1.8 – Classification of sensors as active or passive regarding their inclusion into an electronic circuit.

In short, classifying the sensors as active or passive is related to whether their source of energy is external or not (Figure 1.8). This classification is the same as that used in relation to electronic components in general - resistance and capacitors are passive components while integrated circuits, such as operational amplifiers, are active.

1.4.3 *Regarding Reference Value*

The sensors can also be classified into **absolute** or **relative**. An absolute sensor is responsive to a quantity relative to an absolute scale which is independent of the measurement conditions. A relative sensor, in turn, produces a signal which is proportional to the difference between the value of the measured quantity and a reference value used for that quantity, which is, in general, dependent on the conditions of measurement. A thermistor is an example of an absolute sensor since the absolute value of its electrical resistance depends directly on the Kelvin temperature scale. A thermocouple is an example of a relative temperature sensor since the value of the output voltage depends on the temperature difference between two points.

1.4.4 *Regarding Complexity*

Another way to classify sensors and actuators is into **simple** or **complex**. Simple sensors use a single energy transformation. A thermistor, for example, transforms heat energy into electrical energy. Sensors using more than one transformation of energy are called complex. An example of this type of sensor is a displacement sensor utilizing optical fibers. This sensor has a light-emitting diode (LED) that emits light that is directed onto an optical fiber that focuses it on the object whose displacement is to be measured. The light reflected by the object enters a second optical fiber that directs it to a photodiode which in turn produces an electrical current proportional to the displacement. There is, therefore, conversion of electrical energy into photons (by the LED) and photons into electrical energy (by the photodiode). This makes this device a complex sensor.

1.5 Datasheets

The choice of a sensor or actuator for an application depends on the requirements of that application and the characteristics of the sensors/actuators that are provided by the manufacturer's datasheets. The datasheet of a sensor or actuator is, first and foremost, a marketing document designed to showcase the positive characteristics of a certain sensor/actuator and to highlight all applications where its use may be more advantageous. The less good features are usually left out of the datasheets. It is therefore important that an engineer, who wants to choose a sensor/actuator for a particular purpose, learn from the outset what features are important and what values are acceptable for the desired application.

It is customary to find in datasheets:

- A short list of the most important features and capabilities,
- A general description of how the sensor works,
- A functional block diagram of the sensor,
- A table with the specifications of the sensor,
- A table with the maximum values of some characteristics,
- Information on the packaging of the sensor,
- Indication of the different variants of the sensor.

Note that it is common for the datasheet to cover more than one sensor and to reference different variants of these sensors in terms of packaging and range of operating temperatures.

Next, some specifications that apply to most sensors are discussed.

1.5.1 *Transfer Function*

The transfer function specifies the functional relationship between the input quantity and the electrical output signal in the case of a sensor, or the opposite, in the case of an actuator. This relationship is generally described by a graph, although it may also be presented through a table or a mathematical equation.

In Figure 1.9, one can see, as an example, the transfer function of a relative humidity sensor. The two dashed lines on the graph indicate the manufacturing tolerance. Notice that the thick-line curve is not a perfectly straight line. This is due to variations in the manufacturing process. The sensors leave the production line slightly different. Each will have its transfer function slightly different. What the manufacturer ensures is that most sensors produced will have a transfer function between the two dashed lines shown in the chart. For a given voltage value obtained at the sensor output (e.g., 0.5 V), the user knows from the graph that the relative humidity is within a certain range centered on the straight line (50% relative humidity in the case of the example) and with a half-width of 3%, that is, in the range] 47%, 53% [.

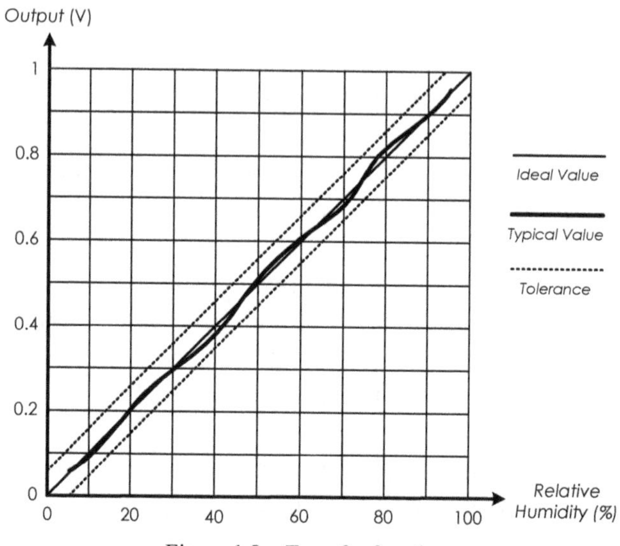

Figure 1.9 – Transfer function.

1.5.2 *Sensitivity*

The parameter "Sensitivity" specifies how large is the variation of the output signal with respect to a variation of the input signal. For example, a thermometer whose output increases 100 mV when the temperature increases by 1°C is more sensitive than one in which the output increases just 50 mV for the same temperature increase. It is defined as the ratio of the variations of the output and input signal. Mathematically it corresponds to the derivative of the transfer function (Figure 1.10).

Typically, the sensors have a linear transfer function, and therefore it is sufficient to specify a single value of sensitivity (the derivative of a straight line is equal in every point along that line). In other cases, it is common for manufacturers to specify the sensitivity via a plot of sensitivity versus the measured quantity or even indicating the maximum sensitivity and the corresponding value of the measured quantity. Even if the transfer function is linear, the sensitivity may vary with other parameters of the sensor. For example, in the case of photodiodes, which are sensors of light intensity, the sensitivity depends on the wavelength of the incident light. It is, therefore, common to find a typical plot of sensitivity versus wavelength as the one shown in Figure 1.11.

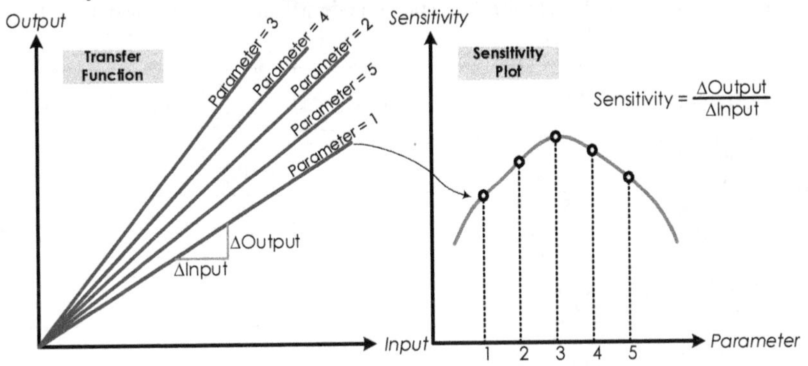

Figure 1.10 – Relationship between the transfer function and the sensibility plot.

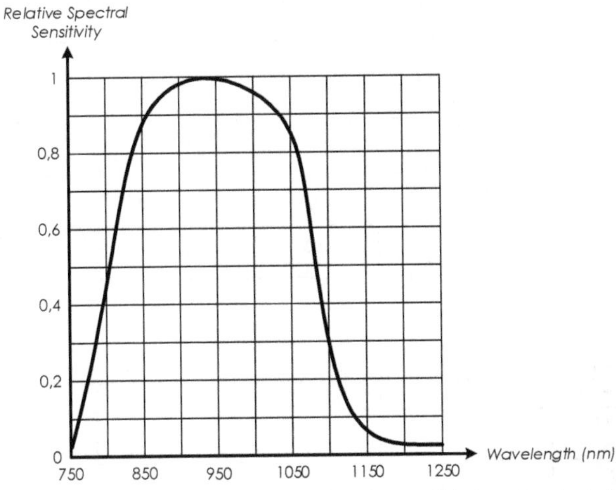

Figure 1.11 – Example of the plot that shows how the sensitivity of a photodiode changes with wavelength. It was obtained from the datasheet of photodiode BPV10NF from Vishay Semiconductors.

In this graph, we can read the sensitivity at a certain wavelength relative to its maximum value. The curve thus takes the value 1 at its peak. Other data present in the datasheet has to be used to obtain the absolute value of sensitivity. As shown in Table 1.2, which presents an example of part of a datasheet of a photodiode, the absolute sensitivity to a wavelength (λ) of 870 nm is 0.55 A/W.

Table 1.2 – Extract from the datasheet of a photodiode which shows the absolute value of spectral sensitivity (0.55 A/W).

PARAMETER	TEST CONDITION	SYM.	MIN.	TYP.	MAX.	UNIT
Noise equivalent power	$V_R = 20$ V, $\lambda = 950$ nm	NEP		3×10^{-14}		W/√ Hz
Forward voltage	$I_F = 50$ mA	V_F		1.0	1.3	V
Breakdown voltage	$I_R = 100$ μA, E = 0	$V_{(BR)}$	60			V
Reverse dark current	$V_R = 20$ V, E = 0	I_{ro}		1	5	nA
Diode capacitance	$V_R = 0$, f = 1 kHz, E = 0	C_D		11		pF
Open circuit voltage	$E_e = 1$ mW/cm², $\lambda = 870$ nm	V_O		450		mV
Short circuit current	$E_e = 1$ mW/cm², $\lambda = 870$ nm	I_K		50		μA
Reverse light current	$E_e = 1$ mW/cm², $\lambda = 870$ nm, $V_R = 5$ V	I_{ra}		55		μA
Detectivity	$V_R = 20$ V, $\lambda = 950$ nm	D*		3×10^{12}		cm/√ Hz/ W
Temperature coefficient of I_{ra}	$E_e = 1$ mW/cm², $\lambda = 870$ nm	TK_{Ira}		−0,1		%/K
Absolute spectral sensitivity	$V_R = 5$ V, $\lambda = 870$ nm	S(λ)		0.55		A/W
Angle of half sensitivity		φ		± 20		°
Wavelength of peak sensitivity		λ_p		940		nm
Nominal current		I_{ra}		60		mA
Range of spectral bandwidth		$\lambda_{(0,5)}$		790 to 1050		nm
Quantum efficiency	$\lambda = 950$ nm	η		70		%

With knowledge of this value and the graph of Figure 1.11, it is possible to determine the absolute value of the sensitivity for any wavelength (from 750 nm to 1150 nm).

Note that, in the case of this sensor, the wavelength for which the absolute sensitivity is specified (870 nm) is not the same for which the sensitivity is maximum, as shown in Figure 1.11 (940 nm).

The sensitivity (and any other characteristics) may depend on various parameters. Usually, the manufacturer presents the sensitivity charts it finds the most important for typical applications of the sensor in question. In the case of photodiodes, one often finds graphs of sensitivity versus wavelength and versus the direction of incidence of the light, as seen in Figure 1.12.

Certainly, in this case, the sensitivity also depends on the operating temperature. However, it must change so little with the temperature that the manufacturer decided it was not worth presenting that information.

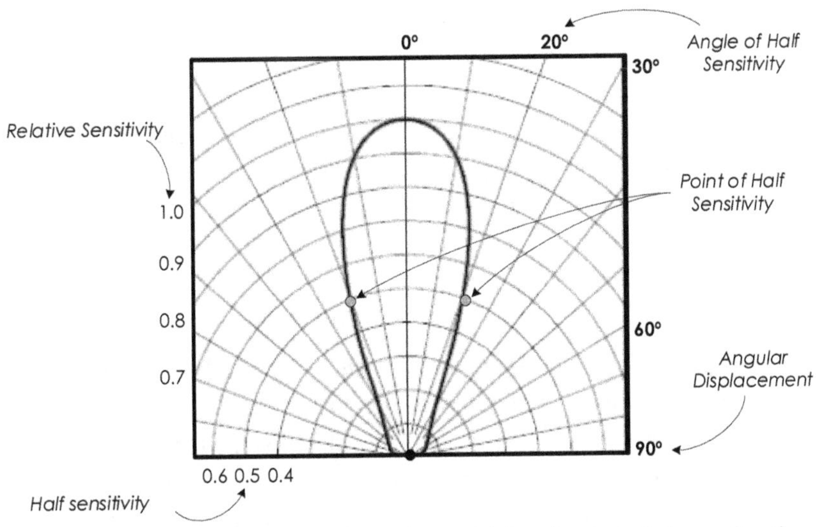

Figure 1.12 – Example of a plot that shows the sensitivity of a photodiode as a function of the angle where light comes from. It was obtained from the datasheet of photodiode BPV10NF from Vishay Semiconductors.

Recall that the sensitivity is the ratio between the variation of the output signal of the sensor and the variation of the quantity to be measured. Therefore, when it is said that the sensitivity varies with wavelength, for example, what is being specified is the dependence of the ratio of the output signal (current) and the measured value (power of light radiation) on the wavelength and not the dependence between the output signal and the wavelength (at least not directly).

1.5.3 *Range*

It specifies the range of values of the quantity being measured that can be converted into an electrical signal. Values outside this range usually lead to large measuring errors.

Note that the operation of a sensor may depend on other characteristics that fall within a certain range of values. For example, in the case of a photodiode, a manufacturer may specify a range of values of the operating wavelength for the sensor. In the case of Fairchild Semiconductor photodiode BPV10NF, this range is 790 nm to 1050 nm, as seen in Table 1.2.

There are cases where it is difficult for the manufacturer to specify clear boundaries for a range of values. In Table 1.2, we can observe that the manufacturer added the value "0.5" to the symbol wavelength (λ). This means that the range of values presented corresponds to a range where the relative sensitivity of the sensor is greater than 0.5, as can be seen in Figure 1.11.

1.5.4 *Accuracy*

Accuracy specifies the maximum error with which the sensor is supposed to make measurements. The error is the difference between the actual value and the measured value. In the example presented in Table 1.3, we can observe, for example, that the relative humidity sensor CHS-UGS from TDK has an accuracy of 5% RH. It should be noted that, in the case of this sensor, because it is a sensor of relative humidity, it represents the absolute value of the error. It does not mean that the error is 5% of the range. Relative humidity expresses itself in percentage (% RH). This is essentially the only case where we encounter this feature.

Table 1.3 – Datasheet extract from a relative humidity sensor (CHS-UGS) from TDK.

PARAMETER	MIN.	TYP.	MAX.	UNIT
Range	5		95	%RH
Nominal accuracy		± 5		%RH

1.5.5 *Precision*

When a sensor performs several measurements of the same quantity under the same conditions, the results obtained are not necessarily the same. This is due to the inevitable presence of noise in the sensor and/or the electronic circuit that accompanies it. The precision specification is used to express this fact quantitatively.

An analogy typically used to illustrate the difference between the concepts of precision and accuracy is that of a target where the goal is to hit the center (perfect measure, zero error). A very accurate sensor "hits" always very close to the target, as seen in the case on the left in Figure 1.13. Each cross means a different measurement. In the case of the other two targets, the measurements are far from the center; that is, they have a big error.

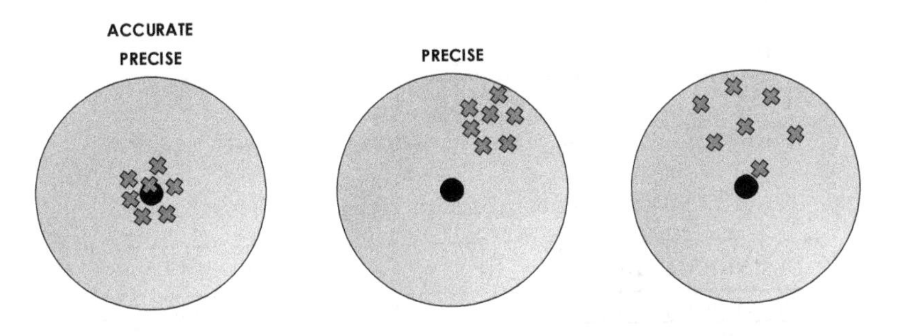

Figure 1.13 – Illustration of the concepts of accuracy and precision.

In this analogy, the concept of precision is observed in how close the crosses are to each other. In the case of the left and the middle, they are quite close, and, therefore, they were produced by a very precise sensor.

1.5.6 *Hysteresis*

Some sensors do not always have the same output for a given value of the measured quantity. This value depends on the derivative of the transfer function, as seen in Figure 1.14. This is called hysteresis.

The hysteresis is expressed as the highest value of the width of the curve of the transfer function divided by the range. Table 1.4 shows the datasheet of a sensor that has 1% hysteresis.

This parameter is more frequently found in magnetic and mechanical sensors.

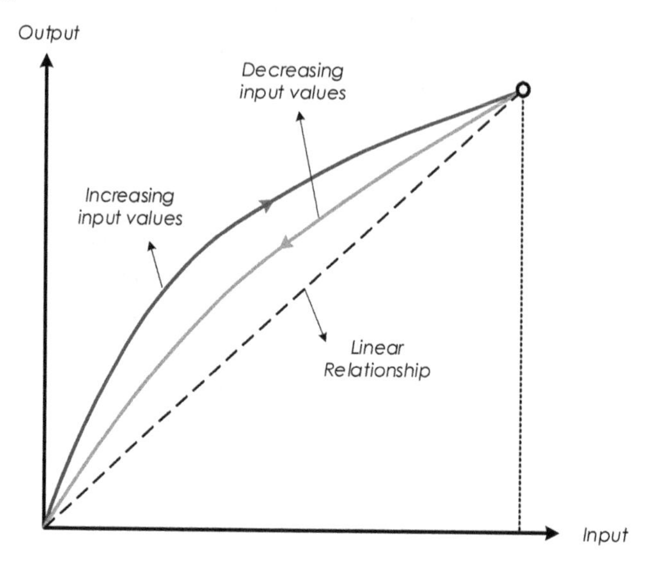

Figure 1.14 – Example of a transfer function that shows hysteresis.

Table 1.4 – Datasheet from a pressure sensor (E8MS) from Omron.

PARAMETER	VALUE
Power supply voltage	12 VDC ±10%. Ripple (p-p) of 5% max.
Current consumption	25 mA max.
Pressure type	Gauge pressure
Applicable fluid	Non-corrosive gas and non-flammable gas
Rated pressure range	0 to 100 kPa
Withstand pressure	400 kPa
Linearity	±1% FS max.
Hysteresis	±1% FS max.

1.5.7 *Nonlinearity*

Ideally, the transfer function of a sensor should be linear. This facilitates the conversion of the value of the output signal into the real value of the quantity measured. On the other hand, this means that the sensor has the same sensitivity for all values of magnitude in its range. One specifies the nonlinearity as the biggest difference between the transfer function and an ideal straight line, divided by the range and generally expressed in % (Figure 1.15). Generally, the ideal straight line used is the one that best fits the transfer function (minimizing the mean squared error, for example) since it leads to a better value of nonlinearity specification.

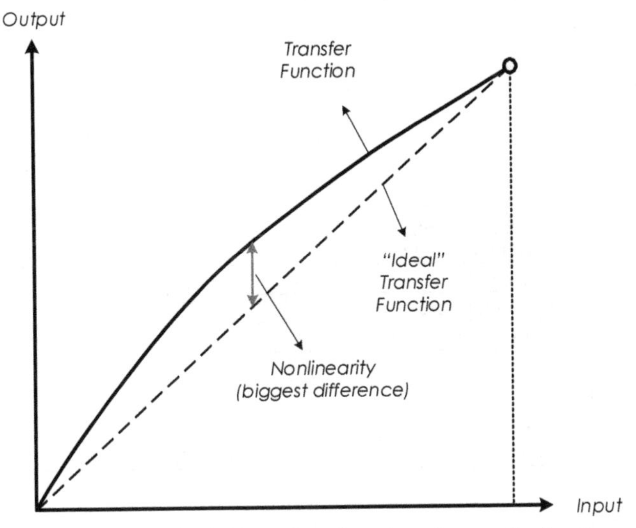

Figure 1.15 – Definition of nonlinearity of a transfer function.

1.5.8 *Noise*

All sensors produce some noise along with the output signal. In some cases, the sensor noise is lower than the noise of the electronic circuitry or the fluctuations of the measured quantity. In such cases, the noise generated by the sensor assumes less importance.

The noise is usually distributed along with different frequencies. Many noise sources produce a "white" spectrum. This means that the noise is distributed equally between all frequencies, such as, for example, with thermal noise in electrical resistances. The voltage noise is typically expressed through its spectral density $V/\sqrt{Hz}$. In the case of sensors, it is expressed in terms of the unit of the measured quantity divided by $\sqrt{Hz}$. In the case of photodiodes, for example, it is expressed in $W/\sqrt{Hz}$, where W corresponds to the unit of radiative power and not to electrical power of a circuit (Table 1.2).

The total amount of frequency-dependent noise depends on the bandwidth of the whole system. There are instances where, instead of specifying the density of noise, the total amount of noise is reported, as in the case of the accelerometer of Table 1.5. This is typical for sensors that have an operating bandwidth that is limited (2 Hz to 1 kHz in the case of that accelerometer).

Table 1.5 – Datasheet of the acceleration sensor 1185 from Monitron.

PARAMETER	VALUE
Frequency response	2 Hz to 1 kHz ±10%
Mounted resonance	18 kHz min.
Measurement range	50 g peak
Isolation	Base isolated
Operating temperature range	−25°C to 80°C
Temperature sensitivity	0,08 %/°C
Electrical noise	0.3 mg max.
Transverse sensitivity	Less than 5%
Supply voltage	10–32 V (smoothed)
Standard cable	5 m armored PVC

There are some cases where the manufacturer presents the spectral distribution of noise, as in the case of accelerometer ADXL150 from Analog Devices (Figure 1.16).

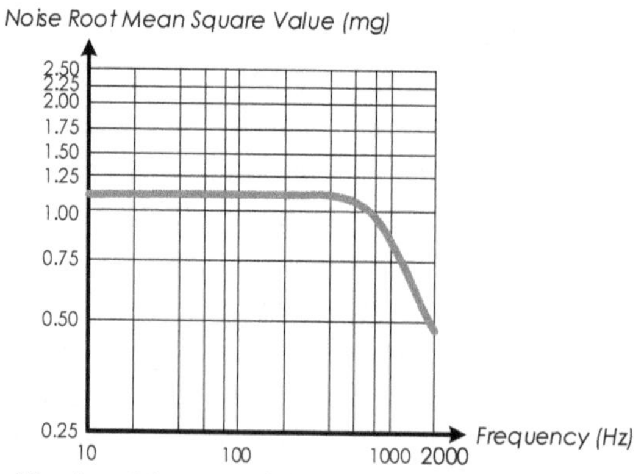

Figure 1.16 – Specification of the spectral distribution of noise in accelerometer ADXL150 from Analog Devices.

Usually, noise is specified in terms of "root mean square" (rms). There are, however, some cases where the value is specified as "peak to peak." Since noise usually has a normal statistical distribution, there is no interval that contains all possible noise values. At most, one can define the probability that the value of noise is within a range (or the opposite). Table 1.6 lists a few examples.

Table 1.6 – Relationship between peak-to-peak noise with normal distribution and percentage of time that it is greater than this value.

PEAK-TO-PEAL VALUE	PERCENTAGE OF THE TIME THAT THE NOISE IS GOING TO BE OUTSIDE THE PEAK-TO-PEAK VALUE
2.0 × rms	32%
4.0 × rms	4.6%
6.0 × rms	0.27%
6.6 × rms	0.1%
8.0 × rms	0.006%

In Figure 1.17, we can see the normal probability density together with the width of three different probability intervals corresponding to three multiples of a standard deviation.

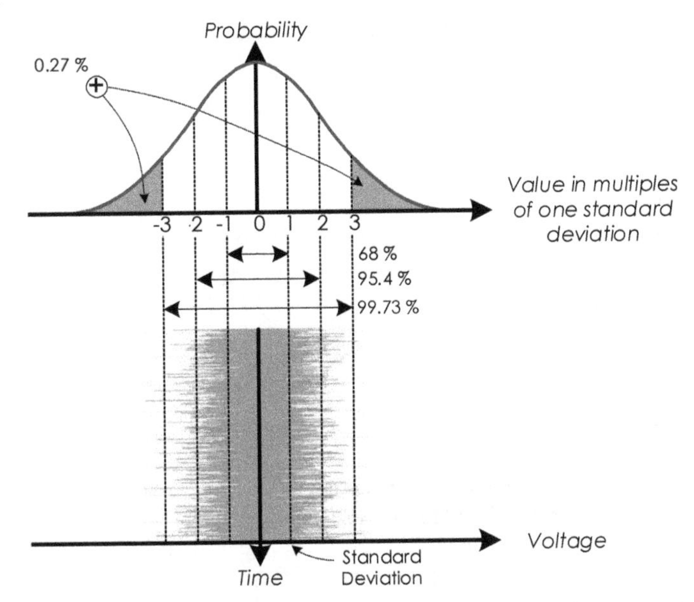

Figure 1.17 – Normal distribution probability with different probability intervals (above). Normal probability density (below).

Remember that when the signal is just random normal noise, the "root mean square" value equals the value of the standard deviation.

1.5.9 *Resolution*

The resolution is defined as the minimum detectable variation of the quantity to be measured. It depends, for example, on the amount of noise or the number of bits of the analog-to-digital converter used (in the case of smart or integrated sensors).

As noted above, the amount of noise present depends on the bandwidth of the sensor and the system in which it is inserted. One way to improve resolution is to decrease the noise, inserting a bandpass filter, covering the frequency range of interest, after the sensor. The narrower the bandwidth of this filter, the less noise goes through, as shown in the graph presented in the specification sheet for the ADXL150 accelerometer from Analog Devices, which can be seen in Figure 1.18.

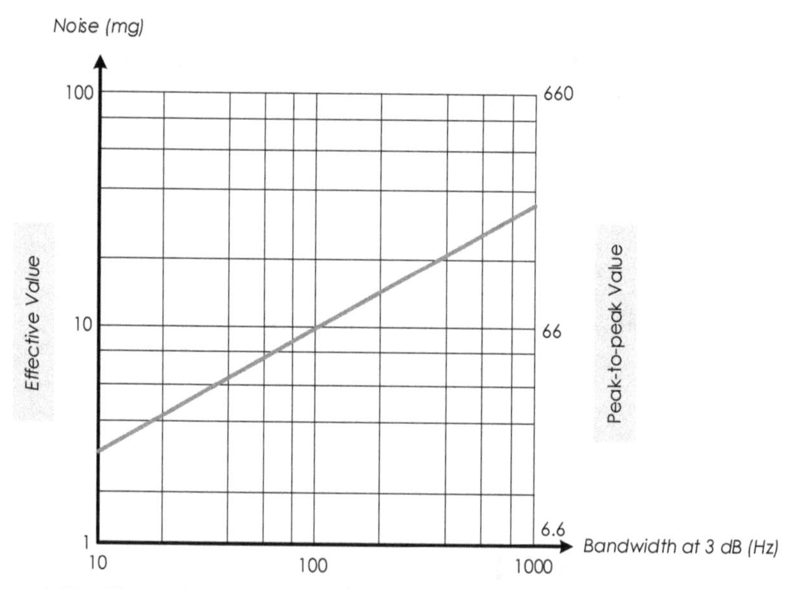

Figure 1.18 – Change in the amount of noise as a function of the bandwidth of a system in the case of accelerometer ADXL150 from Analog Devices.

1.5.10 *Bandwidth*

All sensors have a finite response time to an instantaneous change in the quantity being measured. In terms of sinusoidal variations of this quantity, one can define the limit of the frequency band as the frequency value of these variations that causes a decline to 70.7% (−3 dB)[1] of the output signal of the sensor in relation to its maximum value.

In some cases, the manufacturer provides a graph of the frequency response of the sensor, as seen in Figure 1.19. The frequency response is specified in dB relative to the maximum value, which usually happens at low frequency (systems are typically low pass).

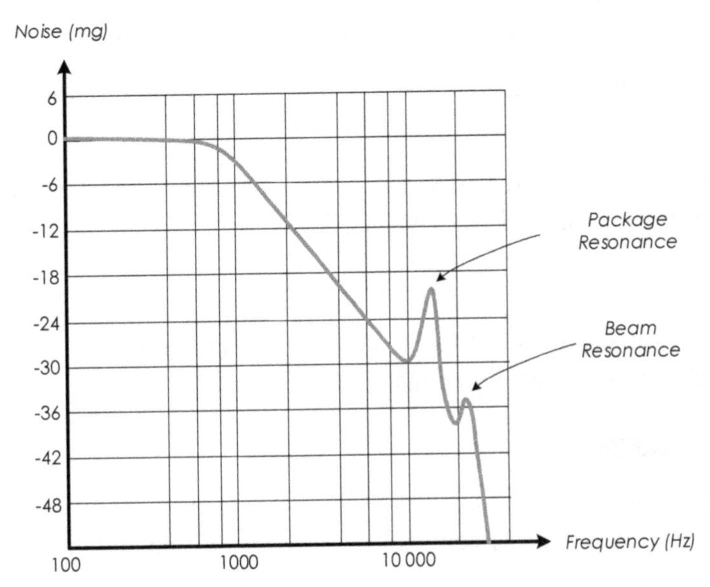

Figure 1.19 – Example of the frequency response specification of a sensor. It was obtained from the datasheet of the accelerometer ADXL150 from Analog Devices.

1.5.11 *Repeatability*

The repeatability error reflects the inability of the sensor to produce the same output when stimulated by a quantity with the same value in the same conditions. It is expressed as the maximum difference between the values of the measured quantity obtained at different calibrations. In Figure 1.20,

[1] $20 \log(0{,}707) \approx -3$ dB.

two transfer functions obtained at two different calibrations of the sensor are presented.

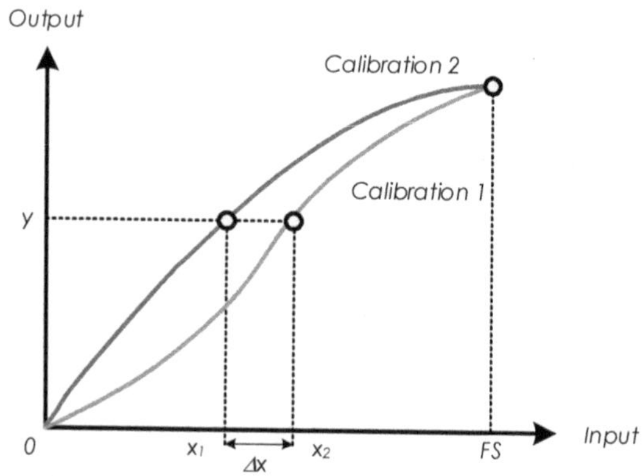

Figure 1.20 – Example of repeatability error: two calibrations lead to two different transfer functions.

The maximum difference, Δx, is used to compute the repeatability error as a percentage of the full-scale value (FS):

$$Repeatability = \frac{\Delta x}{FS}. \tag{1.1}$$

The possible causes of this error are related to thermal noise, charge accumulation, and plasticity of materials, among others.

1.5.12 *Dead Zone*

The dead zone is a range of values of the measurand for which the sensor output varies very little in comparison to the rest of the measuring range (Figure 1.21).

1.5.13 *Saturation*

No sensor is linear for infinite values of the measured quantity. At some point, the output voltage ceases to follow a straight line, as shown in Figure 1.22. This phenomenon is called *saturation*.

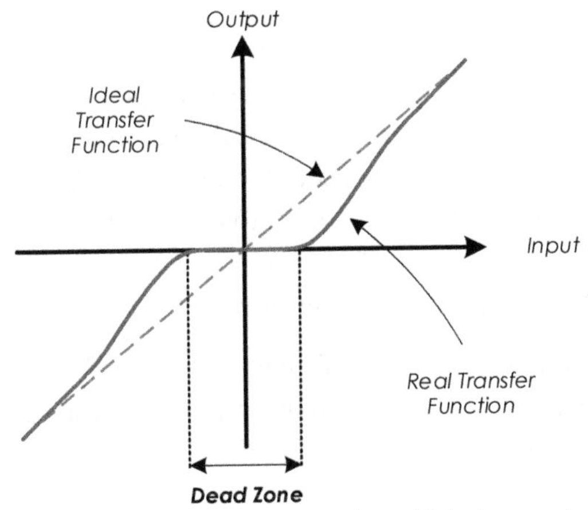

Figure 1.21 – Example of a sensor transfer function which shows a dead zone around 0.

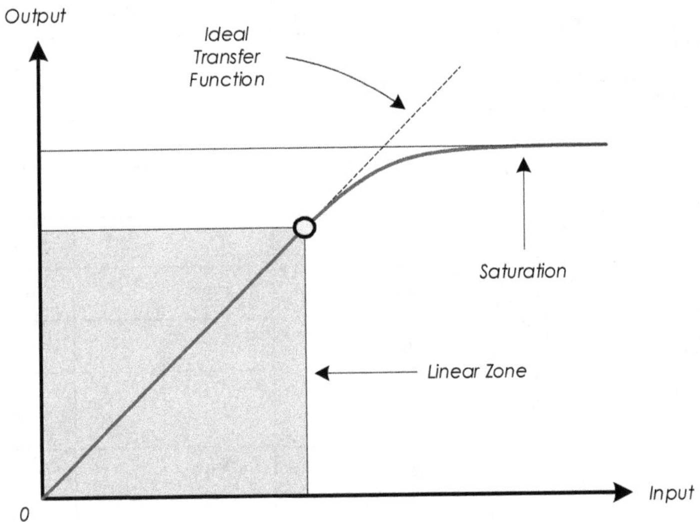

Figure 1.22 – Example of a transfer function presenting saturation (it stops being linear for high input values).

When the input value is too high, the output is not able to keep up. Generally, this maximum output value is close to the power supply value.

1.6 Questions

1. Consider a smart sensor that acquires temperature values at a rate of one per second. The analog/digital converter, used to transform the voltage output of the sensor element into a digital word, has 16 bits. It is only intended to remotely transmit information on measured values once a day. What is the minimum memory size required to store temperature information on the smart sensor?

2. How sensitive is the Vishay Semiconductors BPV10NF photodiode, whose specifications are given, for radiation with a wavelength of 1050 nm?

Table 1.7 – Sensor specifications.

PARAMETER	TEST CONDITION	SYM.	MIN.	TYP.	MAX.	UNIT
Noise equivalent power	$V_R = 20$ V, $\lambda = 950$ nm	NEP		3×10^{-14}		W/√ Hz
Forward voltage	$I_F = 50$ mA	V_F		1.0	1.3	V
Breakdown voltage	$I_R = 100$ µA, E = 0	$V_{(BR)}$	60			V
Reverse dark current	$V_R = 20$ V, E = 0	I_{ro}		1	5	nA
Diode capacitance	$V_R = 0$, f = 1 kHz, E = 0	C_D		11		pF
Open circuit voltage	$E_e = 1$ mW/cm², $\lambda = 870$ nm	V_O		450		mV
Short circuit current	$E_e = 1$ mW/cm², $\lambda = 870$ nm	I_K		50		µA
Reverse light current	$E_e = 1$ mW/cm², $\lambda = 870$ nm, $V_R = 5$ V	I_{ra}		55		µA
Detectivity	$V_R = 20$ V, $\lambda = 950$ nm	D*		3×10^{12}		cm/√ Hz/ W
Temperature coefficient of I_{ra}	$E_e = 1$ mW/cm², $\lambda = 870$ nm	TK_{Ira}		−0.1		%/K
Absolute spectral sensitivity	$V_R = 5$ V, $\lambda = 870$ nm	S(λ)		0.55		A/W
Angle of half sensitivity		φ		± 20		°
Wavelength of peak sensitivity		λ_p		940		nm
Nominal current		I_{ra}		60		mA
Range of spectral bandwidth		$\lambda_{(0,5)}$		790 to 1050		nm
Quantum efficiency	$\lambda = 950$ nm	η		70		%

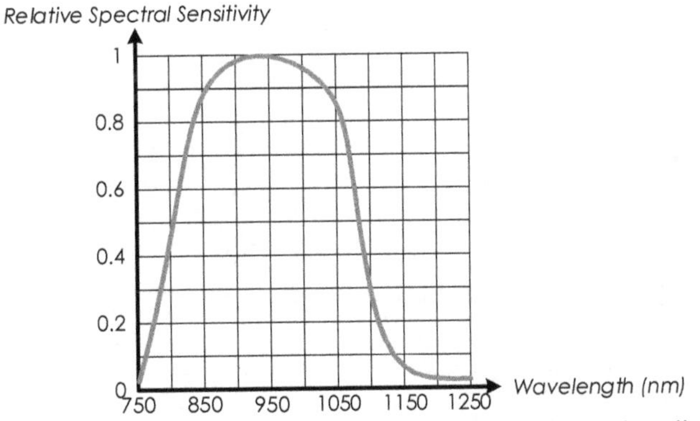

Figure 1.23 – Specification of the relative sensitivity of the photodiode.

3. Considering the specifications provided for the Vishay Semiconductors BPV10NF photodiode (given in the previous problem), calculate the value of the signal-to-noise ratio if it is inserted in a circuit with a bandwidth of 10 kHz and the power of incident monochromatic radiation ($\lambda = 950$ nm) of 3 μW. Consider that the photodiode is inversely polarized with 20 V.

4. Consider a Linear Variable Differential Transformer (LVDT) model SM1 from RS, whose specifications are shown in the table. What is the minimum rms value of the sinusoidal voltage with a frequency of 15 kHz to be used in the primary winding so that the rms output voltage is at least 5 V in the event of a 2 cm displacement?

Table 1.8 – Sensor specifications.

PARAMETER	SM1	SM2
Energizing voltage	1 to 10 V rms	
Energizing frequency (Hz)	1 to 20	1 to 20
Calibration load (W)	100 k	100 k
Primary resistance (W)	102	69
Primary impedance (W) at 1 kHz	120	135
Primary impedance (W) at 2 kHz	160	240
Primary impedance (W) at 5 kHz	310	560
Primary impedance (W) at 10 kHz	620	1000
Primary impedance (W) at 20 kHz	1300	1700
Secondary resistance (W)	204	200
Secondary impedance (W) at 1 kHz	417	295
Secondary impedance (W) at 2 kHz	756	450
Secondary impedance (W) at 5 kHz	1830	930
Secondary impedance (W) at 10 kHz	3646	1880
Secondary impedance (W) at 20 kHz	7280	3900
Sensitivity (mV/V/mm typ.) at 1 kHz	69	118
Sensitivity (mV/V/mm typ.) at 2 kHz	110	134
Sensitivity (mV/V/mm typ.) at 5 kHz	142	136
Sensitivity (mV/V/mm typ.) at 10 kHz	147	130
Sensitivity (mV/V/mm typ.) at 20 kHz	149	128
Zero phase shift frequency (kHz)	14	3.9
Residual voltage at zero (typ.) < %fsd	0.3	0.3

5. What is the current that must be applied to the TLUR6400 LED by Vishay Semiconductors so that the luminous intensity emitted in a 30° direction deviated from the center is 15 mcd?

Table 1.9 – Specification of the actuator.

PARAMETER	TEST CONDITION	SYM.	MIN.	TYP.	MAX.	UNIT
Luminous intensity	$I_F = 10$ mA	I_v	4	15	-	mcd
Dominant wavelength	$I_F = 10$ mA	λ_d	-	630	-	nm
Peak wavelength	$I_F = 10$ mA	λ_p	-	640	-	nm
Angle of half intensity	$I_F = 10$ mA	φ	-	±30	-	deg
Forward voltage	$I_F = 20$ mA	V_F	-	2	3	V
Reverse voltage	$I_R = 10$ μA	V_R	6	15	-	V
Junction capacitance	$V_R = 0$, f = 1 MHz	C_j	-	50	-	pF

6. What is the current that is obtained when Osram's SFH203P photodiode receives radiation with a wavelength of 850 nm and a power of 100 mW at an angle of 90° relative to the center?

Table 1.10 – Sensor specifications.

PARAMETER	TEST CONDITION	SYM.	MIN.	TYP.	MAX.	UNIT
Photocurrent	E_v = 1000 lx, Std. Light A, V_R = 5 V	I_P	5	9.5		μA
Wavelength of max. sensitivity		λ_{Smax}		850		nm
Spectral range of sensitivity		$\lambda_{10\%}$		400 to 1100		nm
Radiant sensitive area		A		1.00		mm^2
Dimensions of radiant sensitive area		L x W		1 x 1		mm x mm
Half angle		φ		± 75		°
Dark current	V_R = 20 V	I_R		1	5	nA
Spectral sensitivity of the chip	λ = 850 nm	$S_{\lambda\ typ}$			0.62	A/W
Quantum yield of the chip	λ = 850 nm	η			0.90	Electrons/s
Open-circuit voltage	E_v = 1000 lx, Std. Light A	V_o	300	350		mV
Short-circuit current	E_v = 1000 lx, Std. Light A	I_{SC}		9.3		μA
Rise and fall time	V_R = 20 V, R_L = 50 Ω, λ = 850 nm	t_r, t_f		0.005		μs
Forward voltage	I_F = 100 mA, E = 0	V_F		1.3		V
Capacitance	V_R = 0 V, f = 1MHz, E = 0	C_0		11		pF
Temperature coefficient of V_o		TC_V		-2.6		mV/K

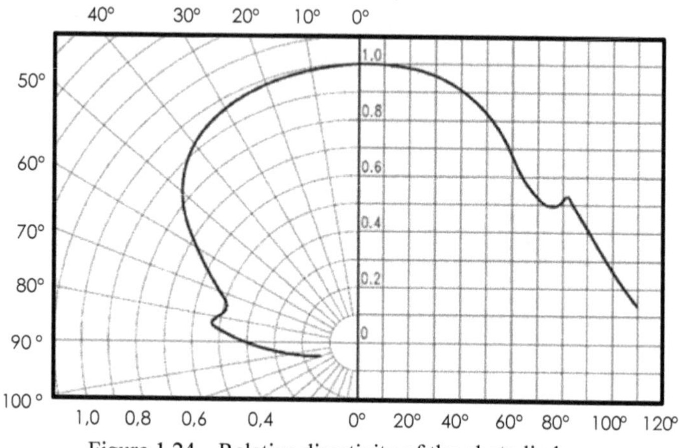

Figure 1.24 – Relative directivity of the photodiode.

7. How do sensors and actuators interact with the Universe around us, and how does this interaction benefit human beings?

8. The Human eye can be considered a sensor created by Nature through evolution. The information generated by this sensor and transmitted to the brain is represented electrically. The physical support of this representation is, however, different from the physical support used by man-made sensors. In which way?

9. What is the difference between sensor, actuator, and transducer? Give examples of each of these devices.

10. How is an integrated sensor built? What is the function of each of its parts?

11. What functions does a smart sensor usually have?

12. What is the difference between an integrated sensor and a smart sensor?

13. What are the possible functions of the signal conditioning circuit used in a sensor?

14. Point out four ways that an electrical signal contains information?

15. What is the difference between a passive sensor and an active sensor? Give an example of each type. What are the advantages of both?

16. The following table presents the specifications of a relative humidity sensor. What is the accuracy when measuring a relative humidity value of 80%?

Table 1.11 – Sensor Specifications.

PARAMETER	MIN.	TYP.	MAX.	UNIT
Range	5		95	%RH
Nominal accuracy		±5		%RH

17. What does the nonlinearity of a sensor translate? In which units is it typically expressed?

18. What does the accuracy of a sensor mean?

19. Give four examples of passive sensors and four examples of active sensors.

20. Which of the following statements is correct?

    a. Sensors and actuators are two examples of transducers.
    b. Sensors and transducers are two examples of actuators.
    c. Transducers and actuators are two examples of sensors.
    d. All transducers are sensors.
    e. None of the above.

21. What type of sensor does a microcontroller or other type of processor have?

    a. Sensor element.
    b. Integrated sensor.
    c. Smart Sensor.
    d. Computational Sensor.
    e. None of the above.

22. Which of the following classifications of a sensor is related to its energy source?

    a. Modulating / Auto-generator.

b. Active passive.
c. Smart / Non-smart.
d. Autonomous / Non-autonomous.
e. None of the above.

23. A classification of a sensor into Active / Passive relates to which aspect?

a. The type of energy source.
b. The operation of the signal conditioning circuit.
c. The existence of moving parts.
d. The way it communicates with the outside.
e. None of the above.

24. Which of the following is the main purpose of a sensor/actuator specification sheet?

a. Publicize the sensor/actuator.
b. Explain the operating principle used.
c. Compare the price of different sensor/actuator variants.
d. Explain how the sensor/actuator should be used and how it behaves in operation.
e. None of the above.

25. Which of the following specifications of a sensor has the definition "maximum error that the manufacturer admits for the sensor"?

a. Precision.
b. Sensitivity.
c. Range.
d. Accuracy.
e. None of the above.

26. What does the sensitivity of a sensor mean?

a. The relationship between the measured quantity and the sensor's output voltage value.
b. How fragile is its handling.
c. What is the smallest value of the quantity as the sensor can detect it.
d. How fast is the sensor reacting to a change in the measured value.
e. None of the above.

27. In specifying how directive the behavior of a sensor is, what would be the advertised value for the sensor whose sensitivity graph is shown?

    a. 0°
    b. 10°
    c. 20°
    d. 30°
    e. None of the above.

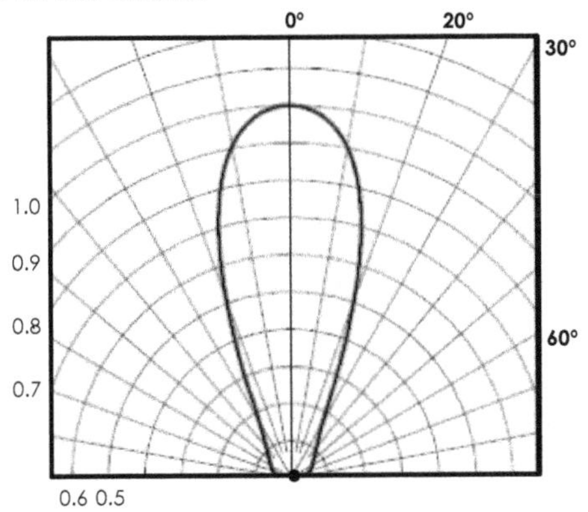

Figure 1.25 – Relative directivity of the photodiode.

28. Which of the parameters specified for a sensor reflects the dependence of the output value with the direction of variation of the measured quantity.

    a. Linearity.
    b. Range.
    c. Saturation.
    d. Hysteresis.
    e. None of the above.

29. How do I specify a sensor's bandwidth?

    a. With the maximum frequency value at which the sensor can operate.
    b. With the frequency value for which the sensor output is 3 dB less than the value at 0 Hz.
    c. With the frequency value for which the output is the highest possible.

d. With the minimum frequency value at which the sensor can operate.
e. None of the above.

Chapter 2
Micro and Nanotechnology

2.1 Introduction

This chapter covers the technologies used to manufacture devices of very small dimensions, in the order of micrometers or nanometers, which are used as sensors or actuators. It addresses the manufacture of MEMS (*Microelectromechanical Systems*).

The main emphasis is to reduce the size of the device. The miniaturization allows an increase of responsiveness, a reduction of the influence of temperature, and the possibility to install sensors in certain places which otherwise would not be able to accommodate them. On the other hand, the energy and materials used in production lead to significant cost reduction, which brings benefits in terms of the cost/performance ratio. The miniaturization allows the construction of batches of identical sensors, which opens the possibility of implementing redundancy in systems. Another advantage is the possibility to integrate the sensor or actuator with electronics for signal conditioning, digitization, and information processing. All in all, we end up with devices that are simpler and less power-consuming. The use of micro-fabrication techniques leads to increased reproducibility, which means higher selectivity and sensitivity, greater dynamic range, and better accuracy.

These advantages, obtained with MEMS devices, have been extended to many quantities and allowed their use in more and more applications in all sectors of activity. Particularly relevant are those applications that involve an interface with a user and the production of many units, such as in the case of mobile phones, smartwatches, and automobiles. The low unit cost of manufacturing MEMS-type sensors allows its numerous uses and the consequent increase in the amount of information that can be easily collected from Nature and presented in a simple, attractive, and undoubtedly useful way to us.

The first MEMS devices, developed in the 1960s, were hydraulic pressure sensors for aircraft. MEMS gained momentum at the beginning of the 1990s with the developments made in the manufacture of integrated circuits and with the possibility of integrating sensors, actuators, and control circuits on the same circuit. The first commercial product appeared in 1993 with the ADXL 50 accelerometer from Analog Devices. The accelerometers are a good example of the advantages that miniaturization can bring. Traditionally they were used in airbags for automobiles and nowadays are also found in cameras and game consoles, for example. The pre-MEMS crash detectors were based on a metal sphere secured by a magnetic field or by a spring. The movement of the ball at the instant of collision closed an electrical circuit that signaled the event. This low-cost and simple method leads to sensors that are susceptible to dirt on the contacts or hindrance to the normal displacement of the sphere that would be impossible to detect. The modern MEMS-based accelerometers may have self-validating systems in which a micro-actuator can simulate the effect of slowing down and thus verify the correct operation of the sensor.

The manufacturers that began producing accelerometers, namely Lucas Nova-Sensors and EG&G IC-sensors, were not able to reduce their price below $10. Only when other manufacturers such as Analog Devices and Motorola managed to produce them at a price of about $3, they began to be adopted on a large scale by automakers leading to a significant increase in sales volume, as shown in Figure 2.1.

Figure 2.1 – Relationship between sales volume and price of accelerometers used in the automotive industry.

This shows the importance of producing a low-cost sensor which is was made increasingly possible through the use of MEMS.

Nowadays, it is possible to find on the market many more MEMS-based devices, which range from traditional print heads used in inkjet printers to micro-bio-analysis systems.

2.2 Manufacture

2.2.1 *Introduction*

Traditional techniques used in manufacturing, such as trimming, punching, or molding, are not suitable for the realization of structures with dimensions less than a millimeter in size. MEMS are manufactured by combining techniques typically used to create integrated circuits, such as lithography, deposition, and etching, and specific techniques of micromachining such as bulk micromachining or surface micromachining.

The techniques used to manufacture chips are essentially two-dimensional, and the creation of structures in three-dimensions is based on the superposition of multiple two-dimensional layers on a substrate. There are, however, practical and economic limitations to the number of layers that can be superimposed, which limits the creation of truly three-dimensional devices. Micromachining techniques allow structures to extend more in the third dimension; however, these structures are simply two-dimensional shape extrusions or depend on the crystalline properties of the material. Truly three-dimensional structures require the possibility to create curved surfaces with arbitrary shapes, which is currently not possible.

MEMS's manufacture tries to use parallel processing techniques to benefit from the economy of scale that has been so successful in reducing the price of integrated circuits. As such, the manufacturing process begins with a wafer of a material (silicon, polymer, glass, etc.) that may play an active role in the final device or can simply be the support in which the device is constructed.

2.2.2 *Use of Silicon*

Silicon is the material most widely used for several reasons:

- Widespread use of electronic integrated circuits in the industry;
- Low cost;
- Well studied electrical properties and of easy control;
- Existence of drawing tools;
- The production of monocrystalline substrates is economic;
- It has good mechanical properties;
- Ease of integration with electronic circuits.

The last item is particularly important in manufacturing microsensors. The monocrystalline silicon is elastic, lighter than aluminum, and has a modulus of elasticity like stainless steel. Note that their properties are anisotropic; that is, they depend on the orientation of the crystalline axis. Table 2.1 shows some properties of silicon and compares them with other materials.

Table 2.1 – Properties of silicon and other materials.

PROPERTY	SI {111}	STAINLESS STEEL	AL	AL_2O_3 (96%)	SIO_2	QUARTZ
Young's Module (GPa)	**190**	**200**	**70**	**303**	**73**	**107**
Poisson Ratio	0.22	0.3	0.33	0.21	0.17	0.16
Density(g/cm^3)	2.3	8	2.7	3.8	2.3	2.6
Yield Strength (GPa)	7	3.0	0.17	9	8.4	9
Thermal Coefficient of Expansion(10/K)	2.3	16	24	6	0.55	0.55
Thermal Conductivity at 300K (W/cm·K)	1.48	0.2	2.37	0.25	0.014	0.015
Melting Temperature (°C)	1414	1500	660	2000	1700	1600

Other materials also used are glass, quartz, ceramics, alloys of various metals, and a wide variety of other materials used in very specific cases.

The silicon wafer is processed with a sequence of processes that add, modify, or remove materials according to well-defined patterns (Table 2.2).

Table 2.2 – Classification of processes used in the manufacture of MEMS.

ADDITIVE PROCESS	MODIFYING PROCESS	SUBTRACTIVE PROCESS
Evaporation	**Oxidation**	**Wet Etching**
Sputtering	Doping	Dry Etching
Chemical Vapor Deposition	Annealing	Sacrificial Etching
Spin-coating	Ultraviolet Exposure	Development
...	...	...

The most used materials in the case of MEMS are silicon dioxide (SiO_2), polycrystalline silicon, or polysilicon (poly-Si), and aluminum. The silicon dioxide serves as an insulator between conductive layers. The polysilicon is a conductive material used to make the structures themselves. The aluminum is used for connections to electronic circuits.

2.2.3 *Creation of a silicon dioxide layer by thermal oxidation*

A thin layer of silicon dioxide is placed on top of the silicon substrate using thermal oxidation. This is achieved by heating the silicon in the presence of oxygen which may be done by inserting the wafers into a furnace at a temperature between 900°C and 1200°C, where oxygen is circulated as illustrated in Figure 2.2.

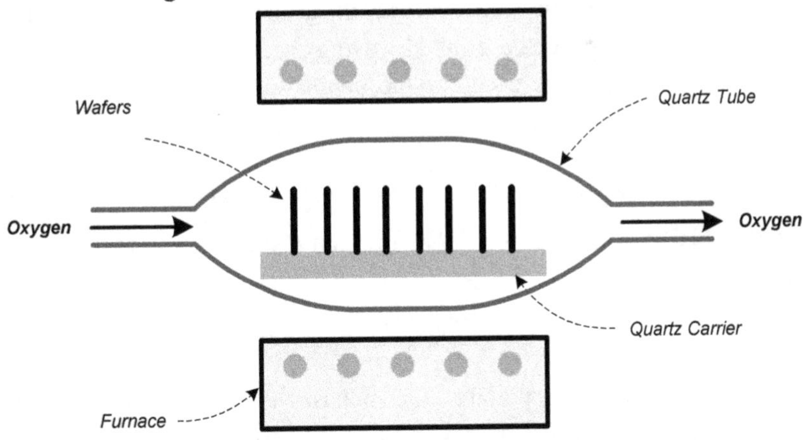

Figure 2.2 – Illustration of the process of growing a layer of silicon dioxide on a silicon wafer by thermal oxidation.

The chemical reaction that occurs is:

$$Si_{(s)} + O_{2(g)} \xrightarrow{900-1200ºC} SiO_{2(s)}. \quad (2.1)$$

Instead of oxygen, water vapor can be used with the same results:

$$Si_{(s)} + 2H_2O_{(g)} \xrightarrow{900-1200ºC} SiO_{2(s)} + 2H_{2(g)}. \quad (2.2)$$

The greater the time the process takes, the thicker is the layer of silicon dioxide created. The use of oxygen or water vapor at high pressure allows a higher speed and the possibility of working at lower temperatures. From the density of silicon and silicon oxide, it is possible to show that a layer of silicon dioxide with height x consumes a silicon layer with a height of $0.44x$. In practice, layers of silicon dioxide with a thickness of up to 2 μm can be created.

2.2.4 *Chemical Deposition by Vapor*

Another technique, which allows creating layers of different materials, including silicon dioxide or polysilicon, is by using chemical vapor deposition (CVD). The process is like thermal oxidation, but there is no reaction with the materials existing on the wafer. The material to be deposit is created within the furnace in gaseous form through the introduction of appropriate gases and by using an appropriate temperature.

Silane and oxygen are used for the deposition of a layer of silicon dioxide. The chemical reaction is

$$SiH_4 + O_2 \xrightarrow{300-500ºC} SiO_{2(s)} + 2H_2. \quad (2.3)$$

The polysilicon is created by using only silane at a temperature between 600 and 650°C:

$$SiH_4 + O_2 \xrightarrow{600-650ºC} Si + 2H_2. \quad (2.4)$$

The silane at room temperature exists in gaseous form. Using a pressure between 25 Pa and 150 Pa, a growth rate of between 10 and 20 nm per minute can be achieved.

2.2.5 *Photolithography*

The creation of patterns on a material consists of two steps:

- Deposition and creation of a pattern on a preliminary layer;
- Transfer of the pattern created in this layer to the final material.

In the most widely used process, photolithography, the preliminary layer is made of a photo resistive material. This material, a polymer sensitive to ultraviolet radiation, still in liquid form, is deposited on the substrate, which is rotated around its center to evenly spread the polymer (Figure 2.3) and then baked in an oven to solidify and adhere to the material on which it rests. The thickness depends on the concentration, viscosity, angular velocity, and process duration.

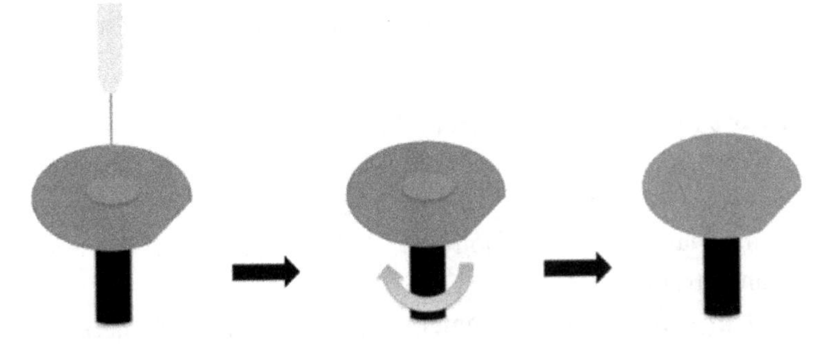

Figure 2.3 – Process of placing the photo resistive on the wafer.

Figure 2.4 illustrates, briefly, the different steps in creating an integrated circuit. As previously mentioned, the silicon substrate (Figure 2.4(a)) is covered by a layer of silicon dioxide (SiO_2) which will constitute the electronic circuit (Figure 2.4(b)). At this point, the temporary layer of the photoresist material is placed on the top (Figure 2.4(c)).

Next, a mask with the desired pattern is placed on top, and ultraviolet light is shined on it (Figure 2.4(d)). The light that goes through the mask, in the area where the pattern is not drawn, strikes the wafer, and the photoresist material, changing its properties (Figure 2.4(e)). The wafer is then placed in a solution that removes the photo resistive material in a developing process (Figure 2.4(f)). Next, the unprotected parts of the silicon dioxide layer are removed in a process known as corrosion (Figure 2.4(g)).

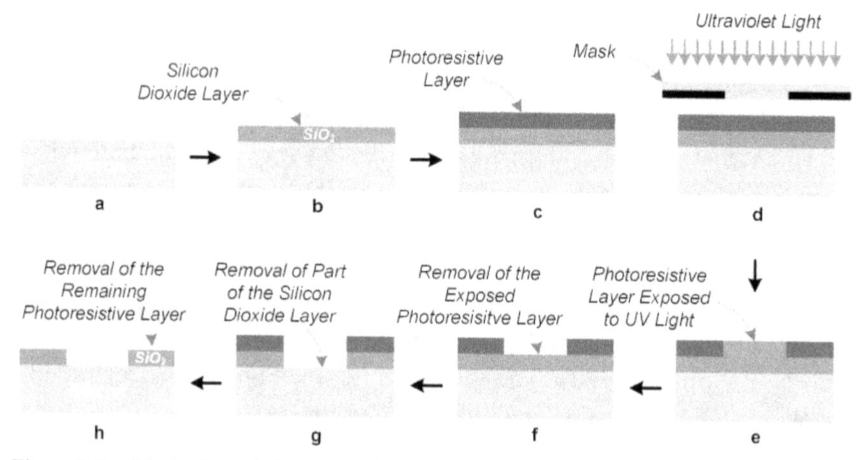

Figure 2.4 – Illustration of photo resistive polymer applied over a layer of silicon dioxide on which you want to create a pattern.

There are two types of corrosion:

- Wet — uses chemicals in the liquid phase. It is highly selective; that is, it removes only the desired material leaving behind other materials. It may be isotropic, that is, equal in all directions, or anisotropic.
- Dry — uses ion bombardment. It is anisotropic and, therefore, less susceptible to unwanted side removal. It is not, however, selective and occurs at a slower rate than wet corrosion.

In the case of isotropic wet corrosion, there is always extra material underneath the material that serves as a mask (photo resistive material, for example) which is also removed. Therefore, it is not a suitable technique for manufacturing MEMS unless one wants to use it for the creation of suspended beams, for example.

For amorphous and homogeneous materials, corrosion must always be isotropic. In crystalline materials, such as silicon, corrosion can be anisotropic. This is possible because the corrosion rate is different depending on the plane of the crystal. Figure 2.5 shows the different planes typically used and their description.

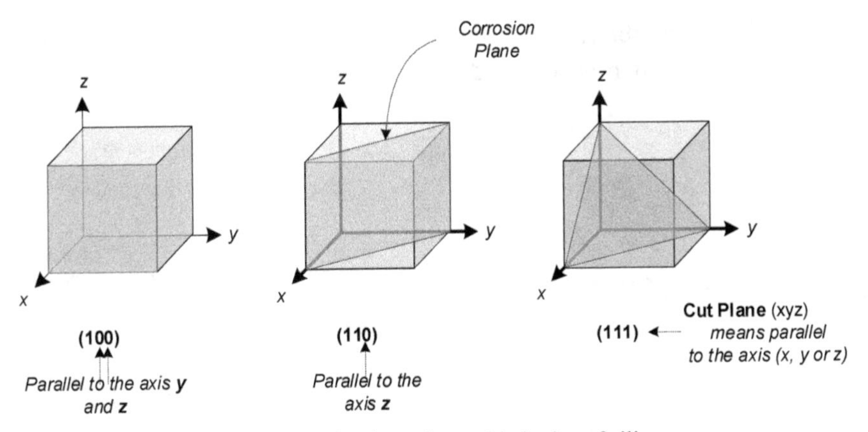

Figure 2.5 – Main planes in a cubic lattice of silicon.

Solutions with different compounds have different corrosion rates depending on the plane. It is possible to create quite sophisticated structures as cavities, slots, beams, bridges, and holes using different combinations of substrate orientation and mask patterns.

In the case of dry corrosion, a series of subtractive methods in which the material surface to corrode is removed by different gases are used. Plasma is often used to increase the rate of corrosion. This corrosion can happen physically by ion bombardment, chemically through a chemical reaction on the surface of the material, or with a combination of the two processes. In general, physical corrosion produces more vertical boundaries than chemical corrosion, which on the other hand, has the advantage of being more selective to the material corroded.

The last step in the lithography process is the removal of photo resistive materials (Figure 2.4(g)). The deposition process, the creation of patterns, and pattern transfer may be repeated several times to create complex structures. In conventional integrated circuits, tens of layers are used, while in MEMS, nowadays, the process does not usually consist of more than five layers.

2.2.6 *Bulk Microfabrication*

The creation of MEMS structures can be made with different lithography-based methods. The most typical methods are bulk microfabrication and superficial microfabrication.

In bulk microfabrication, used since the 60s, is the substrate itself, usually silicon, which is corroded to create the desired structures (Figure 2.6).

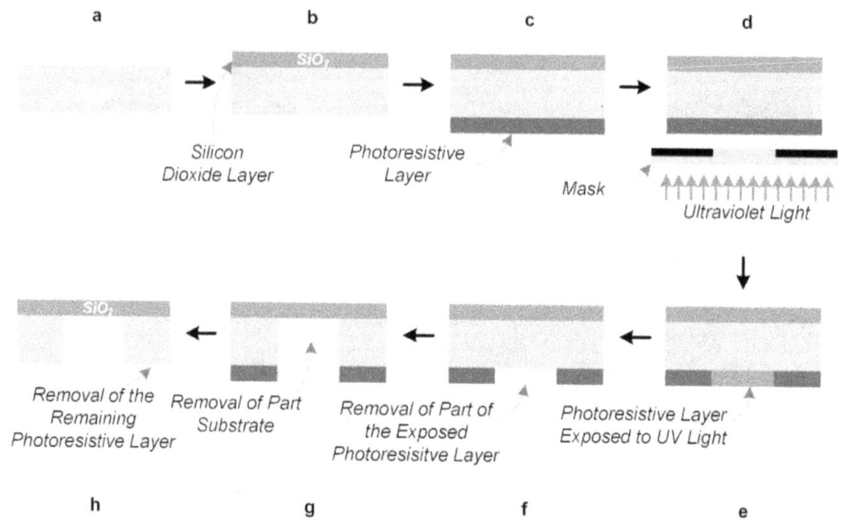

Figure 2.6 – Illustration of bulk microfabrication.

It is possible to create structures with a height less than 1 μm up to the height of the wafer, which can reach 500 μm. Horizontally structures can be created ranging from a few μm to the diameter of a wafer, which may reach 200 mm.

2.2.7 *Superficial Microfabrication*

Since the 80s, the method of superficial microfabrication can also be used. With this method, structures that are an order of magnitude smaller than the density of the bulk microfabrication method can be created. Another advantage of this method is the possibility of easy integration with electronic circuits that can be created on the same wafer.

Unlike bulk microfabrication, where the substrate material is selectively removed to form the structures, on superficial micro-fabrication, these structures are created on top of the substrate by depositing provisional layers, which end up being removed, originating suspended structures. Figure 2.7 illustrates the steps in the superficial microfabrication method. The initial state is, for example, something like the final stage of the process illustrated in Figure 2.4. In that example, a layer of silicon dioxide with the desired pattern (circuit) had been created on top of the silicon substrate. This layer of silicon dioxide (or other material) is temporary because, at the end of the process, it will be removed. It only serves to temporarily support the other upper layers that will be built.

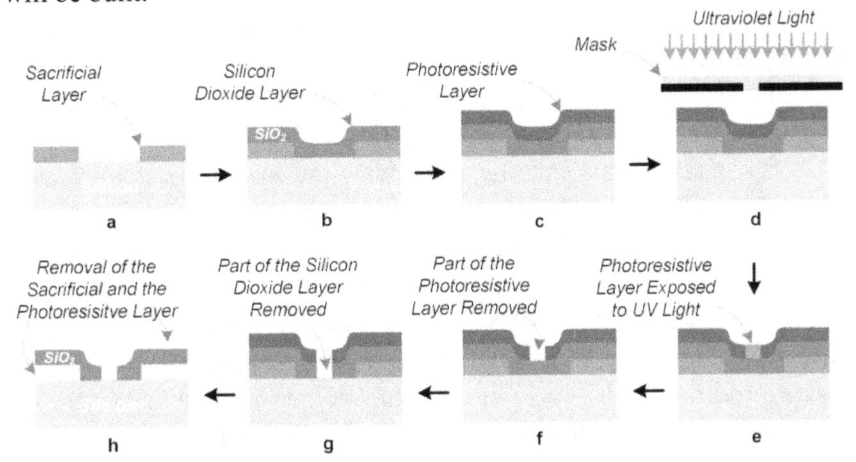

Figure 2.7 – Illustration of the method of superficial microfabrication.

2.3 Application Examples

The advantages that are obtained with MEMS devices have been extended to more and more quantities and allowed their use in more and more applications in all sectors of activity. Particularly relevant are those applications that involve an interface with a user and the commercialization of many units, such as in the case of mobile phones, smartwatches, and automobiles. The low unit cost of manufacturing MEMS-type sensors allows its numerous uses and the consequent increase in the amount of information that can be easily collected from Nature and presented in a simple, attractive, and undoubtedly useful way, to the user.

In the case of automobiles, for example, in addition to the collection of information for presentation to the driver, there is a huge amount of information collected from all parts of the vehicle and which allows for an optimization of its operation and the diagnosis of problems.

One of the quantities that are fundamental in this area is pressure. A common application in modern automobiles is the measurement of air pressure in the tires. Since it is not possible to connect electrical cables between the wheel and the rest of the car, due to its movement, solutions were developed that in addition to the sensor element that measures the pressure, there is also a microcontroller and a radio frequency transmitter that sends the pressure data to the car. This entire system is powered by a battery whose charge level is also monitored.

In the combustion car engine itself, there are several MEMS sensors that measure not only pressure but also flow. These sensors measure the flow of air and gasoline entering and leaving the engine to obtain greater combustion efficiency. The pressure inside each cylinder of the engine is also measured, which allows regulating, for example, the gas injection time and the recirculation of the exhaust air.

Another area where MEMS sensors are used in automobiles is security. The best known is the accelerometer that is used to trigger the driver and passenger airbag in the event of a crash. Another example is the use of gyroscopes to trigger side airbags when they detect when a rollover occurs or even before that rollover starts. These sensors can even be used in the dynamic control of the car, automatically reducing its speed in curves, for example.

Yet another area where continuous monitoring is important is in terms of the gases released. The most important are carbon monoxide (CO), carbon dioxide (CO_2), and nitrous oxides (NO, NO_2 and N_2O). It is important to measure and limit the production of these gases due to the negative effect they have on the planet and on humans. Carbon monoxide, for example, binds with the iron in hemoglobin and affects the absorption of oxygen in our lungs. The carbon dioxide absorbs infrared radiation, which leads to a change in the planet's temperature. In humans, it also causes damage leading to rapid breathing, increased heart rate, dizziness, etc. Nitrous oxide creates smog that causes eye irritation, for example.

There are MEMS-type sensors that help us monitor our health status. Among them, we can highlight, for example, the pedometer that allows the number of steps taken, the speed of travel, the distance covered, and, indirectly, the calories expended. Another example is that of a contact lens for placement in the eye, which allows early diagnosis of glaucoma by measuring deformations in the cornea caused by variations in intraocular pressure.

The most prevalent consumer device today is the mobile phone. It works as an interpersonal communication device, as a source of information, and as an object of entertainment. There is no application that has so strongly benefited from the miniaturization of sensors and actuators. Some examples of sensors are:

- **Accelerometer** — Measures the acceleration to which the mobile phone is subjected. The original function was to measure the acceleration of gravity to orient the image on the screen when the phone was oriented differently. Other applications have emerged, such as measuring the number of steps taken or estimating speed.
- **Gyroscope** — Measures the angular speed of the phone. This information allows the accelerometer to help determine the orientation of the mobile phone. It can also be used in games where the mobile phone simulates the steering wheel of a vehicle.
- **Magnetometer** — Determines the magnetic field around the phone. The determination of the direction of the earth's magnetic field allows the real-time orientation of the maps presented in navigation applications. This sensor is also used in applications that make it possible to detect the presence of metals as they disturb the magnetic field around them.
- **Global Positioning System (GPS)** — The mobile phone receives information from several satellites in orbit around the Earth that allows the estimation of the position of the mobile phone on the Earth's surface. This is used in navigation applications.
- **Barometer** — Measures atmospheric pressure and can be used to determine the altitude at which the device is located as well as providing information used for weather forecasting.
- **Proximity Sensor** — Located next to the microphone, it allows you to determine when the phone is close to the ear and turning off the screen, preventing unwanted commands from being executed through the touch panel.
- **Ambient Light Sensor** — Determines the amount of ambient light and is used to decrease the brightness of the phone's screen in low light situations to save battery.

- **Fingerprint Sensor** — Determines the user's fingerprint and compares it with prints stored in a local database to allow access to the mobile phone.
- **Iris Sensor** — Allows identifying the user from the image of the iris in the user's eye.
- **Temperature sensor** — Used to determine the air temperature.
- **Camera** — Used to take pictures.
- **Touch Screen** — Used as an interface between the user and the device.
- **Microphone** — Captures the user's voice for phone calls or the surrounding sound when recording videos.

There are also other external sensors that can be connected to mobile phones and allow you to measure, for example, airspeed or blood sugar level.

The number of MEMS sensors currently in existence is truly endless. Every day more are created, making our lives easier through the features offered and the low price that allows almost universal access.

2.4 Questions

1. What does the acronym MEMS mean?

2. Name four advantages of micro and nanosensors?

3. Why is silicon used to build MEMS sensors?

4. How are integrated circuits built using lithography?

5. How does the bulk microfabrication technique work in the construction of MEMS sensors?

6. How does the superficial microfabrication technique work in the construction of MEMS sensors?

7. What is the chemical vapor deposition of polysilicon on a silicon substrate?

8. There are two ways to create a layer of silicon dioxide on a silicon substrate: by thermal oxidation and by chemical vapor deposition. What is the difference between these two techniques?

9. What is the technique of creating MEMS in which several layers are deposited on a substrate, and subsequently, some of the lower layers are completely eliminated when making suspended elements?

   a. Bulk micromachining.
   b. Surface micromachining.
   c. Lithograph.
   d. Milling.
   e. None of the above.

Chapter 3

Devices Based on the Electric Field

3.1 Introduction

Electricity is nothing more than the use of the electron. This small particle has led to a technological development never seen in human history. In engineering, for example, it allows Man to solve many problems: How to inform a driver in real-time about the state of the brakes on your car? How to make the image on a mobile phone, PDA, or camera display stay always oriented correctly relative to the user even when the device is rotated? These two problems and many more are solved based on the concept of capacity, which is directly related to the accumulation of electrons. The same concept is also fundamental in the creation of motion within an integrated circuit, allowing, for example, commanding micro-mirrors to switch light signals used in optical fiber communications.

3.2 Force, Electric Field, and Voltage

All matter is composed of atoms which, in turn, are composed of protons, neutrons, and electrons. These particles have a property that we call electric charge. Nowadays, there is not yet any theory that explains what the electric charge at a fundamental level is. There is no theory that can explain "where it comes from," why there are particles that have an electric charge and others do not and why the value of the electric charge of fundamental particles (electrons, quarks, etc.) has the value that it has. What we know is that the charge of a certain type of particle is the same for all particles of that type. The charge of an electron, for example, is 1.602×10^{-19} C (coulomb), whatever the electron. In fact, there is no property that distinguishes one electron from another electron. Actually, the current consensus is that all the electrons are excitation in one and the same field (the electron field).

Another fact that we also know about the electric charge is that it determines the interaction of a particle with other particles that also have this property. This interaction takes the form of a force because it can change the movement of the particles involved (Figure 3.1).

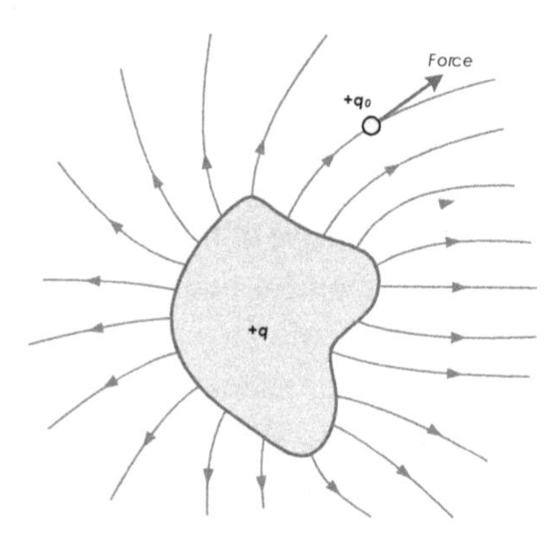

Figure 3.1 – Illustration of the force (*f*) that acts on a particle with electric charge ($+q_0$) when immersed in an electric field created by another electric charge ($+q$).

The force at which a particle with electric charge is subject when in proximity to another electrical charge is dependent on the relative position of the two particles. The Coulomb Law allows the calculation of this force using the charge of the two particles (q and q_0) and the distance between them (r):

$$f = \frac{1}{2\pi\epsilon_0} \times \frac{q \cdot q_0}{r^2}, \tag{3.1}$$

where ϵ_0 is the dielectric constant of vacuum ($8{,}854\ 187\ 817 \times 10^{-12}\ \mathrm{Fm}^{-1}$). If the two electric charges have the same algebraic sign, the force is repulsive, and if they have opposite signs, it is attractive.

To conceptualize this phenomenon, we use the concept of Electric Field — "Field" because it is something that is spatially distributed and "Electric" because it depends on the electrical charge of the particles. Mathematically we define the electric field as the ratio of the force that is subjected to a particle immersed in this field and its electric charge (the Lorentz Force Law):

$$\vec{E} = \frac{\vec{F}}{q_0}. \tag{3.2}$$

Imagine a small particle at point A immersed in an electric field created by another particle at B with a charge of the opposite sign. As noted, this particle will be under the action of a force that "pulls" the particle in the direction of B. If we want to move this particle to another position away from A, it is necessary to carry out work to overcome this force. In other words, it takes energy to move the particle at A. The work/energy that is necessary depends on the distance that you want to move the particle – the greater the distance, the greater the work. As was done in the previous paragraph, in which the concept of the electric field was introduced from the force to which a particle is subject, we can also introduce the concept of Electric Potential (V) from the amount of work (W_{AB}) that it is necessary to perform to move a particle between two points (A and B):

$$V_B - V_A = -\frac{W_{AB}}{q_0}. \tag{3.3}$$

The electrical potential at every point in space depends on the value of the electric field at that point. We "voltage" to the potential difference between two points. Mathematically it can be calculated through the electric field using the line integral:

$$V_B - V_A = -\int_{\gamma} \vec{E} \cdot \overrightarrow{d\gamma}. \tag{3.4}$$

Note that, in many situations, the value of the potential difference does not depend on the path you choose.

3.3 Concept of Capacity

If we have two conductive objects, for example, two parallel plates as shown in Figure 3.2, and we put positive charges in one plate and negative charges in the other, they will remain there because they are immersed in a non-conducive environment. We can say there is an electric field in the space between the plates. Between the two plates, we can say that there is a potential difference (V) since, to move a test charge from the right plate to the left plate, we have to do work, given by (8).

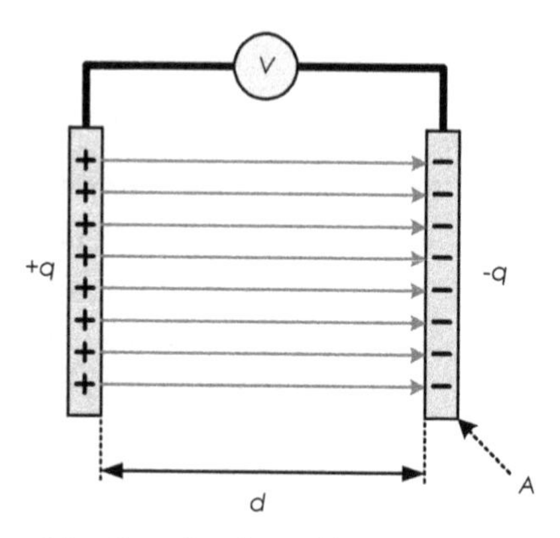

Figure 3.2 – Capacitor formed by two parallel plates.

The relationship between the charge on each plate (q) and the potential difference that exists between them (V) is called **Capacity** (C):

$$C = \frac{q}{V}. \tag{3.5}$$

The value of the capacity of this structure, which we call a capacitor, depends on its dimensions and shape. In the case of the geometry of Figure 3.2, the capacity is given by

$$C = \epsilon_0 \frac{A}{d}. \tag{3.6}$$

This expression is valid if the plates are in a vacuum (and are infinitely large). If there is a material between them (dielectric), this capacity is greater:

$$C = \epsilon_0 \epsilon_r \frac{A}{d}. \tag{3.7}$$

The constant ϵ_r is the relative dielectric constant (in vacuum) of the material between the capacitor plates. This increased capacity is due to the molecular polarization of the dielectric. In certain materials, such as water, the molecules have a permanent dipole, while in others, it required an electric field to create the dipoles. In any case, the molecules of the material tend to align with the direction of the electric field — Dielectric Polarization — as illustrated in Figure 3.3.

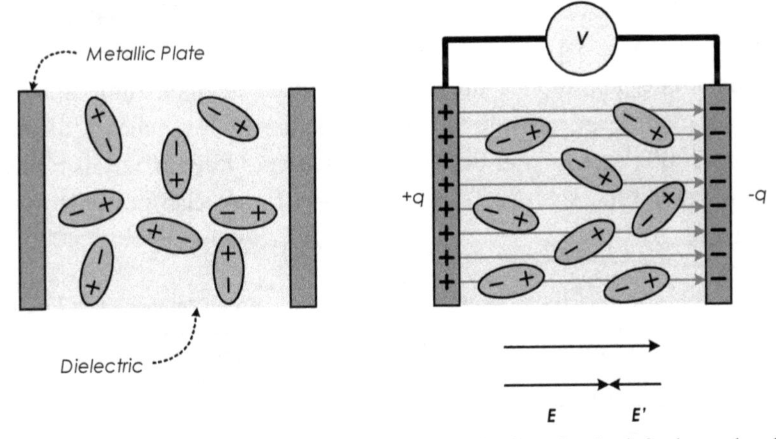

Figure 3.3 – Illustration of the effect of dielectric polarization. On the left, the molecules are randomly oriented. On the right, the molecules are aligned (imperfectly) according to the applied electric field.

In the left figure, the molecules are randomly aligned, while in the right figure, they are more or less aligned according to the applied electric field (E_0) caused by the charges present on the capacitor plates. Note that, due to thermal agitation and molecular organization of the material, the molecules in the latter case are not perfectly aligned with the field. Nevertheless, the positive side of the dipoles is facing the plate that has a negative electrical charge (right plate). The dipoles having an electric charge can give rise to an electric field. Given that the electric field has the orientation of positive charges to negative charges, the field created by the dipoles of the dielectric (E') is opposed to the field created by the charges in the capacitor (E_0) plates. The resulting field (E) is, therefore, smaller than the field that we would have if there were no dielectric (E_0) i.e., if the plates were in a vacuum. Weaker electric fields lead to a lower voltage between the plates. The voltage that was V_0 in vacuum becomes $V = V_0/\epsilon_r$. Inserting this into (10) leads to

$$C = \epsilon_r \frac{q}{V_0} = \epsilon_r C_0, \tag{3.8}$$

where C_0 is the capacity value in a capacitor without the dielectric. That is how you get the expression (3.7). In short, the presence of the dielectric decreases the voltage for the same charge. It means that, with a dielectric, it is possible to have more charge with the same voltage, that is, greater capacity. In Table 3.1, we can read the value of the dielectric constant of different materials. Notice the compound used in thick-film capacitors, which is made especially for capacitors and which reaches a dielectric constant 5000 times greater than that of vacuum. Other materials typically used in capacitors, such as ceramics (14–110), are also presented in the same table.

Table 3.1 – Dielectric constant of different materials at room temperature (25°C).

MATERIAL	DIELECTRIC CONSTANT
Air	1.00054
Acrylic	2.5 to 2.9
Ceramic	14 to 110
Dielectric for Capacitors	300 to 5000
Diamond	5.5
Paper	3.5
Porcelain	6.5
Silicon Rubber	3.2
Water	78.5

The dielectric constant of a material depends on the temperature and operating frequency. The thermal agitation of the dielectric molecules due to an increased temperature will affect the alignment of their dipoles. The dependence on frequency is since the dipoles must reorient themselves in each period of the applied voltage because the applied electric field inverts its direction. Different crystalline structures of the dielectric make that reorientation easier or more difficult. Ideally, to make a capacitor, a dielectric whose dielectric constant does not vary much with frequency, as is the case with polyethylene, is desirable.

The capacitors can have many different forms. The most typical shapes are flat plates, as shown in Figure 3.2, and whose capacity is given by (3.7) and the circular plates as illustrated in Figure 3.4 have a capacity given by

$$C = 2\pi\epsilon_0\epsilon_r \frac{l}{ln(b/a)}, \tag{3.9}$$

where l is the length of the capacitor plates, a is the inner plate radius, and b is the outer plate radius.

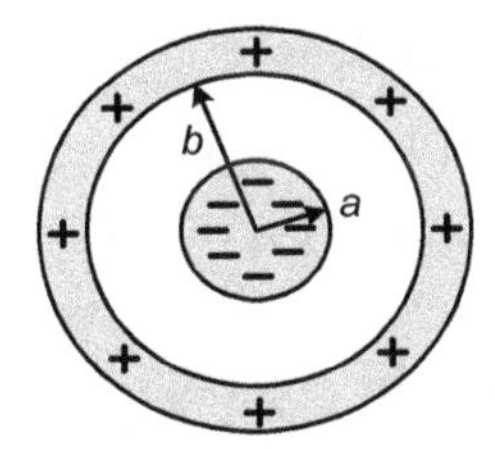

Figure 3.4 – Illustration of a cylindrical capacitor.

In a "good" capacitor, the dielectric constant and the geometry must be stable. Ideally, they should not vary with temperature, humidity, pressure, or any other environmental factor. "Good" capacitors are fundamental components of electronic circuits. However, to create sensors, it is desirable to have a "bad" capacitor, i.e., a capacitor whose capacity varies considerably with the physical parameter we are interested in measuring.

In order to build a sensor that is based on the capacitive effect, it is only necessary to look at the analytical expression that determines your capacity. In the case of the plane capacitor, whose capacity is given by (3.7), it is possible to build a sensor if the physical quantity to measure can cause a change in:

- the value of the area of the plates (A).
- the value of the distance between them (d).
- the value of the dielectric constant of the dielectric (ϵ).

Note that the expression of the capacity of a capacitor is an approximation used for condensers with infinite plates. In the real case where the length of plates of one flat conductor is w, the exact expression of the capacity is

$$C = \epsilon\frac{A}{d} + \frac{2w}{\pi}\epsilon\ln\left(\frac{\pi w}{d}\right). \tag{3.10}$$

The second term leads, in the final sensor, to a nonlinearity that may, however, be greatly reduced using guard electrodes that completely involve the capacitor plates, as will be seen later.

In the following, some examples of sensors that used the capacitive effect are addressed.

3.4 Capacitive Displacement Sensor

Automobiles increasingly allow the driver to control different aspects of cars for greater comfort and safety of the passengers. Examples are the activation of the airbags in the event of a collision, seat heating, or mirror adjustment. Another aspect is the amount of information that is available to the driver, from the water temperature and engine oil to the consumption of gasoline and tire pressure.

In line with these advances in terms of available information, there is a desire to monitor aspects related to the mechanics of the automobile, such as the state of wear of the brake discs. For that purpose, the system requires a sensor capable of measuring the thickness of a brake disc that can be permanently mounted in the car so you can make continuous monitoring of the status of the disks. The brake discs wear out due to the friction with the brake pads when slowing down the car. It must be a sensor that can, for example, measure the distance from a fixed point to the surface of the disk. The greater the wear, the greater is the distance. Given that the sensor is supposed to be fixed permanently to the car, it must be able to measure this distance even with the disc moving. What solutions exist, and which could be used?

- The sensors of resistive displacement, wherein the displacement is determined by the position of a cursor potentiometer, are not adequate because there can be no contact between the sensor and the disc.
- The inductive displacement sensors in which displacement is measured by varying the thickness of an air gap of a magnetic circuit could be used if the disk brake would itself be part of the magnetic circuit. This, however, would only work when the disk is made of iron and thus would not work in the case of ceramic discs because they are not ferromagnetic materials.
- The optical displacement sensors that work by measuring the time of flight of a light beam emitted by a LED would not be the best choice because the surface of a disc is not a very good reflector. Another problem with this type of solution has to do with the fact that this type of sensor is likely to be affected by dirt which is a very real possibility in the case of an automobile wheel.

- The ultrasonic displacement sensors, which work by measuring the time of flight of an acoustic wave, are also not suitable due to noise and multiple reflections present.
- The best solution for this application would be the use of a capacitive sensor. This type of sensor discussed in detail below works as a capacitor in which one of the plates is the brake disc itself. This sensor can be used because the brake disc is conductive. We can thus build a non-contact sensor that is immune to this dirt in the measurement environment.

Let us now look at a displacement sensor that uses the capacity variation with the distance between the plates of a flat capacitor.

Keeping one of the plates fixed and associating the other plate to the object whose displacement is to be measured, we can have a capacity whose distance between plates varies with displacement. This operating principle is illustrated in Figure 3.5.

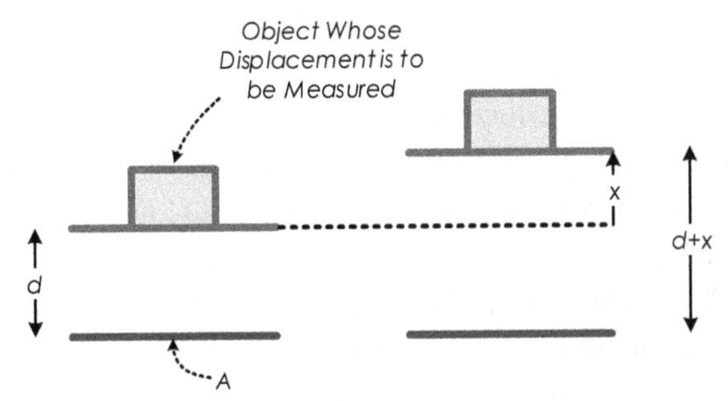

Figure 3.5 – Illustration of the operating principle used by a capacitive sensor which translated the displacement to measure into a change in distance between the two plates.

In this example, the capacity is

$$C = \epsilon_0 \epsilon_r \frac{A}{d+x}, \tag{3.11}$$

where d was replaced by $d + x$ where x is the displacement to be measured. The area of the plates, A, is equal to the product of its width l_1 by its length l_2.

This sensor is a contactless displacement sensor that measures the displacement of a conductive body. The body itself is the movable plate of the flat capacitor. The dielectric of this sensor is air. In Figure 3.6, there is an illustration of the constitution of the sensor. Note that besides the fixed plate of the plane condenser (measuring electrode), there is a guard ring electrode which helps to reduce the effect of the boundary of the condenser plate.

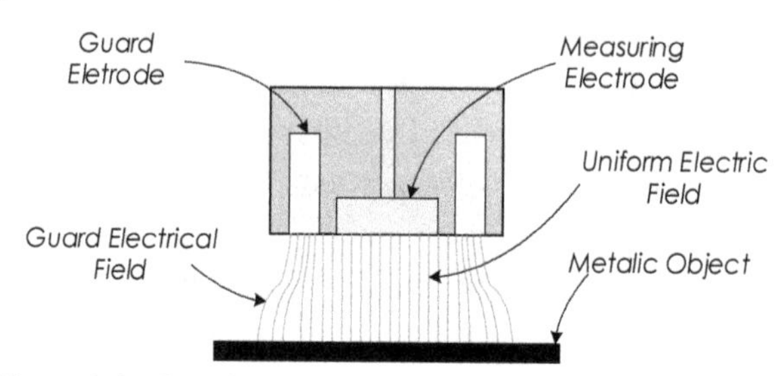

Figure 3.6 – Constitution of the capacitive displacement sensor.

From the concept of the sensor, shown in Figure 3.5 and supported by Eq. (3.11), to the construction of a functional product, there are other aspects to consider. One of them is the fact that a capacitor alone does not produce an electrical signal. It is necessary to include the sensor into an electronic circuit to produce an electrical signal proportional to the quantity to be measured. Another aspect that must be considered has to do with the linearity of the sensor. Equation (3.11) shows clearly that the relationship between capacity and displacement is not linear. It is, therefore, necessary to design the circuit taking this into account to obtain at the end a signal that varies linearly with displacement. The typical linearity values of existing sensors are 1%.

The solution adopted by Micro-Epsilon in its sensor was to use the electrical property of a capacitor that is its impedance. The impedance relates the voltage and current of a device in a sinusoidal regime. In the case of a capacitor, it is given by

$$\bar{Z} = \frac{\bar{V}}{\bar{I}} = \frac{1}{j\omega C}, \tag{3.12}$$

where ω is the angular frequency. Looking at (3.12) we realize that the module of the impedance is inversely proportional to the capacity value, as is the relationship between capacity and displacement. Inserting (3.11) into (3.12) leads to the following expression for the module of the impedance,

$$Z = \frac{d+x}{\epsilon_0 \epsilon_r \omega A}. \tag{3.13}$$

Note that since the dielectric is air, the relative dielectric constant is almost equal to 1 (see Table 3.1).

The relationship between modulus of the impedance (Z) and displacement (x) is therefore linear. Thus, if a sinusoidal current with amplitude I is applied to the capacitor, you get a sinusoidal voltage drop with amplitude directly proportional to the displacement:

$$v(t) = x \frac{I}{\epsilon_0 \epsilon_r \omega A} \cos(\omega t) + d \frac{I}{\epsilon_0 \epsilon_r \omega A} \cos(\omega t). \tag{3.14}$$

An RMS/DC converter, for example, can be used to determine the amplitude of this sinusoid (Figure 3.7).

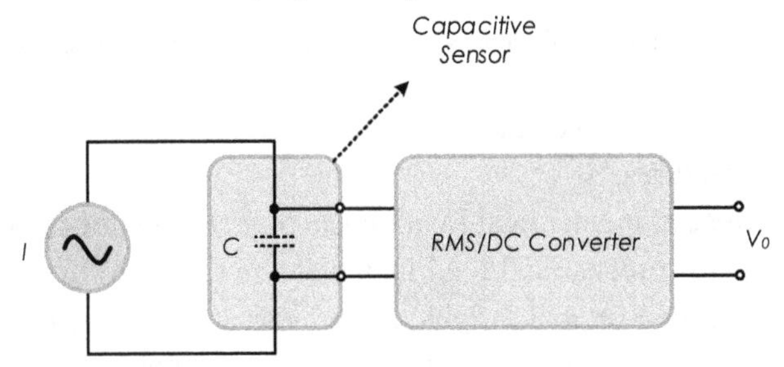

Figure 3.7 – Example of a signal conditioning circuit for a capacitive sensor that used an RMS/DC converter.

The solution adopted is one of many possible. Another solution would be to use an oscillator whose oscillation frequency depends on the value of a capacitor, as is the case of a Wien bridge oscillator, whose output can be connected to a frequency-to-voltage converter (Figure 3.8).

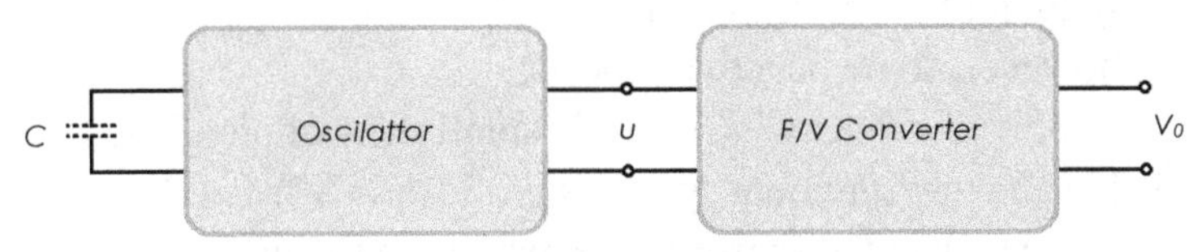

Figure 3.8 – Example of a signal conditioning circuit for a capacitive sensor using an oscillator and an F/V converter.

The voltage output of the oscillator is a sine wave whose frequency depends on the value of the capacitor.

$$v(t) = A\cos(2\pi f_c t). \tag{3.15}$$

The example shown here is only one possible configuration of capacitive displacement sensors. In addition to varying the distance between the plates (Figure 3.9(a)), it is possible to vary its area by sliding one plate relative to the other (Figure 3.9(b)) or vary the dielectric constant sliding the dielectric relative to the capacitor plates (Figure 3.9(c)).

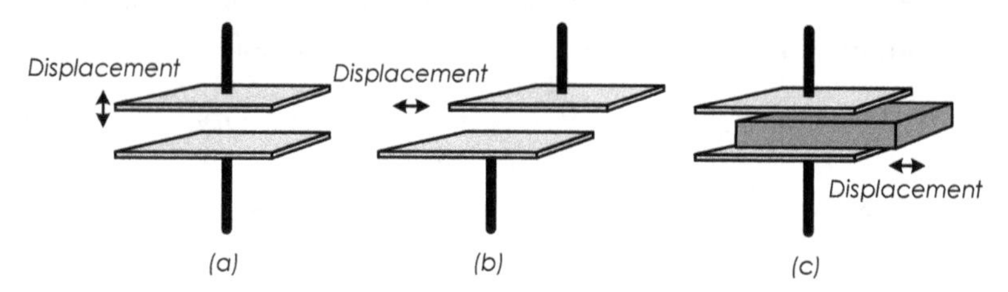

Figure 3.9 – Illustration of three ways to build capacitive displacement sensors.

In general, capacitive displacement sensors have the following advantageous features:

- The effect of mechanical load on the object to be measured is very small. This mechanical load is due to the Coulomb force between the charges of opposite signs that appear at the capacitor plates.
- There is no friction because there is no contact between the capacitor plates and the dielectric.
- There is no hysteresis. The hysteresis typically arises in sensors that use magnetic fields and magnetizable material, which is not the case with capacitive sensors.
- Have good stability and reproducibility.
- Have an excellent resolution of up to picometers.
- Easy to implement in silicon.

It has, however, some negative aspects:

- The value of capacity is generally small (tens of picofarad), which makes the measurement system significantly affected by the cable used to connect the sensor to the rest of the system.
- A capacitive sensor is susceptible to electrical fields that may be present in the sensor area.
- In practice, the analytical expressions used for the capacity do not take into account the fact that the electric field between the plates

does not have parallel field lines at the edges. They are approximate expressions, which leads to a transduction error if appropriate measures with the use of guard ring electrode or the use of calibration and linearization techniques are not used.

- Capacitive sensors with air dielectric are especially sensitive to moisture because of the effect it has on the value of the dielectric constant of air.

3.5 Capacitive Acceleration Sensor

On the market, there are also capacitive acceleration sensors. They work intrinsically as displacement sensors in which the displacement is caused by the acceleration acting on a body that is free to move in the direction of the acceleration to be measured. Figure 3.10 illustrates the principle of operation of such sensors.

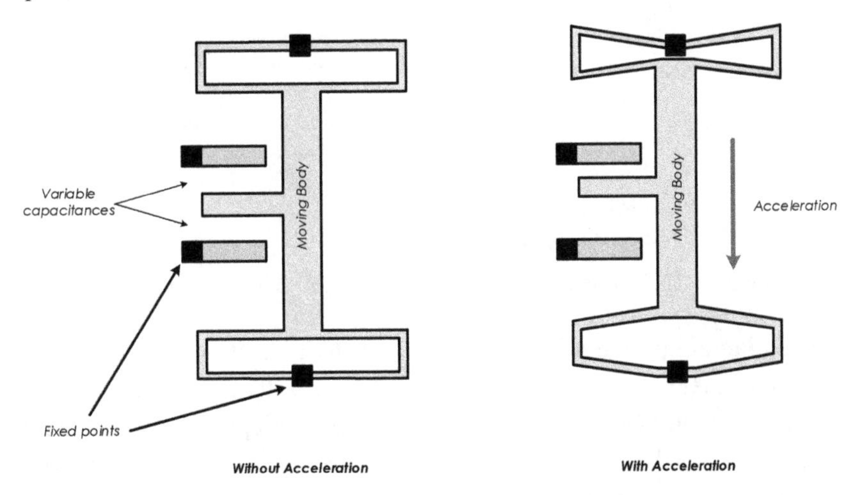

Figure 3.10 – Illustration of the operation of a capacitive acceleration sensor. It was extracted from the datasheet of sensor ADXL150 from Analog Devices.

If m is the mass that is free to move and a is *the* acceleration in the direction of movement allowed, the force exerted on the mass is (Newton's Second Law of Motion),

$$F = ma. \tag{3.16}$$

The mass, which is secured by springs on both ends, suffers a displacement (x) which depends on the deflection constant of the spring (k):

$$x = \frac{F}{k}. \tag{3.17}$$

Inserting (3.17) into (3.16) leads to

$$x = \frac{m}{k}a. \tag{3.18}$$

We obtain a displacement that is proportional to the acceleration imposed on the sensor. The rest of the operation of the sensor is just like a displacement sensor, where in practice, the displacement is very small (no more than a few μm).

There is a plate attached to the moving body, which, together with the other two, forms two capacitors connected in series, as illustrated in Figure 3.11.

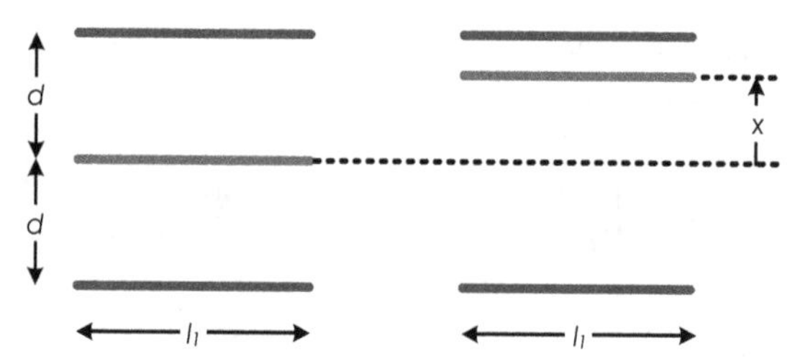

Figure 3.11 – Illustration of a differential assembly of plate capacitors.

The diagram on the left side corresponds to a situation where there is no displacement, so we have two capacitors with the same distance between the plates (*d*). On the right side, there is an upward displacement by *x*, which leads to a different distance between the plates of both capacitors. The upper capacitor (C_1) has a distance between plates of $d - x$ and the lower capacitor (C_2) has a distance of $d + x$. The capacities depend therefore on the displacement x as follows:

$$C_1 = \frac{\epsilon_0 A}{d-x} \text{ and } C_2 = \frac{\epsilon_0 A}{d+x}. \tag{3.19}$$

This configuration is implemented entirely within an integrated circuit with approximately 10 mm per side. This is achieved by building it as a micro-electromechanical device (MEMS).

A Wheatstone bridge is used to produce a voltage proportional to the displacement, as illustrated in Figure 3.12. This bridge is fed with a sinusoidal voltage with a certain frequency (100 kHz in the case of the Analog Devices ADXL150 sensor).

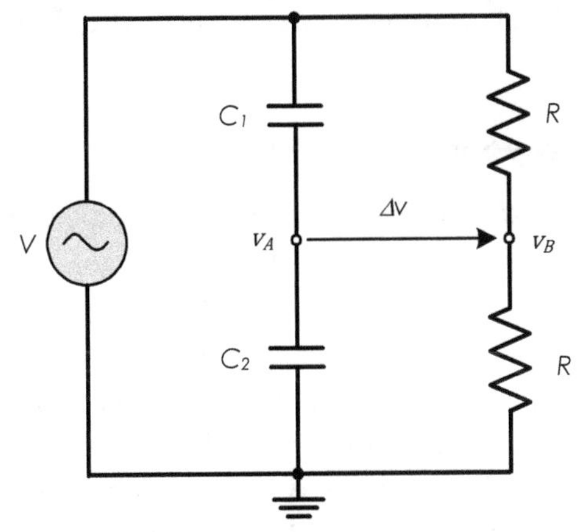

Figure 3.12 – Part of the electronic circuit used to produce the output voltage of the capacitive sensor.

If the supply voltage is V the voltage drop across the capacitor C_2 is

$$v_A = \left(\frac{Y_1}{Y_1+Y_2}\right)V = \frac{C_1}{C_1+C_2}V. \tag{3.20}$$

The voltage at B is $V/2$. Thus, the difference between the two voltages is

$$\Delta v = v_A - v_B = \frac{C_1}{C_1+C_2}V - \frac{V}{2} = \frac{1}{2}\left(\frac{2C_1}{C_1+C_2} - \frac{C_1+C_2}{C_1+C_2}\right)V = \frac{1}{2}\left(\frac{C_1-C_2}{C_1+C_2}\right)V. \tag{3.21}$$

Introducing (24) we have

$$\Delta v = \frac{v}{2}\frac{\epsilon_0 A\left(\frac{1}{d-x}-\frac{1}{d+x}\right)}{\epsilon_0 A\left(\frac{1}{d-x}+\frac{1}{d+x}\right)} = \frac{v}{2}\frac{\epsilon_0 A(d+x-d+x)}{\epsilon_0 A(d+x+d-x)} = \frac{v}{2d}x. \tag{3.22}$$

This voltage is directly proportional to the displacement x and, therefore, to the acceleration.

The biggest advantages of capacitive accelerometers are

- the ability to measure low accelerations (less than 2 g),
- work at low frequencies (to DC) of acceleration,

- be able to withstand large shocks (typically more than 5000 g).

Some of the disadvantages of this type of accelerometer are

- does not work with high-frequency acceleration,
- a large phase delay,
- high noise levels when compared with a piezoelectric acceleration sensor.

3.6 Angular Velocity Sensor (Gyroscope)

A gyroscope is a sensor that measures angular velocity (in radians per second). This is a three-dimensional vector that contains information about the axis of rotation as well as the speed of rotation.

There are two types of gyroscopes: rotary, which is based on conservation of angular momentum, and vibratory, which are based on conservation of linear momentum. The first ones have a mass in continuous rotation (a wheel) whose axis of rotation is independent of the orientation of the sensor. The second has a mass in continuous vibration whose direction of vibration is independent of the orientation of the sensor. This second type is the most commonly found in consumer electronic devices because they can be built using integrated circuit manufacturing techniques (MEMS) with the consequent decrease in production costs. It is this second type that will now be covered in detail.

A vibrating gyroscope has one (or more) test masses that are vibrated by an actuator, as shown in Figure 3.13. The test mass thus performs a periodic linear movement.

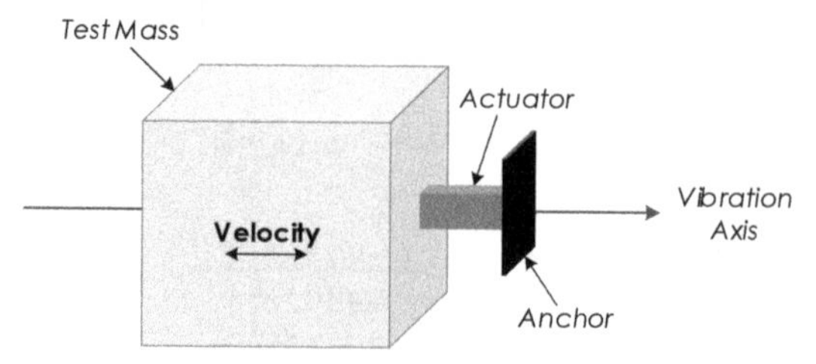

Figure 3.13 – Illustration of the constitution of a vibrating gyroscope with a single test mass.

In practice, two test masses are used instead of one to double the sensitivity and make the response more linear.

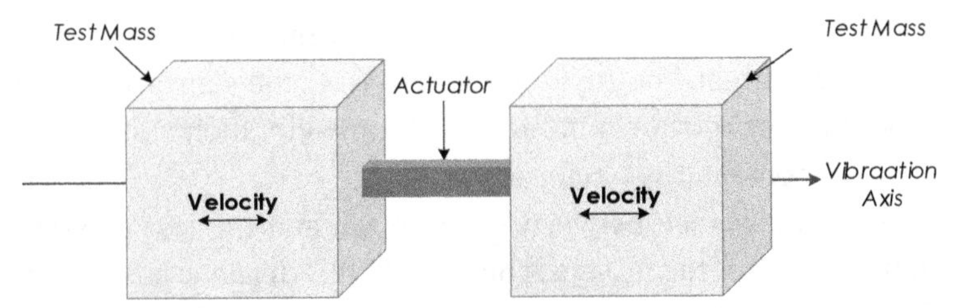

Figure 3.14 – Illustration of the constitution of a vibrating gyroscope with two test masses.

Figure 3.15 also shows part of the sensor frame. Between the test masses and the frame, there are springs that serve to balance the fictitious force that appears when the sensor rotates, as shown below.

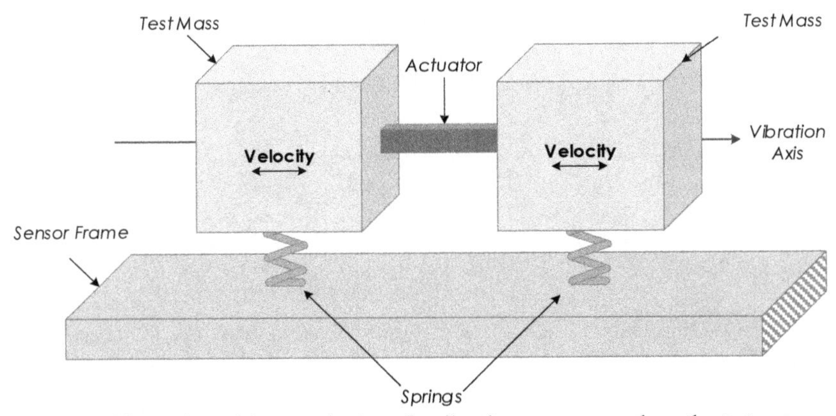

Figure 3.15 – Illustration of the constitution of a vibrating gyroscope where the test masses and part of the sensor frame are visible.

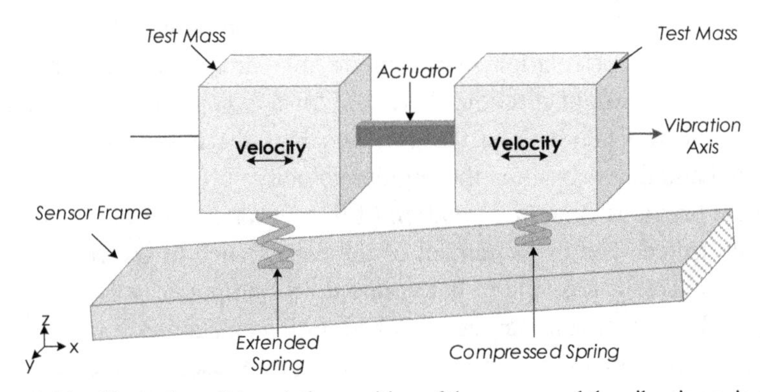

Figure 3.16 – Illustration of the relative position of the sensor and the vibration axis of the test mass when the sensor rotates about the y-axis.

Inertia determines that the test masses keep their movement always along the same straight line (axis of vibration). If the sensor rotates in a different orientation, as shown in Figure 3.16, the vibration axis of the test mass remains (momentarily) in the same place.

As noted, the distance between the package and the test masses has changed. In the case of the test mass on the left, this distance has increased, and in the case of the test mass on the right, this distance decreases. These distances are measured indirectly by measuring the capacities of the flat capacitors whose armatures are attached to the frame and the test masses as is done in the accelerometer already covered.

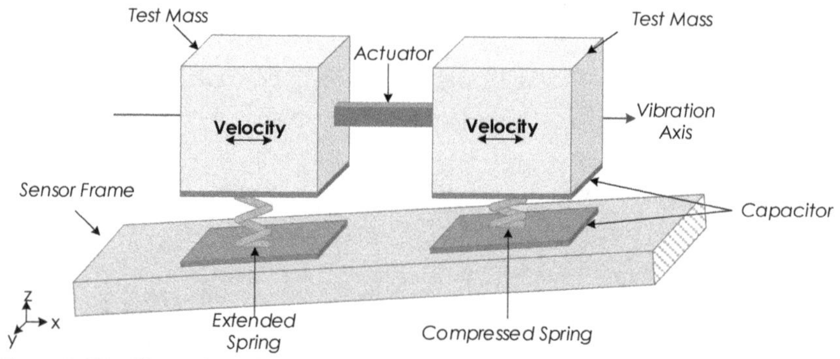

Figure 3.17 – Illustration of flat capacitors formed between the sensor frame and the test masses.

If the sensor rotates a certain amount and then stops, the test masses, due to the springs that connect them to the package, will reposition themselves parallel to the package as they were initially. This sensor is not, however, an orientation sensor but an angular velocity sensor. If the package is in constant rotation, there will always be a different distance between the test masses and the package. It is these distances that are measured and that determine the angular velocity.

In the previous figures, a rotation of the sensor around the y axis has been illustrated. The measurement of the capacitance of the represented capacitor makes it possible to determine the y component of the angular velocity. Two other test masses, with a vibration axis along y and another pair of capacitors (and restitution springs), are used to measure the component according to x.

For the measurement of the component according to z of the angular velocity, the same test masses are used as in the case of the measurement of the component according to y (masses to vibrate along the x-axis), but the displacement between the test masses and the frame occurs along the y-axis and is measured with capacitors whose armatures are perpendicular to that axis as shown in Figure 3.18.

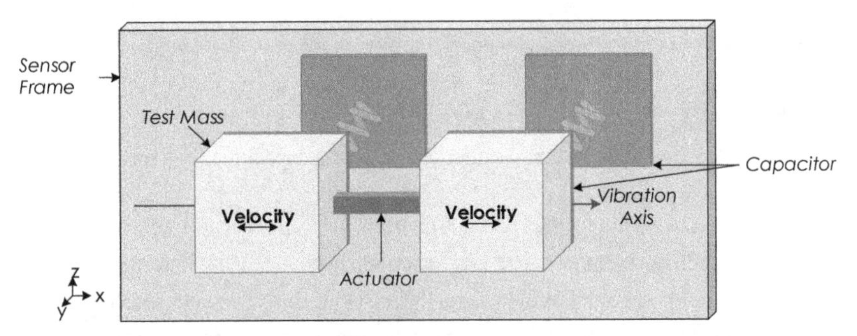

Figure 3.18 – Illustration of the flat capacitors formed between the frame and the test masses to measure the z component of the angular velocity.

There are sensors in the market with ranges from ±250°/s, ±500°/s and ±2000°/s with measurement frequencies from 100 Hz to 800 Hz. Its size is extremely small, with 4 mm on the side and 1.1 mm in height.

3.7 Capacitive Fingerprint Sensor

A fingerprint sensor allows you to recognize the fingerprint of a human finger. There are several techniques, including the use of light, the use of ultrasound, and the use of electrical capacity. The latter will now be covered. The sensor has a two-dimensional array of capacitors whose dielectric is formed partly by the finger and partly by air. This creates a 2D "image" of the dielectric constant of the material leaning against the sensor surface.

Figure 3.19 shows a simplified diagram of the sensor's electronic circuit. There are only six capacitor plates and an example of the circuit used to produce an electrical voltage that depends on the dielectric constant of the condenser and which can be inferred from the presence of air or a finger near the sensor surface.

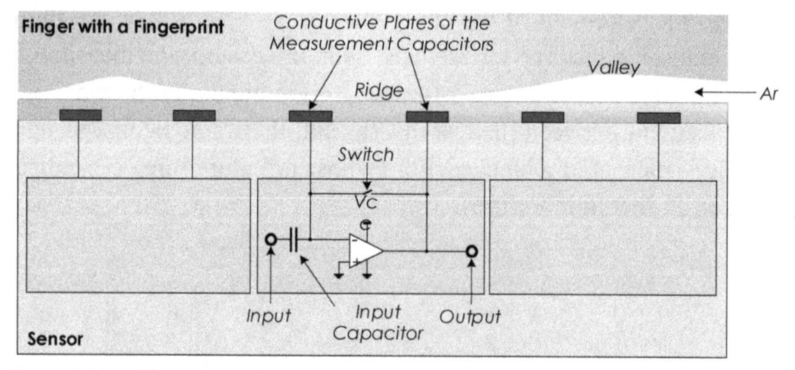

Figure 3.19 – Illustration of the electronic circuit of a capacitive fingerprint sensor.

The size and spacing between the plates of the capacitors are smaller than the spacing between ridges and valleys of the fingerprint. It is common to have sensors with more than 300 capacitors per inch.

Each electronic circuit has a switch that allows the capacitors to be discharged at the end of the measurement to be ready for the next measurement. The output of each circuit is stored and forms an image.

3.8 Electrostatic Loudspeaker

A loudspeaker is an actuator that converts an electrical signal into sound waves. It is the opposite of a microphone. An electrostatic speaker uses the fact that equal charges repel, and different charges attract. It consists of a diaphragm between two conductive grids separated by a small air layer (Figure 3.20).

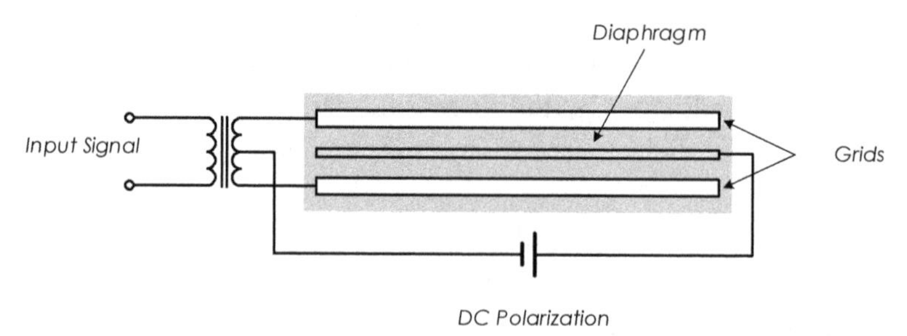

Figure 3.20 – Illustration of an electrostatic speaker. The vibration of the diaphragm causes the sound waves.

The diaphragm is made of a polyester film having a thickness between 2 and 20 μm. This film is an electrical conductor because it must receive the electrical charges that are going to be attracted or repelled by the electrical charge present in the grids and controlled by the signal applied to the loudspeaker. The electrical charges of the diaphragm must therefore exist but cannot change rapidly over time (compared to the audio signal). For this to happen, the diaphragm must be made with a material of high resistivity so that it is almost an insulator.

The diaphragm is kept at a constant DC potential of several kV relative to the grid. An audio signal is applied to the grids. This signal will cause an electric field between the grids that will attract the diaphragm towards one or the other grid, depending on the polarity of the audio signal. The movement of the diaphragm causes pressure variations of the surrounding air creating sound waves. Note that the signal applied to the grids has to be high voltage, and therefore a transformer is used.

The main advantage of an electrostatic loudspeaker is its excellent frequency response because of the low weight of the diaphragm. The main drawbacks have to do with a limited capacity to produce low-frequency sounds and their sensitivity to the relative humidity of the air.

3.9 Electrostatic MEMS Actuator

An electrostatic actuator uses the electric field to create a force between two conductive parts: a fixed and a moving one. It is possible to create electrostatic actuators within an integrated circuit made of silicon.

The application of a voltage between two electrodes causes the attraction of opposite charges due to the Coulomb force (Figure 3.21).

The value of the exerted force can be calculated, as with any mechanical system, by varying the energy stored in the system to displacement caused by the force

$$F = \frac{\partial W}{\partial x}, \tag{3.23}$$

where x is the direction in which the comb is free to move. Energy in a condenser is given by

$$W = \frac{1}{2} C V^2. \tag{3.24}$$

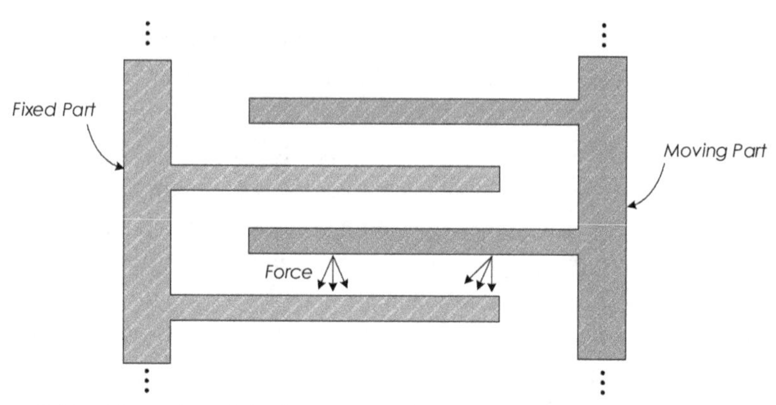

Figure 3.21 – Illustration of the force exerted by the "stator" on the "rotor" of a comb actuator.

The force is therefore proportional to the square of the voltage and to the derivative of the capacity in relation to x,

$$F = \frac{1}{2}\frac{\partial C}{\partial x}V^2. \tag{3.25}$$

Given the expression for the capacity of a flat capacitor as a function of its geometry, Eq. (3.7) we have in the case of a comb actuator, like the one illustrated in Figure 3.21, a force

$$F \approx n\epsilon_0 \frac{h}{g}V^2, \tag{3.26}$$

where n is the number of teeth of the comb, this is an approximate value because the expression (3.7) is only accurate in the case of a plate capacitor with an infinite area.

The force that acts upon each tooth of the comb with a height (h) equal to its spacing (g) for a 15 V voltage is 10^{-9} N.

It is found that the smaller the distance between the teeth, the greater the force. On the other hand, the force is proportional to the square of the applied voltage between the fixed comb and the movable comb. Note that the force does not depend on the displacement x; that is, the force is always the same throughout the travel of the movable comb.

In the case of the "gap-closing" actuator (Figure 3.21), where the displacement of one plate is in the direction of the other plate, the force is given by

$$F \approx \epsilon_0 \frac{A}{2x^2} V^2. \tag{3.27}$$

As noted, the force, in this case, is not the same throughout the path of the plate — the closer they are, the greater the force is.

3.10 Questions

1. Is a capacitive displacement sensor based on the variation in distance between the plates linear? Justify.

2. Determine the relationship between the variation in the sensor capacity shown in the figure and the displacement. The sensor consists of 3 distinct parts: a flat plate that moves, a cylindrical tube open at both ends, and a cylindrical tube coaxial with the first open only at one end.

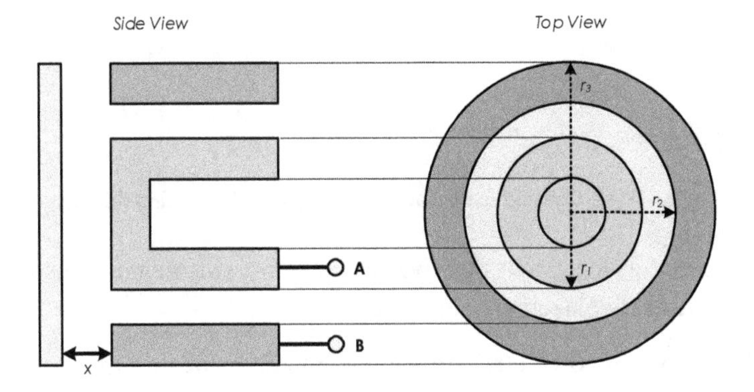

Figure 3.22 – Cylindrical capacitive sensor.

3. Consider a capacitive displacement sensor with a variable dielectric constant. Obtain the expression of the capacity as a function of the x-displacement. Is this function linear? Help: Look at this geometry as being made up of two capacitors in parallel.

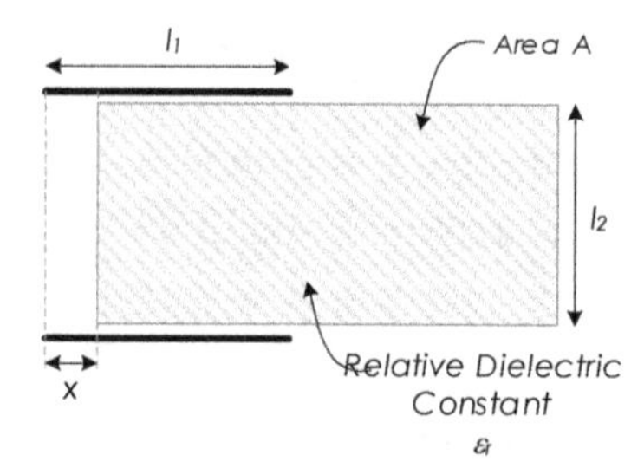

Figure 3.23 – Capacitive displacement sensor with variable dielectric.

4. Consider the capacitive displacement sensor with the variable area. Obtain the expression of the capacity as a function of the x-displacement.

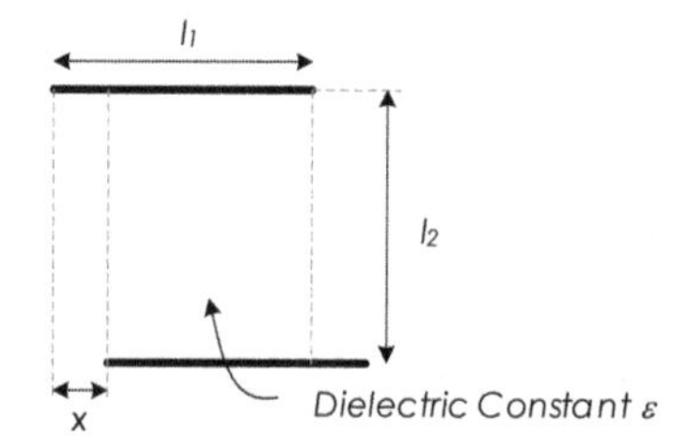

Figure 3.24 – Capacitive displacement sensor with the variable area.

5. Point out at least two positive aspects and two negative aspects of capacitive displacement sensors.
6. What does the dielectric constant of material mean in the context of a capacitor?
7. How can you build a displacement sensor based on the capacity phenomenon?
8. Give three examples of signal conditioning circuits capable of producing a continuous voltage dependent on the capacity value of a sensor element.
9. Explain how a capacitive acceleration sensor works.
10. Point out the advantages and disadvantages of capacitive accelerometers.
11. Explain how an electrostatic loudspeaker works.

Chapter 4

Devices Based on Electrical Resistance

4.1 Introduction

The movement of electrons within a material is also used to make sensors that solve daily problems like knowing the temperature of the air or ensuring our safety by monitoring the bridges and the buildings where we live and work with the use of strain gauges that measure the deformation of the materials. All this is possible due to the resistance that the materials present to the movement of electrons and how this resistance depends on some physical variables such as temperature, humidity, or mechanical strain.

4.2 Definition of Electric Resistance

In any material, the electrons move around randomly as if they were like a gas in a closed container. There is no preferred direction, and the average electron concentration is the same in any part of the homogenous material.

Imagine a bar of a certain material with a length L as illustrated in Figure 4.1.

When a battery is connected to the ends of a material rod, an electric field arises within it. The battery causes an increase of negative charges on the negative terminal (right side of the figure).

Inside the bar, the electric field is uniform and has intensity given by

$$E = \frac{V}{L}. \tag{4.1}$$

Due to the electric field, the free electrons in the material feel a force that makes them move in the opposite direction of the field (from right to left, in the figure). An electrical current in the direction of the electric field arises because the charge of the electrons is negative, which is defined as

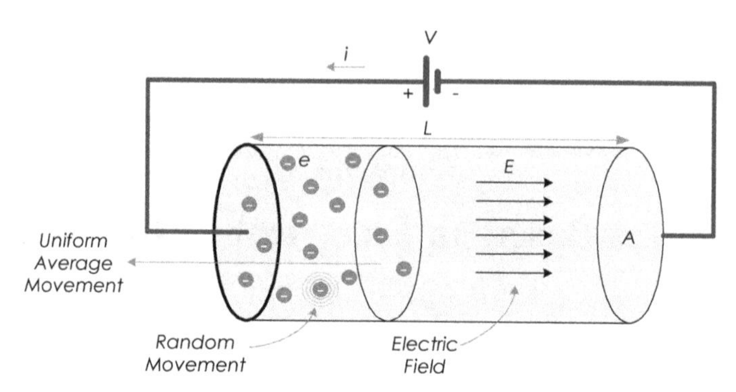

Figure 4.1 – Illustration of the electric field in a material due to the potential difference promoted by a battery. The electron movement is in the opposite direction from the electric field due to their negative electric charge.

$$i = \frac{dq}{dt}, \tag{4.2}$$

which means that the electric current through a given surface is the amount of electric charge that flows through it per unit of time. The concept of electric current is used to represent the motion of electric charges.

The electric current through a section in the material is always the same regardless of the section or form of the section. It is like what happens with the water flow into a series of pipes with different diameters. The water flows more rapidly in the narrower tubes and slower in the wider tubes but the total quantity of water passing through a section per unit time is the same. This is because water does not appear or disappear. The same happens with the electric charge. There is, therefore, conservation of charge. Everything that enters on one side of the wire must leave through the other side.

A conductive material, such as copper can be modeled as a semi-rigid periodic structure of positive ions of copper. These ions are "linked" together by strong electromagnetic forces. Each copper atom has a free electron that can move freely throughout the structure of the material. When establishing the electric field inside the conductor, each electron experiences a force $(-qE)$ that makes it move. This movement is, however, very short because it ends up colliding with a neighboring copper atom that is in constant motion due to the material's temperature. When the electron transfers moment to the structure is often caught by an ion releasing another positive ion which moves along the structure, by the action of the electric field, until it collides with another atom. The average time between collisions in a conductor of pure copper at room temperature is typically 2.5×10^{-14} s, covering an average distance of 0.04 μm. Note that the electrons entering at one end of a conductor are not the same coming out the other end. A constant flow of electrons through the material is, however, maintained. The collision of electrons with atoms increases the agitation and, therefore, the temperature of the material, which is then released as heat (Joule effect).

If we apply the same voltage to two different materials, the flowing electric current will be different. The ratio of the applied voltage (V) and the resulting electrical current (I) is called electrical resistance:

$$R = \frac{V}{I}. \tag{4.3}$$

This law is known as Ohm's Law.

The electrical resistance of a device not only depends on the material type as well as on its geometry. The material itself can be characterized with the resistivity parameter (ρ), which is defined by the ratio between electric field (E) and current density (j):

$$\rho = \frac{E}{j} \tag{4.4}$$

The inverse of the resistivity is named conductivity,

$$\sigma = \frac{1}{\rho}. \tag{4.5}$$

The resistivity of a material is determined by the mean time between collisions (τ), the electric charge (q), the electron mass (m) and the number of conduction electrons per unit volume (n):

$$\rho = \frac{m}{nq^2\tau}. \tag{4.6}$$

The electrical resistance of a uniform conductor with section A and length l is given by

$$R = \rho\frac{L}{A}. \tag{4.7}$$

The resistance depends on the cross-section area, the length, and the resistivity of the material from which the conductor is made of.

4.3 Potentiometric Displacement Sensors

Using the fact that the resistance of a conductor depends linearly on the length (Eq. (4.7)), It is possible to make a displacement sensor since the object whose displacement is to be measured can cause a change in the length of the conductor. In practice, what is done is to have a driver of fixed length and a cursor that moves along that conductor (Figure 4.2). There are also angular displacement sensors that are similarly constructed as well as displacement sensors of this type using conductive films instead of a coiled thread.

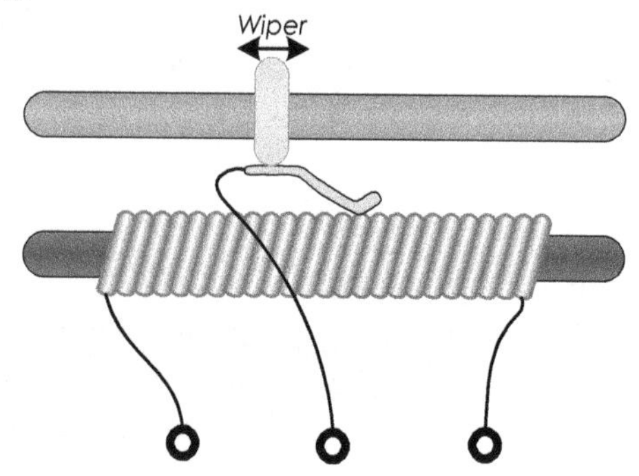

Figure 4.2 – Construction of a linear resistive displacement sensor.

To produce a voltage proportional to the displacement, a voltage (E) is applied to the complete conductor (Figure 4.3).

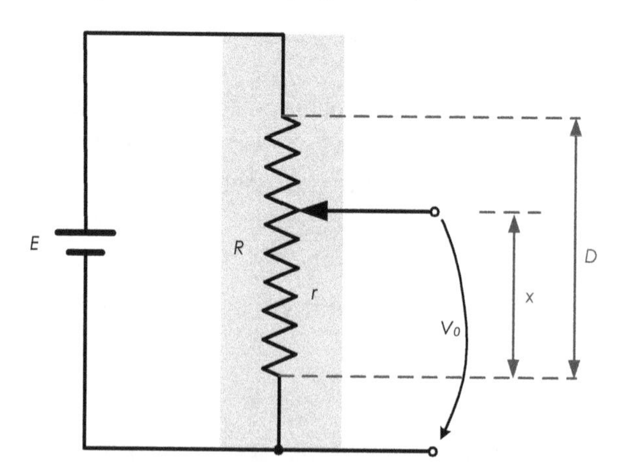

Figure 4.3 – Part of a signal conditioning circuit for a resistive displacement sensor.

The voltage between an end of the conductor and the slider (V_0) varies linearly with the position of the slider (x).

$$v_0 = \frac{r}{R}E = x\frac{E}{D}. \tag{4.8}$$

In order not to electrically load the potentiometer, a buffer circuit is usually used.

Another type of signal conditioning is to use a constant current source and to measure the voltage between one end of the conductor and the slider.

The sliders used are made of a material that is resistant to wear caused by friction. Examples of some materials are:

- alloys of precious metals;
- hardened copper alloys;
- Iron-bronze;
- Beryllium-copper.

The coiled sensors may have resolutions of 10 μm depending on the diameter of the wire used. Due to the contact between the cursor and the wire conductor, the output voltage has a variation with displacement by steps (Figure 4.4).

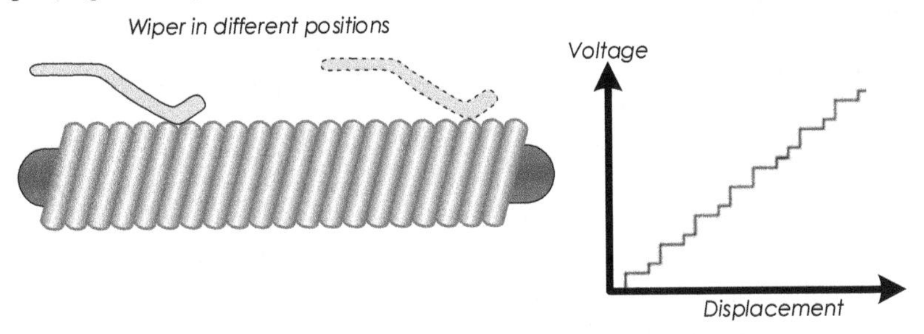

Figure 4.4 – Illustration of the nonlinear operation of a coiled resistive displacement sensor.

Film sensors may have resolutions of 0.1 μm. The resolution, which is theoretically infinite, is, however, limited by the noise in the electronic circuit and the film's uniformity. The films used are, in general, plastic, carbon, metal, or a mixture of ceramic and metal known as cermet. These types of potentiometric displacement sensors are more expensive than the coiled ones but have a longer life.

The potentiometric sensors have the following disadvantages:

- Mechanical load due to friction;
- Need for a physical connection with the object of measurement;
- Low speed;
- Heating due to friction and the applied voltage;
- Low environmental stability (sensitive to temperature and humidity).

Have, however, some advantages:

- Economic;
- Robust;
- Lightweight;
- Small.

Table 4.1 presents some typical specifications of the potentiometric displacement sensors.

Table 4.1 – Typical specifications of a resistive displacement sensor.

SPECIFICATION	LINEAR SENSOR	ANGULAR SENSOR
Range	2 mm to 8 m	360°
Linearity	0.002% to 0.1% of full scale	
Resolution	50 μm	0.2° to 2°
Power	0.1 to 50 W	
Temperature Coefficient	20 ppm/°C to 100 ppm/°C	
Maximum Operating Frequency	3 Hz	
Lifetime	400 million cycles	

This type of displacement sensor is, for example, used in electronic throttles in modern cars. These throttles, which operate without a mechanical connection between the throttle pedal and the engine, have a DC motor that controls the opening of the valve used to let fuel into a combustion engine.

4.4 Dependence of Resistivity with Temperature and Moisture

The resistivity of a material depends on its temperature (T). This ratio is, in general, not linear. However, for a small temperature range, it may be considered linear:

$$\rho = \rho_0[1 + \alpha(T - T_0)], \tag{4.9}$$

where ρ is the resistivity at a given temperature T_0 (typically 0°C or 25°C) and α is the temperature coefficient. Figure 4.5 shows the variation of the resistivity of tungsten with temperature.

Metals have a positive temperature coefficient, while some semiconductors and oxides have a negative temperature coefficient. In electronic circuit components, it is desirable that the temperature coefficient has a small value. On the other hand, a high value of temperature coefficient allows the construction of good temperature sensors.

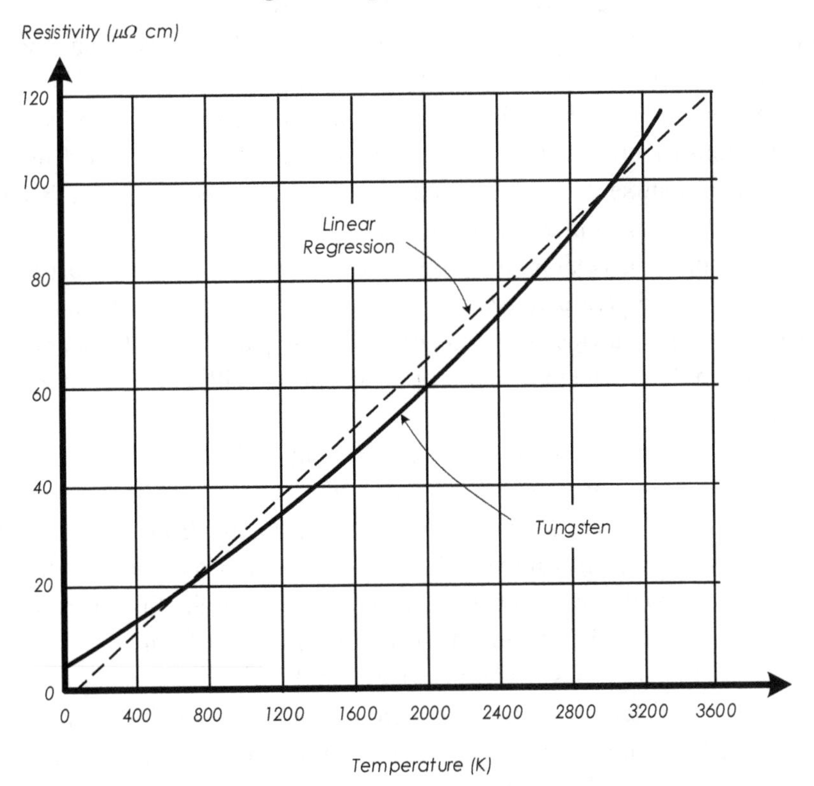

Figure 4.5 – Graph of the resistivity of tungsten according to temperature.

At a temperature of 0 K, the resistivity of a material is due to the collision of electrons that form the current with the structure of the material (the material's atoms) and other electrons.

When the temperature rises, the structure of the material begins to vibrate, and therefore, the probability of an electron colliding with the frame (and other electrons) increases, resulting in an increase in resistivity. This is the main phenomenon responsible for the increase in resistivity in metals.

The molecules and atoms of a material are connected to each other (by forces of electrical nature) and cannot vibrate independently. These vibrations take the form of waves that propagate through the material at the speed of sound in that material. Also, the energy of these waves is quantified (not all energy values are allowed), which causes an increase or decrease of their energy to be in multiples of a fixed quantity. That amount is called "phonons" by analogy with photons.

In semiconductors, in addition to this phenomenon, there is an increase in the number of available electrons within the material, caused by the increase in temperature. This means that there is a decrease in resistivity (because there is an increase in current).

Some materials exhibit the property to stop having electrical resistance below a certain temperature (superconductivity). This is because the electrons form pairs that are "connected." These pairs, called Cooper pairs, are formed due to the interaction of the electrons with the structure of the material.

Imagine an electron moving from the right to the left, as illustrated in Figure 4.6. Due to the attraction between the electron and the ions of the material, it will move. The passage of the electron, therefore, leaves a trail. Any other electron that moves close to it will suffer an attraction by the structure's ions that were displaced by the passage of the first electron.

Therefore, a link between these two electrons is formed as a slight attraction that opposes their electrical repulsion. These electron pairs behave very differently than the electrons that are isolated in such a way that the effects of a collision with the structure (and other electrons) stop happening, eliminating the resistivity source.

The resistivity of materials is also dependent on humidity. There are materials that can absorb more moisture than others. As more water is absorbed, the material becomes more conductive, and thus its resistivity is lower. Because of this, it is possible to produce humidity sensors called hygristors.

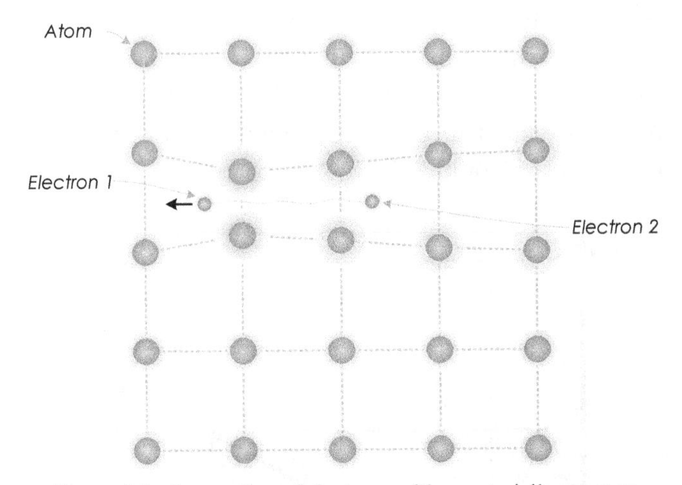

Figure 4.6 – Interaction of electrons with a material's structure.

4.5 Resistive Temperature Detector

Resistive temperature detectors (RTD) use an element whose electrical resistance varies with temperature. The most common material used in this type of sensor is platinum. The resistive element may be a coiled platinum wire or a film deposited on a platinum substrate. In both cases, the resistance/temperature relation is very linear, as one can see from Figure 4.7 with a non-linearity in the –20°C to 120°C range of less than 0.25%.

The most used standard for this kind of temperature sensors (IEC 751:1983) specifies a resistance value of 100 Ω for a temperature of 0°C and a temperature coefficient of 0.385 Ω/Ω/°C:

$$R = 100 + 0.385 \cdot T. \tag{4.10}$$

The non-linearity is very predictable and repeatable and is thus easily corrected. A second-order polynomial is a function that best describes the relation resistance/temperature

$$R = 100 + 0.391 \cdot T - 5.56 \cdot 10^{-5} \cdot T^2. \tag{4.11}$$

This type of temperature sensor is very accurate but expensive due to the use of platinum. Another disadvantage of this sensor is self-heating. To obtain a voltage proportional to the temperature, you must pass a current through the platinum resistance. This current will artificially increase the temperature leading to a measurement error. For example, if a current of 5 mA with a platinum RTD having a resistance of 100 Ω (Figure 4.8) is used, the power dissipated is 2.5 mW. If the relationship between power dissipation and temperature is 1 mW/°C, then the self-heating causes an error of 2.5°C.

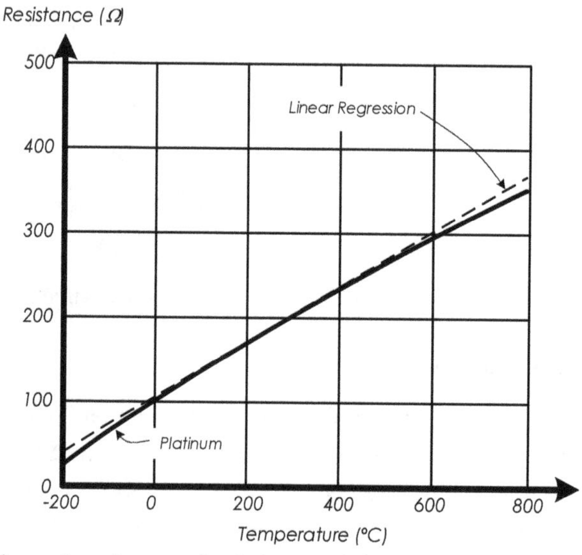

Figure 4.7 – Change in resistance of a platinum resistive temperature detector (solid line). The dashed line represents a line for comparison.

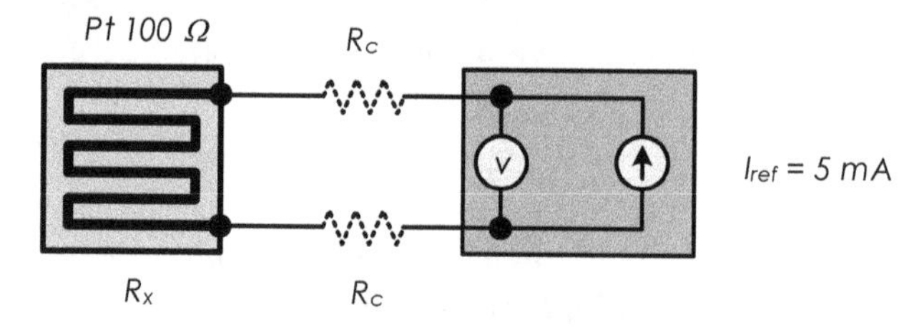

Figure 4.8 – Example of signal conditioning used in RTD circuit.

Yet another disadvantage is their low resistance which causes the resistance of the connecting cables to have a great influence. In the example of Figure 4.8, if the resistance of each cable connector is 1 Ω (R_c), the temperature error, assuming a temperature coefficient of 0.385 Ω/Ω/°C is 5.195°C:

$$\Delta T = T_{R_c \neq 0} - T_{R_c = 0} = \frac{(R_x + 2 \cdot R_c) - 100}{0.385} - \frac{R_x - 100}{0.385} = \frac{2 \cdot R_c}{0.385} = 5.195\ ºC. \tag{4.12}$$

One should therefore use low current and short cables with the lowest possible resistance. Another solution is to use a 4-wire connection, as illustrated in Figure 4.9. The cables used to connect the current source to the RTD are not the same ones used to connect the voltmeter. As the current drawn by the voltmeter is usually very small, the voltage drop in the cables will be minimal.

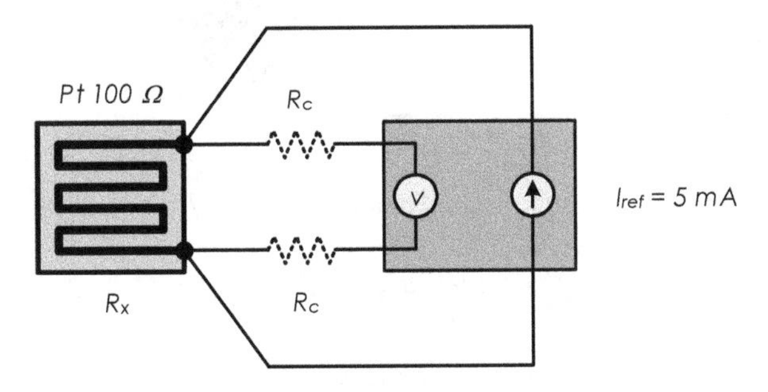

Figure 4.9 – Example of signal conditioning used in RTD circuit.

Whenever a conductor is heated, there is the possibility of a voltage being generated. This voltage of thermal origin can be compensated by performing the measurement of the RTD voltage with and without applied current. The voltage difference is then divided by the value of the applied current.

Voltmeters are more sensitive when measuring voltages close to 0 because it is possible to use a lower range. To take advantage of this, one can use a bridge signal conditioning circuit, as shown in Figure 4.10.

The voltage measured by the voltmeter is given by

$$\Delta v = V_A - V_B. \tag{4.13}$$

The voltage measured in B is half of the supply voltage of the bridge

$$V_B = \frac{E}{2}. \tag{4.14}$$

while the voltage in A depends on the value of the thermistor's resistance (R_x),

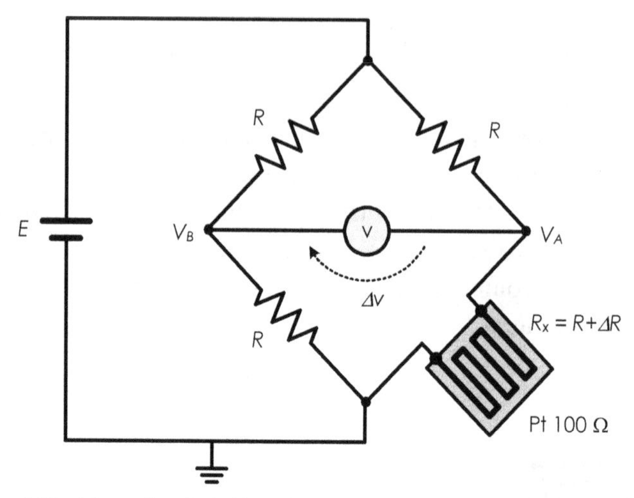

Figure 4.10 – Example of a bridge signal conditioning circuit to be used with an RTD.

$$V_A = \frac{R_x}{R_x+R}E = \frac{R+\Delta R}{2R+\Delta R}E. \tag{4.15}$$

Inserting (45) and (46) into (44) leads to

$$\Delta v = \left(\frac{R+\Delta R}{2R+\Delta R} - \frac{1}{2}\right)E = \frac{E}{4} \times \frac{\Delta R}{R+\frac{\Delta R}{2}}. \tag{4.16}$$

In this case, the disadvantage is that the RTD is very close to the bridge's resistors; therefore, their value will be variable due to an increase in the temperature of measure. One alternative is to keep the RTD away from the bridge and use the third wire to connect the voltmeter, as shown in Figure 4.11.

This method requires one less wire than the 4-wire method but is generally less accurate.

In summary, RTDs are almost linear, accurate, very stable, and work in a wide temperature range but are large, expensive, and have slow responses when compared with thermistors.

The bridge's unbalance voltage is generally small and needs to be amplified. The most common way is to use an instrumentation amplifier, as illustrated in Figure 4.12.

The amplifier must be powered by a bipolar DC source in order to be able to amplify the positive and negative values of the imbalance voltage.

The gain (G) is usually adjusted through the choice of an external resistor. The output voltage of the amplifier is given by, considering (4.16),

$$v_s = \frac{E \cdot G}{4} \times \frac{\Delta R}{R + \frac{\Delta R}{2}}. \tag{4.17}$$

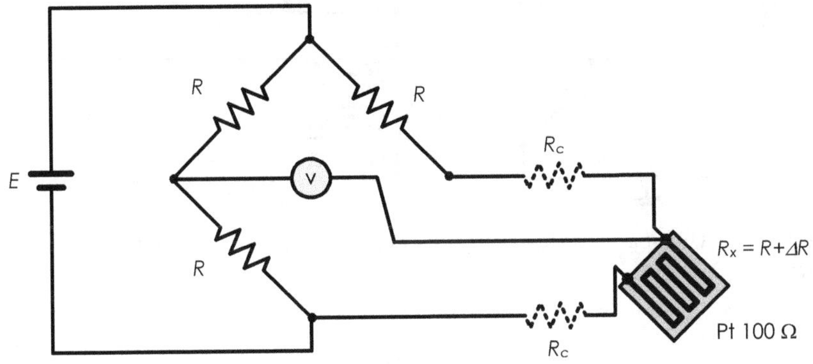

Figure 4.11 – Example of signal conditioning bridge with three wires to be used with an RTD circuit.

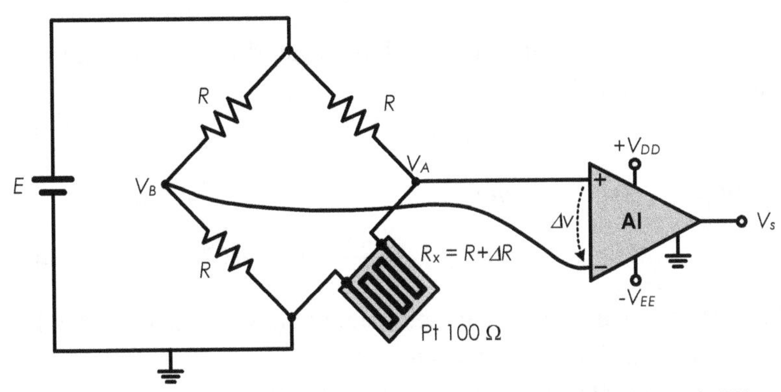

Figure 4.12 – Use of an instrumentation amplifier to amplify the imbalance of a Wheatstone bridge voltage.

That non-linear relation between the resistance's variation and the output voltage can be eliminated by using the circuit in Figure 4.13 [4].

By considering the behavior of the operational amplifier as ideal, one can see that the imbalance voltage is forced to zero. The currents that go through the fixed resistors are

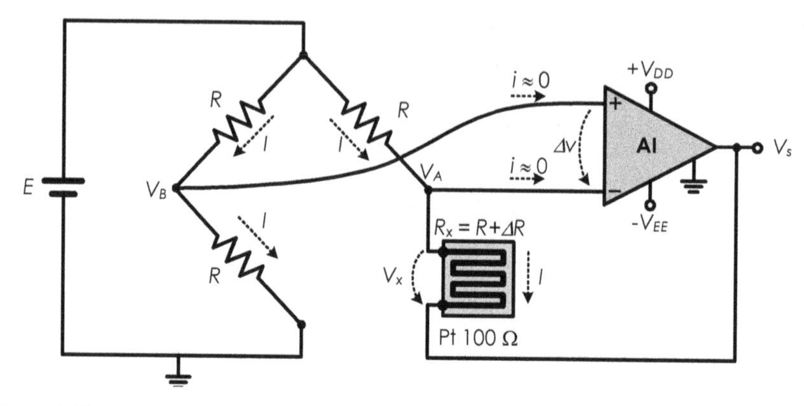

Figure 4.13 – Use of an operational amplifier to linearize the operation of the Wheatstone bridge.

$$I = \frac{E}{2R} \tag{4.18}$$

and it is independent of the variable resistor's value. That same current is the one that goes through the variable resistor, creating a voltage of

$$V_x = \frac{E}{2R}(R + \Delta R). \tag{4.19}$$

The voltage in the negative terminal of the amplifier is equal to the one in the positive terminal:

$$V_+ = V_- = \frac{E}{2}. \tag{4.20}$$

The output voltage is, therefore,

$$V_S = V_- - V_x = \frac{E}{2} - E\left(\frac{R+\Delta R}{2R}\right) = -E\frac{\Delta R}{2R}. \tag{4.21}$$

This way, a linear relation can be obtained between the output voltage and the variation of the resistor's value.

There are other ways to obtain a linear relationship between the output voltage and the measurand. For instance, the non-linearity caused using the Wheatstone bridge can be corrected by using an electronic circuit after the bridge, such as the one in Figure 4.14.

At the input of the multiplier, we have the input voltage (V_S) and the output voltage (V_o). Therefore, the voltage at the output of the multiplier is

$$V_W = \frac{V_S V_O}{C}. \tag{4.22}$$

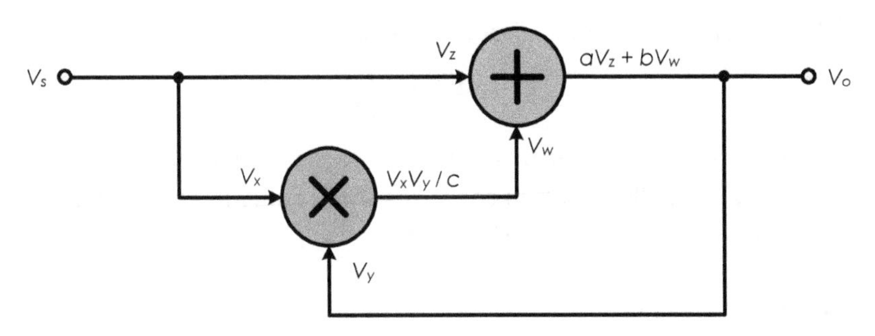

Figure 4.14 – Use of an electronic circuit to correct the nonlinearity introduced by the Wheatstone bridge.

Considering that this voltage is one of the inputs of the adder and that the other is the input voltage of the circuit (V_S), the output voltage's expression is given by

$$V_o = aV_s + b\frac{V_S V_O}{c}. \tag{4.23}$$

Solving for V_o one gets

$$V_o = \frac{aV_S}{1-\frac{b}{c}v_S}. \tag{4.24}$$

By inserting the expression of the output voltage of the instrumentation amplifier used in Figure 4.12, given by (4.17), one can obtain

$$V_o = \frac{a\left(\frac{EG}{4}\times\frac{\Delta R}{R+\frac{\Delta R}{2}}\right)}{1-\frac{b}{c}\left(\frac{EG}{4}\times\frac{\Delta R}{R+\frac{\Delta R}{2}}\right)}. \tag{4.25}$$

By multiplying the numerator and the denominator by $R + \Delta R/2$, one obtains

$$V_o = \frac{aEG}{4} \times \frac{\Delta R}{R+\frac{\Delta R}{2}-\frac{b}{c}\times\frac{EG}{4}\Delta R}. \tag{4.26}$$

The b constant can be chosen in such a way that both terms in ΔR in the denominator cancel each other.

$$b = \frac{2c}{EG}. \tag{4.27}$$

The output voltage of the circuit is, then,

$$V_o = \frac{aEG}{4} \times \frac{\Delta R}{R}, \tag{4.28}$$

which demonstrates the linear relation that can be obtained between the output voltage and the variation of the resistor's value in the RTD (ΔR).

The main advantages of RTDs are [8].

Advantages:

- Good accuracy;
- Wide range of measurement;
- Stable;
- Good linearity;

and the main disadvantages are [8]:

- More expensive than thermistors and thermocouples;
- Auto-heating;
- Bigger than thermistors and thermocouples;
- Less robust than thermocouples.

4.6 Thermistor

A thermistor is a type of sensor used to measure the temperature, like an RTD, but which is typically made of a metal oxide semiconductor (manganese, nickel, cobalt, iron, copper, titanium, etc.) and is encapsulated in glass or epoxy.

The word "Thermistor" comes from the combination of two words, "thermal" and "resistor." The main advantage of thermistors is their high sensitivity (Figure 4.15). A thermistor with a resistance of 2252 Ω has a sensitivity of −100 Ω/°C at room temperature.

There are two types of thermistors:

- **NTC (*negative temperature coefficient*)** — Thermistors with a negative temperature coefficient.
- **PTC (*positive temperature coefficient*)** — Thermistors with a positive temperature coefficient.

The NTCs are the ones used more often. Sensors of this type have the advantage of having high impedance, and therefore it will not be necessary to use a 4-wire connection, as in the case of RTD. Another advantage is that its resistance varies a lot with temperature, which allows it to be a very sensitive sensor.

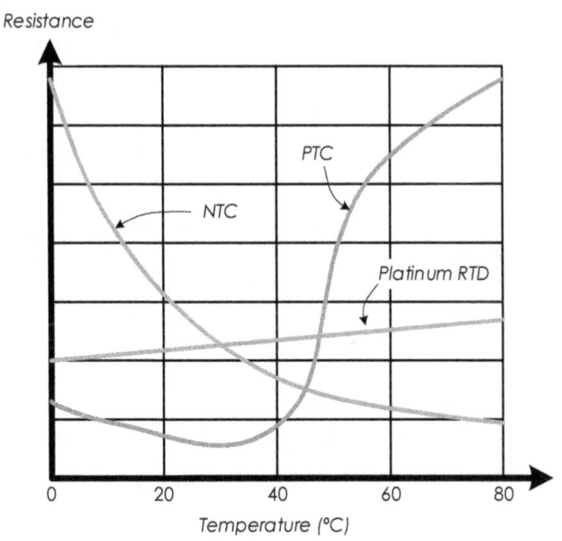

Figure 4.15 – Comparison of resistance/temperature of a thermistor and RTD characteristics.

Another important advantage is that this type of sensor is generally small and therefore has a small thermal mass. This implies that heat will not load the object of measurement. This feature, however, also brings a disadvantage — it is susceptible to self-heating.

Since the characteristic of a thermistor is nonlinear, you must use a signal conditioning circuit to correct it. One of the easiest ways to do this is to add a fixed resistance in parallel, as illustrated in Figure 4.16.

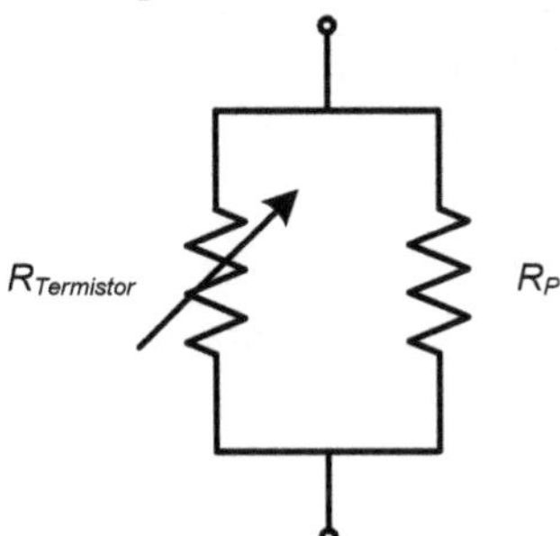

Figure 4.16 – Linearization circuit of a thermistor using a parallel resistor.

An approximate way to calculate the value of the resistor R_p is using

$$R_p = \frac{R_2(R_1+R_3)-2R_1R_3}{(R_1+R_3)-2R_2}, \tag{4.29}$$

where R_1, R_2 and R_3 are the resistance values of the thermistor for three equally spaced temperature values. For example, using the values of Table 4.2, one obtains a resistance R_p of 12.29 kΩ.

Table 4.2 – Example of resistance values for a thermistor.

QUANTITY	TEMPERATURE (°C)	RESISTANCE (kΩ)
R_1	−15	23
R_2	25	10
R_3	65	4

In Figure 4.17, one can observe the thermistor's resistance and the value of the parallel between the resistor (R_p) and the thermistor. It is shown that the characteristic function became more linear, as desired, although the sensitivity has decreased.

This kind of linearization is called **Linearization by Sensor Circuit Change**.

The main advantages and disadvantages of the thermistor are listed below [8]. The advantages are:

- Low Cost;
- Short Reaction Time;
- High Sensitivity;
- High resistance value, which makes it less critical the value of cable resistance,

and the disadvantages are:

- Limited range of measurement;
- The characteristic may vary according to the manufacturer;
- Self-heating;
- Non-linearity;
- It needs many additional components, which compromises its reliability.

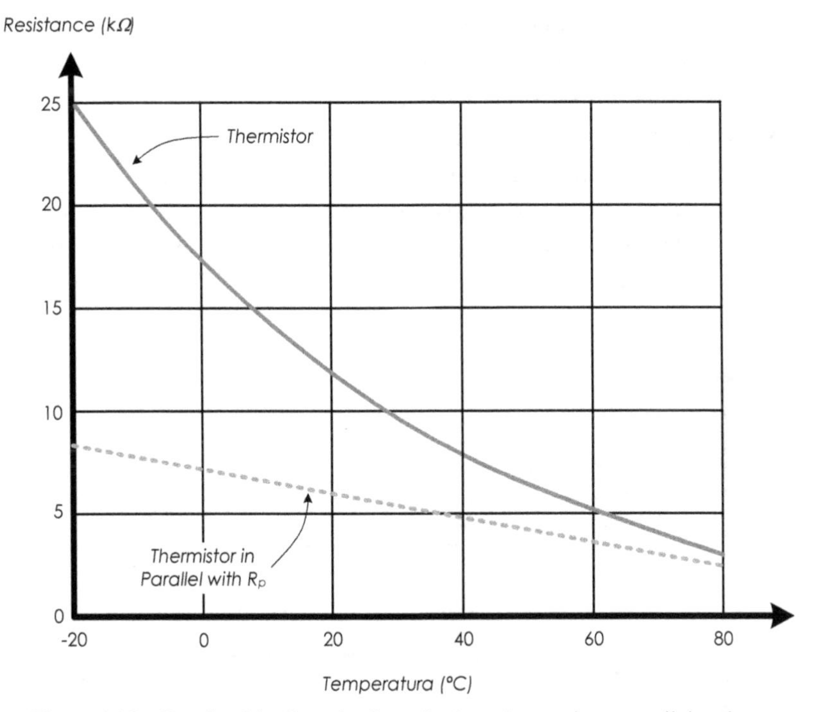

Figure 4.17 – Result of the linearization of a thermistor using a parallel resistor.

4.7 Integrated Temperature Sensor

The voltage in a P-N junction biased with a constant current depends on temperature with a sensitivity of about −2mV/°C.

It is possible to use a junction of this type to make an integrated temperature sensor. This allows the correction of non-linearity and the production of a high voltage output with the use of integrated amplifiers.

This type of temperature sensor is cheap and has good accuracy for temperatures near room temperature. It has the disadvantage of having a limited measurement range.

Such a sensor is manufactured, for example, by the National Semiconductors company that produces the LM35 model. It has the following specifications:

- Works from −55°C to 150°C;
- Sensitivity of 10 mV per °C;
- Continuous supply voltage between 4 V and 20 V;

- Current consumption below 60 μA;
- Self-heating limited to 0.08°C (in still air);
- Accuracy of 0.5°C (at 25°C);
- Non-linearity below 0.25°C;
- Low output impedance (0.1 Ω for a load of 1mA);
- No additional components required.

Figure 4.18 shows the temperature error specification as a function of the temperature for this sensor.

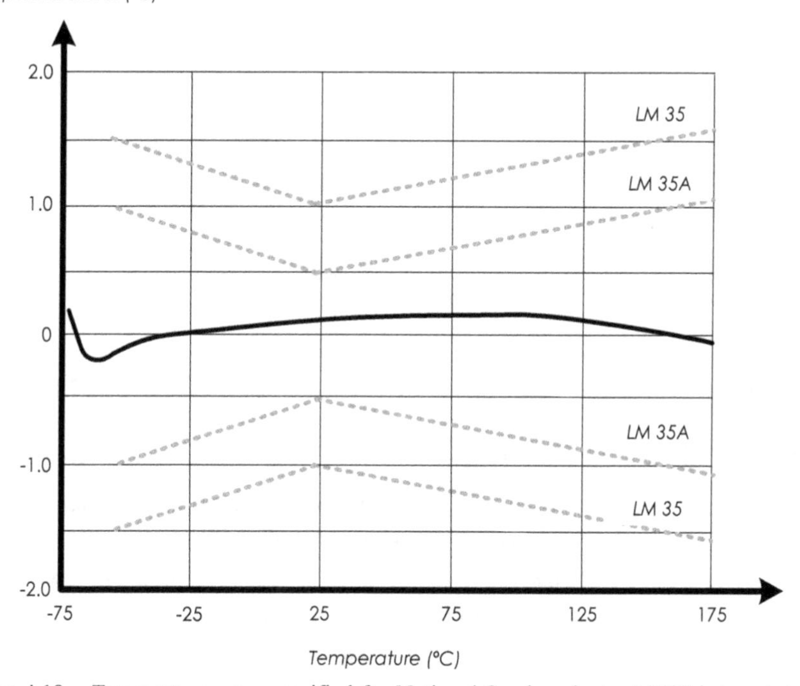

Figure 4.18 – Temperature error specified for National Semiconductor LM35 integrated temperature sensor.

Each sensor produced will have a slightly different transfer function. The manufacturer guarantees that the measurement error does not exceed the limits dotted in the figure. At 25°C, the maximum guaranteed error is less. There are two models represented in the figure. The LM35A model is sold with a guarantee of a maximum error below 0.5°C (for 25°C), while the LM35 model is sold with a guarantee of a maximum error of 1°C (at 25°C).

The main advantages and disadvantages regarding integrated temperature sensors are [8]:

Advantages

- Cheaper than other RTDs
- More linear than thermistors
- Contain incorporated signal conditioning

Disadvantages

- Not as linear as other RTDs
- Low Precision
- More expensive than thermistors and thermocouples
- Reduced measuring range
- Slower response time

4.8 Dependence of Resistivity with Deformation

The electrical resistance of materials changes when they are mechanically deformed (piezoresistive effect). This effect may, in some cases, be a source of error but can also be used to make sensors of mechanical deformation.

The electrical resistance of a conductor is given by (4.7) and can be written, multiplying and dividing by the length of the conductor, as follows

$$R = \rho \frac{l^2}{V}, \tag{4.30}$$

in which V is the volume of the conductor ($A \times l$). By differentiating with respect to the length,

$$\frac{dR}{dl} = 2\rho \frac{l}{V}, \tag{4.31}$$

the sensitivity of the resistance value of the deformation (dl) of the conductor is obtained. Note that the sensitivity increases with the length of the wire.

Combining (61) and (62) leads to

$$\frac{dR}{R} = 2\frac{dl}{l}. \tag{4.32}$$

The ratio dl/l is usually called "strain," and is usually represented with the letter ϵ:

$$\epsilon = \frac{dl}{l}. \tag{4.33}$$

The factor multiplying the strain in Eq. (4.32) is called *gauge factor* and represented by GF. Equation (4.32) takes the form

$$\frac{dR}{R} = GF \cdot \epsilon. \tag{4.34}$$

In metal wires, the gauge factor has a typical value of 2. There are materials, however, that have gauge factors of up to 6, such as platinum. There are semiconductor materials manufactured with the purpose of having a high gauge factor, as seen in Table 4.3.

Table 4.3 – Some characteristics of materials that are used to build deformation sensors.

MATERIAL	SENSITIVITY	RESISTANCE	TEMPERATURE COEFFICIENT ($°C^{-1}$ X 10^{-6})
Copper and Nickel Alloy (57/43)	2	100	10.8
Platinum	4 to 6	50	2160
Silicon	−100 to 150	200	90000

The reasoning that led to the derivation of (63) is assuming that the volume of the material remained constant before deformation. This is not always true.

The typical behavior of materials is that they do not deform only along the axis where the force is applied but also according to other directions. To the ratio of symmetrical deformation in the perpendicular direction and the deformation in the axial direction (direction in which the force is applied) is given the name of Poisson's ratio:

$$\nu = -\frac{\epsilon_{transversal}}{\epsilon_{axial}}. \tag{4.35}$$

Table 4.4 shows Poisson's ratio of some materials.

Typically, Poisson's ratio value is between 0 and 0.5. The value of 0.5 corresponds to a material that maintains its volume when it undergoes compression or traction. The relative volume change is given by

$$\frac{\Delta V}{V} = (1 - 2\nu)\frac{\Delta l}{l}. \tag{4.36}$$

Table 4.4 – Poisson's ratio for some materials.

MATERIAL	POISSON COEFFICIENT
Rubber	0.5
Aluminum	0.33
Steel	0.3
Glass	0.25
Cork	0
Special Foam	−0.6
Material	Poisson Coefficient

A value of 0 for the Poisson's ratio corresponds to a material that does not alter its transverse dimensions when subjected to mechanical stress, as is the case of cork. There are some materials, typically foams made from polymers, which have a negative value of Poisson's ratio (Figure 4.18).

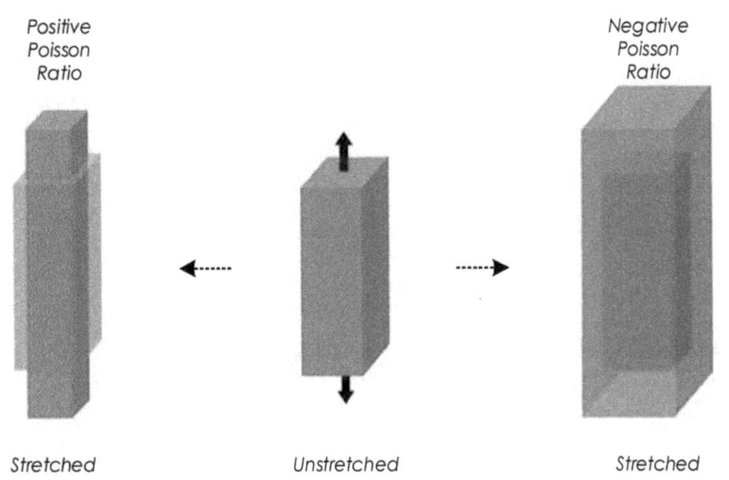

Figure 4.18 – Example of the deformation caused on materials with positive (a) and negative (b) Poisson's ratio. It was adapted from a drawing by Abhishek Ghosh.

The resistance of a material varies not only with length, as considered previously (Eq. (62)) but also with its volume and resistivity. The variation of the resistance of a material with the dimensions of Figure 4.19 is given by

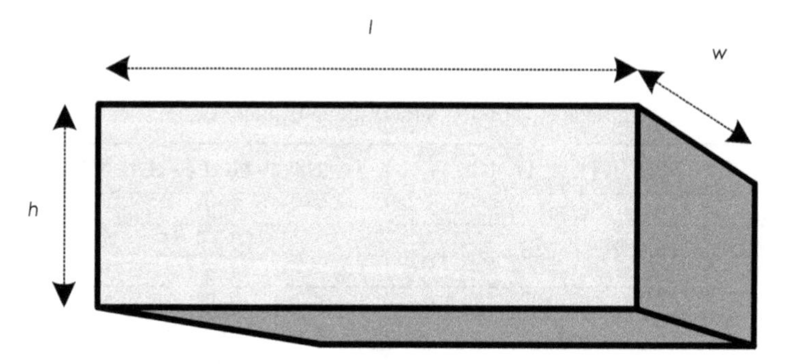

Figure 4.19 – Dimensions of a given slab of material.

$$\frac{\Delta R}{R} = \frac{\Delta l}{l} + \frac{\Delta \rho}{\rho} - \frac{\Delta A}{A}. \tag{4.37}$$

When there is a variation of the length l, there is also a variation of the area, given by

$$\frac{\Delta A}{A} = \frac{\Delta h}{h} + \frac{\Delta w}{w} = -\nu\frac{\Delta l}{l} - \nu\frac{\Delta l}{l} = -2\nu\frac{\Delta l}{l}. \tag{4.38}$$

Likewise, the volume will vary

$$\frac{\Delta V}{V} = \frac{\Delta l}{l} + \frac{\Delta h}{h} + \frac{\Delta w}{w} = \frac{\Delta l}{l} - \nu\frac{\Delta l}{l} - \nu\frac{\Delta l}{l} = (1 - 2\nu)\frac{\Delta l}{l}. \tag{4.39}$$

The volume change is directly related to the variation in resistivity. This relationship is given by the Bridgman constant (C):

$$\frac{\Delta \rho}{\rho} = C\frac{\Delta V}{V} = C(1 - 2\nu)\frac{\Delta l}{l}. \tag{4.40}$$

Inserting (4.38) and (4.40) in (4.37) leads to

$$GF = \frac{\frac{\Delta R}{R}}{\frac{\Delta l}{l}} = 1 + C(1 - 2\nu) + 2\nu. \tag{4.41}$$

Therefore, the gauge factor depends on Bridgman's constant and Poisson's ratio of the material. Usually, the Bridgman constant has a value of 1, leading to a sensitivity of 2 as a typical value between most metals.

4.9 Strain Gauge

Using the fact that the electrical resistance of a material depends upon its mechanical deformation is possible to make sensors that measure the strain — Strain Gauges.

There are several kinds of Strain Gauges:

- Metallic
- Wired
- Film
- Helicoidal
- Thin Film
- N-Type Semiconductor
- P-Type Semiconductor

A Strain Gauge of this kind uses the fact that the electric resistance of a conductive wire is given by

$$R = \rho \frac{l}{A}. \tag{4.42}$$

If a copper wire, for example, is stretched, its length increases and its section decreases, leading to increased resistance. If it is compressed, the opposite happens. A metallic conductor is arranged as a coil to increase its sensitivity. For example, if 16 conductive portions are laid side by side. A variation in length (Δl) of the material to which the Strain Gauge is glued causes a variation in the length of the wire extensometer of approximately $16\Delta l$. The relative length variation leads to a relative change of the resistance given by

$$\frac{\Delta R}{R} = GF \frac{\Delta l}{l}. \tag{4.43}$$

The maximum strain that a Strain Gauge can measure is typically between 0.4 and 0.5%, i.e., for a Strain Gauge with a nominal resistance of 120 Ω this corresponds to 0.96 to 1.2 Ω. This small resistance change can be detected by inserting the Strain Gauge into the appropriate signal conditioning circuit, which is generally a Wheatstone bridge.

There are also Strain Gauges made with thin films. There are the same cases in which this film is deposited directly on the material under test as with the extensometer made of Palladium and Chromium developed by NASA to measure deformations in jet engines. This Strain Gauge can operate at temperatures up to 1100°C and is highly stable and reproducible. Reproducibility is 200 μm compared to 1000 μm of film strain gauges.

Strain Gauges made of wire, foil, and thin films are metallic strain Gauges. There is also semiconductor Strain Gauges whose main advantage is that they have a much higher sensitivity (50 to 200). These Strain Gauges are made with the same techniques used in integrated circuits, is possible to create very small sensors.

Semiconductor Strain Gauges have, however, the disadvantage of being less linear and more dependent on temperature.

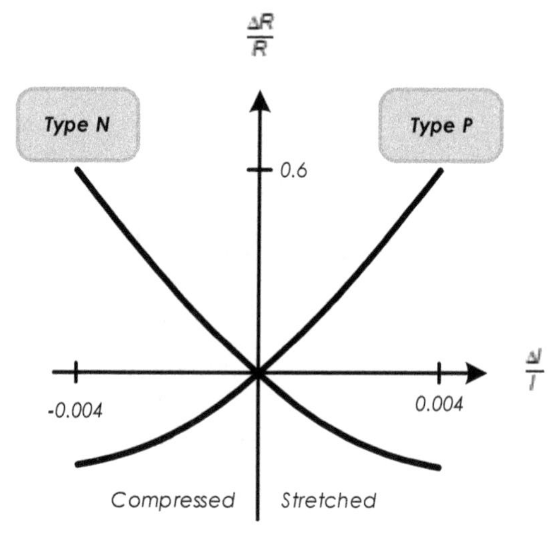

Figure 4.20 – Typical Characteristic curves of semiconductor Strain Gauges.

Table 4.5 presents a comparison between the main characteristics of Strain Gauges (metallic and semiconductors).

In practice, metal Strain Gauges are still the most widely used. Semiconductor ones are used for specific applications where sensitivity is important and where the temperature variations are small.

Table 4.5 – Comparison of the main characteristics of metallic and semiconductor strain gauges.

PARAMETER	METÁLICO STRAIN GAGE	SEMICONDUTOR STRAIN GAGE
Measurement Range	0.1 με to 40000 με	0.0001 με to 3000 με
Gauge Factor	2 to 4.5	50 to 200
Resistance	120, 350, 600, …, 5000 Ω	1000 to 5000 Ω
Resistance Tolerance	0.1% to 0.2%	1% to 2%
Size	0.4 mm to 150 mm	1 mm to 5 mm

A Wheatstone bridge is typically used for signal conditioning, as shown in Figure 4.21.

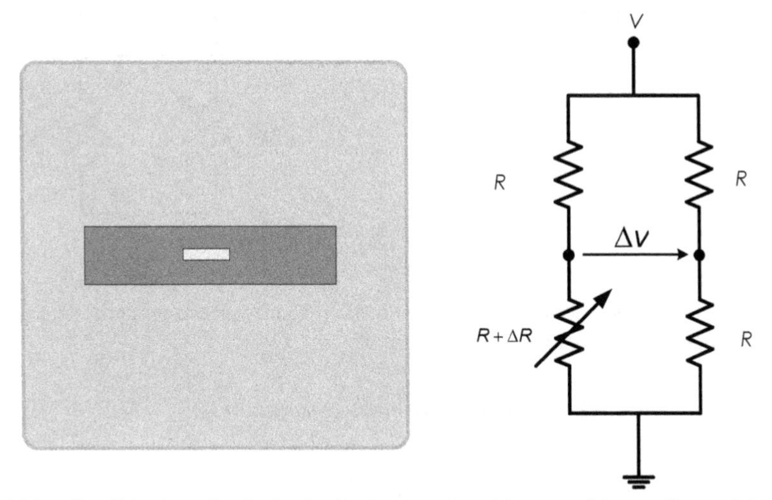

Figure 4.21 – Conditioning circuit for Strain Gauges placed in one of the surfaces of the material of which the deformation is to be measured.

For this circuit,

$$\frac{\Delta v}{v} = \frac{GF\frac{\Delta l}{l}}{2\left(2+GF\frac{\Delta l}{l}\right)} \approx \frac{GF}{4}\frac{\Delta l}{l}. \tag{4.44}$$

The sensitivity is, therefore $GF/4$. If we consider a variation in the resistance with temperature, $\Delta R(T)$, we obtain

$$\frac{\Delta v}{v} = \frac{GF\frac{\Delta l}{l}+\frac{\Delta R(T)}{R}}{2\left(2+GF\frac{\Delta l}{l}+\frac{\Delta R(T)}{R}\right)} \approx \frac{GF}{4}\frac{\Delta l}{l} + \frac{1}{4}\frac{\Delta R(T)}{R}. \tag{4.45}$$

The voltage imbalance, therefore, depends on the operating temperature, which is undesirable. If it is possible to use two Strain Gauges, one on each face of the material whose strain is to be measured (one in compression and one in tension), can use the assembly of Figure 4.22.

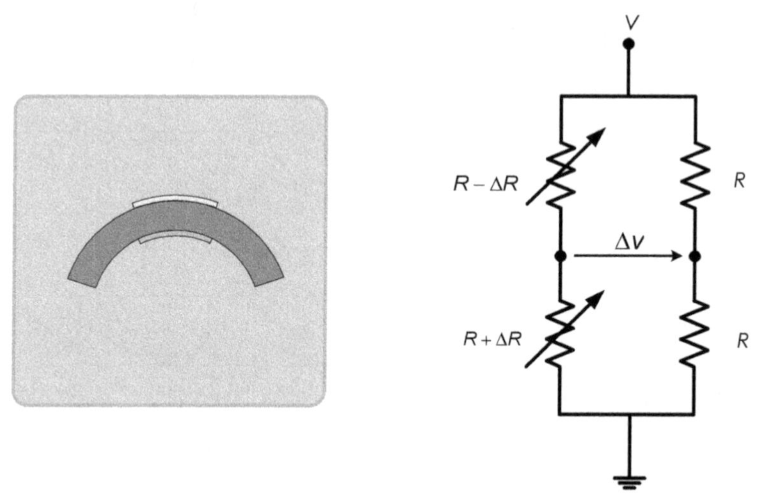

Figure 4.22 – Conditioning circuit for two strain gauges placed on opposite sides of the material which deformation is to be measured.

In this case, the imbalance voltage is given by

$$\frac{\Delta v}{v} = \frac{GF}{4} \cdot \frac{\Delta l}{l} \tag{4.46}$$

This circuit has twice the sensitivity of the assembly with a single Strain Gauge. It is also more linear. The use of two sensors with complementary behaviors is one form used to linearize the input/output relation.

Considering the same resistance change caused by temperature in both strain gauges, one obtains

$$\frac{\Delta v}{v} = \frac{\Delta R}{2(R+\Delta R(T))} \approx \frac{GF}{4} \cdot \frac{\Delta l}{l}, \tag{4.47}$$

which shows that this circuit is also less sensitive to temperature variations.

In the case where it is possible to use four Strain Gauges, two should be connected for compression and the remaining for traction, as shown in Figure 4.23.

By connecting the Strain Gauges in the bridge as shown in Figure 4.23, it is possible to obtain a sensitivity of GF which is four times greater than in the case of using a single Strain Gauge:

$$\frac{\Delta v}{v} = GF \cdot \frac{\Delta l}{l}. \tag{4.48}$$

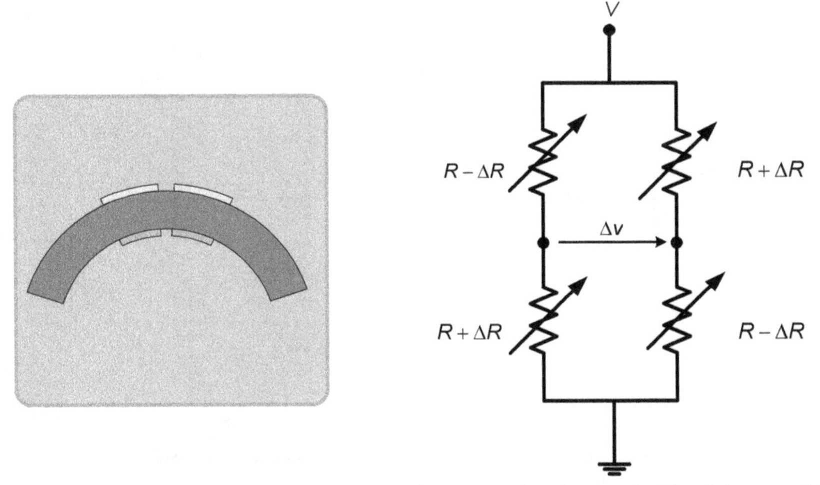

Figure 4.23 – Conditioning circuit for two strain gauges placed on each side of the material in which deformation is to be measured.

For the same variation of resistance with temperature, an imbalance voltage can be obtained, also not very sensitive to temperature changes:

$$\frac{\Delta v}{v} = \frac{\Delta R}{R+\Delta R(T)} \approx GF \cdot \frac{\Delta l}{l}. \tag{4.49}$$

When there is only the possibility of placing Strain Gauges on one side of the material, one can use the same strain but four perpendicularly arranged as illustrated in Figure 4.24.

The variation of the resistance of the two Strain Gauges arranged perpendicularly to the direction of deformation has a resistance change that depends on the Poisson's ratio of the material, that is, a variation of $-\nu\Delta R$. In this case, the voltage is given by imbalance is

$$\frac{\Delta v}{v} \approx \frac{1+\nu}{2} GF \frac{\Delta l}{l}. \tag{4.50}$$

The sensitivity is not as high as in the case of four strain gauges placed at opposite sides. This assembly, however, is also insensitive to temperature:

$$\frac{\Delta v}{v} \approx \frac{1+\nu}{2} \frac{\Delta \mathrm{R}}{R+\Delta R(T)} \approx \frac{\Delta v}{v} \approx \frac{1+\nu}{2} GF \frac{\Delta l}{l}. \tag{4.51}$$

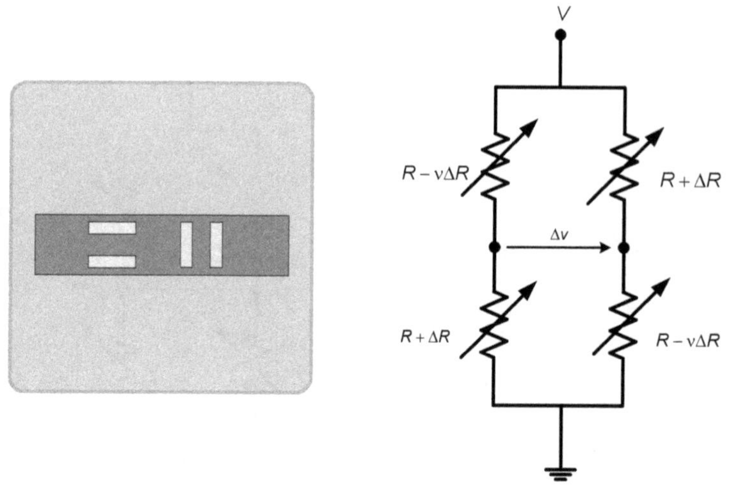

Figure 4.24 – Conditioning circuit for four strain gauges placed in one side of the material, which deformation is to be measured.

In summary, the more Strain Gauges used, the higher the sensitivity. Also, the use of more than one Strain Gauge allows better linearity and greater insensitivity to variations with temperature (Figure 4.25).

Strain Gauges are often used to measure other quantities indirectly as the force, torque, pressure, mechanical stress, flow, etc.

Figure 4.26 illustrates a way of measuring the flow of a fluid using a flexible rod which is associated with one or more strain gauges. The higher the flow, the greater the deformation. The precise relationship depends on the system geometry and the type of liquid.

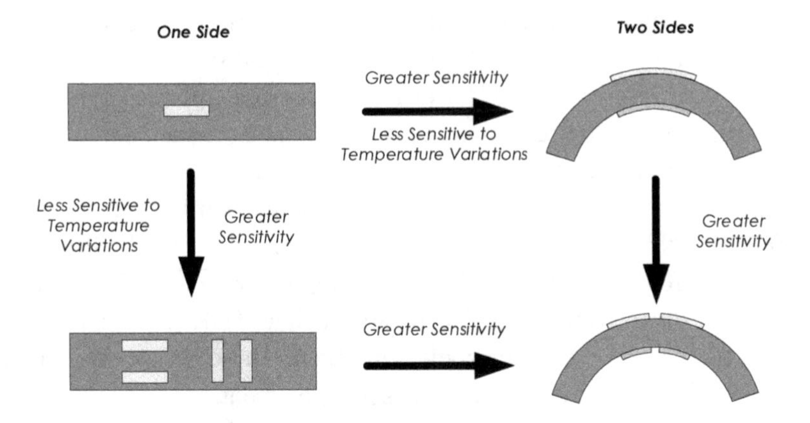

Figure 4.25 – Comparison of the different Strain Gauge assemblies.

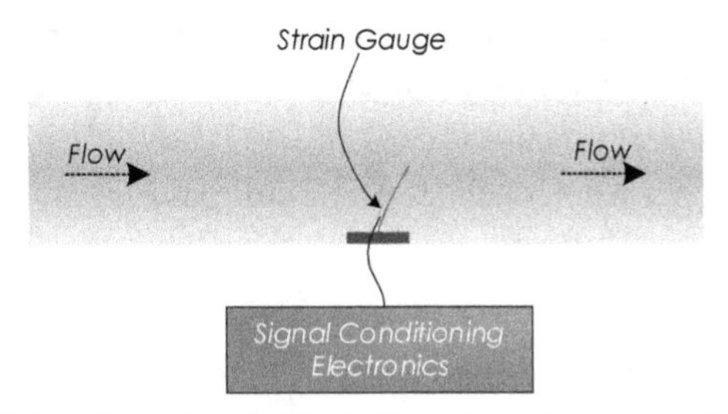

Figure 4.26 – Example of a Strain Gauge used to measure the flux.

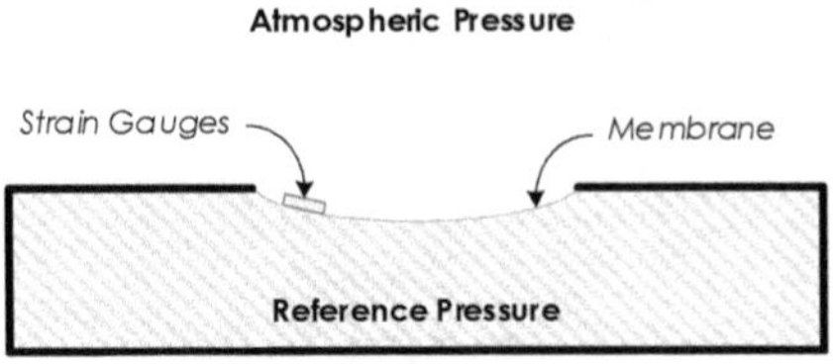

Atmospheric Pressure

Strain Gauges

Membrane

Reference Pressure

Figure 4.27 – Constitution of a barometric pressure sensor.

Another application that uses strain gauges indirectly is the measurement of barometric pressure. A very thin membrane is used that separates the inside of a closed cavity from the outside environment. The pressure inside that cavity does not vary. The variation of the outside atmospheric pressure causes the membrane to deform due to pressure imbalance. The deformation of this membrane is related to this pressure difference and can be measured with one or more strain gauges included in a Wheatstone bridge in order to create an electrical signal (Figure 4.27).

This type of sensor is used, for example, in mobile phones to provide weather information to users, in drones to maintain a constant altitude, and in cars to measure tire pressure.

There are several manufacturers that produce barometric pressure sensors of this type, such as Bosch (model BMP388) and STMicroelectronics (LPS 22HB). There are other manufacturers that produce pressure sensors that use capacity, such as TDK (ICP-10100 and ICM 20789) or that use the piezoelectric effect. These sensors are smart sensors that include, in addition to the sensor element, a microcontroller, and a digital interface, usually I^2C and/or SPI. In addition, they have very small energy consumption with a power of the order of µW.

These sensors can detect very small pressure differences in the order of 1 Pa. Note that an altitude difference of 8.5 cm corresponds to a difference of 1 Pa in atmospheric pressure. It is possible to determine, with a sensor of this type, when each step of a staircase is traversed.

4.10 Questions

1. Calculate the differential gain of an instrumentation amplifier to be used in a signal conditioning circuit to obtain a continuous voltage proportional to the strain in a bar using four metal strain gauges with a Gauge Factor (GF) of 2 to obtain a sensitivity of 1000. Use a Wheatstone bridge powered by 500 mV.

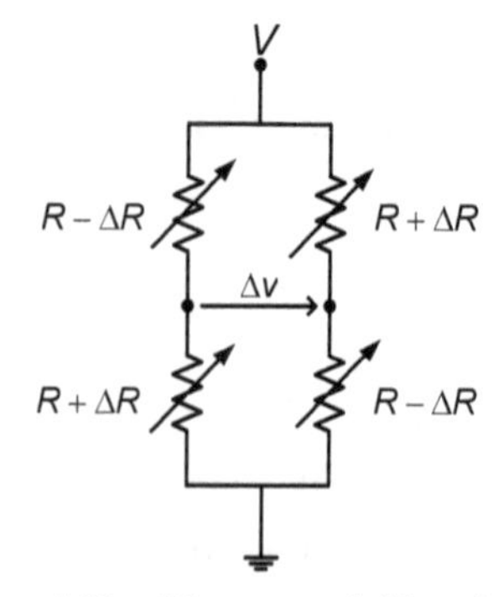

Figure 4.28 – Wheatstone bridge circuit.

2. The error caused by a finite CMRR value of the instrumentation amplifier is intended to be less than 1% when used together with a Wheatstone bridge and four strain gauges, with a Gauge Factor of 2, to measure a strain of 100 με. What is the minimum value of CMRR to use?

3. What is the value of the output voltage of a strain transducer using four metal strain gauges with a Gauge Factor (GF) of 2 and an instrument amplifier with a differential gain of 2000 when measuring a strain of 30 με? Consider that the transducer is powered with 500 mV.

4. Consider a strain transducer using four metallic strain gauges with a Gauge Factor (GF) of 2 and an instrumentation amplifier with a differential gain of 2000. Consider that the transducer is powered with 500 mV. What is the error value in the strain measurement caused by the amplifier's CMRR being 120 dB and not infinite?

5. How does the electrical resistance of metals vary with temperature? Why is that?

6. Describe one of the physical phenomena responsible for the decrease in the electrical resistance of semiconductor materials.

7. Point out at least two positive aspects and two negative aspects of potentiometric displacement sensors.

8. What is the difference between an NTC thermistor and a PTC thermistor?

9. What does a material's resistivity depend on?

    a. The dimensions of the material.
    b. The hardness of the material.
    c. The type and dimensions of the material.
    d. The type of material.
    e. None of the above.

Chapter 5

Devices Based on the Magnetic Field

5.1 Magnetic Properties of Materials

Magnetic properties were discovered in prehistoric times in some species of the iron mineral known as magnetite (Fe_3O_4). It was also discovered that some pieces of iron, when rubbed on a magnetic material, acquired the same properties as the magnetic materials themselves, attracting other pieces of iron and magnets.

There are many similarities between electricity and magnetism. Just as there are two types of electric charges that attract and repel each other, there are two types of magnetic poles that attract and repel (the North Pole and the South Pole). The magnetic poles always come in pairs contrary to electric charges.

5.2 Concept of Magnetic Field

In the same way that the concept of Electric Field was introduced to represent the phenomenon of the existence of forces between particles with electric charge, also the concept of **Magnetic Field** was created to represent the forces that exist between magnetic and magnetizable materials. By placing an iron piece in the proximity of a magnet, this piece will suffer the action of a force that does not exist if the magnet was not present.

The forces applied to the magnetic materials also have a direction (from the North Pole to the South Pole) and, because of that, it is said that the magnetic field is a vector field. If a small magnet is placed near a large magnet, the north pole of this small magnet will be attracted to the south pole of the large magnet and vice versa. Hans Christian Oersted discovered that magnetic fields could also be created by electrical currents, i.e., electric charges in motion. One of the things that this discovery allowed was the explanation of the reason why there are materials that are magnetic and others that are not and where does the magnetic field of these materials came from. The electrons of an atom are continuously in movement. The movement of each electron can be seen as constituting a small electric current around the atom nucleus. These currents are associated with a magnetic field. The combination of the magnetic fields of the currents of several electrons of the atom results in a considerable magnetic field if these atoms have a concordant orientation. This happens in materials that have an appropriate chemical composition and crystalline structure — ferromagnetic materials.

5.3 Magnetic Force

A charged particle (q) in a magnetic field ($\vec{B}$) moving with a certain velocity ($\vec{v}$) feels a force given by

$$\vec{F} = q \cdot \left(\vec{v} \times \vec{B}\right), \tag{5.1}$$

which is perpendicular to both the velocity and the magnetic field, as shown in Figure 5.1.

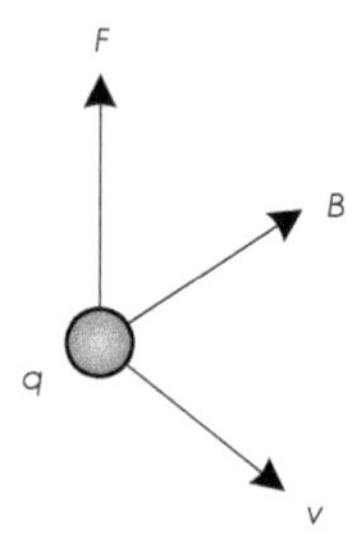

Figure 5.1 – Magnetic force felt by an electric charge moving in a magnetic field.

The force that acts on an electric charge in the presence of both an electric and a magnetic field is

$$\vec{F} = q \cdot \left(\vec{E} + \vec{v} \times \vec{B}\right) \tag{5.2}$$

which is known as "Lorentz's Force." This force is used, for instance, in the case of the Hall effect.

5.4 Stepper Motor

5.4.1 *Introduction*

A stepper motor, like other motors, is used to make things (loads) move. The goal of a movement can be to change the position of the load or give it a certain velocity or acceleration. There are several forces that naturally oppose the change in position or speed of a body, namely inertia, friction, gravity, springs, etc. It is, therefore, necessary that a motor can exert enough force to overcome these other forces that may be acting on a given body.

In general, the motors used are rotary motors, i.e., the bodies are rotated around a fixed axis. When we want a linear displacement, we often use a rotary motor and a mechanical device that transforms this rotation into linear movement. An example of this is the automobile. The rotation of the car motor makes the wheels rotate, which, on the other hand, and due to the friction that they have with the pavement, makes the car move linearly.

A stepper motor is different from other motors in as much as the movement of rotation is not uniform. In electric motors of DC or AC type, for example, the applied voltage is proportional to the speed of rotation. If the voltage stays constant, the rotational speed will stay constant too. If we wish to apply a given amount of acceleration to a body, we just increase or decrease this voltage (positive or negative acceleration). If the goal, however, is to position a body in a certain location using these types of motors, a complex control circuit is required to adjust the voltage applied to the motor. Necessarily, a position sensor is required in these cases. The closed-loop control circuit uses this information to increase or decrease the applied voltage of the motor until it reaches the desired position (Figure 5.2).

It is exactly in applications that require an easy positioning of a body that stepper motors are used. The control of those motors is simple since they can operate in an open loop (Figure 5.3).

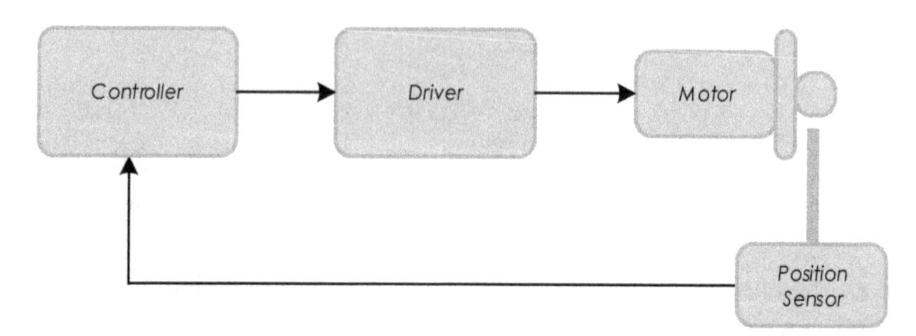

Figure 5.2 – Closed-loop control of an AC or DC motor in positioning applications.

Figure 5.3 – Open-loop control used with stepper motors in positioning applications.

One of the most common applications for this type of actuator is in robotic arms. It is possible to perform the same movements repeatedly using stepper motors. Another application is in milling machines that need to move the cutting element (drill, for example) to very exact locations with high precision to mold or perforate a certain part, whether in metal, wood, or even plastic. Other applications are, for example, welding, textiles, traditional paper-based printing, and 3D printing.

Stepper motors are constructed such that the motion of its axis occurs by fixed amounts (steps); that is, each time "a command is sent," the motor rotates a fixed amount. For instance, in a stepper motor where the step amount is 7.5°, and one wants to rotate it by 30°, it is necessary to issue that "command" four times for the motor to advance four steps.

An analogy that can be made to illustrate the difference of motion between the stepper motor and continuous motors is, for example, ladder climbing by a person versus the ascending of a ramp by an automobile (Figure 5.4). A person climbs the ladder one step at a time. An automobile, in turn, easily ascends a ramp with constant velocity. It becomes more difficult, however, for an automobile to travel on the ramp an exact distance while a person easily does it on a ladder.

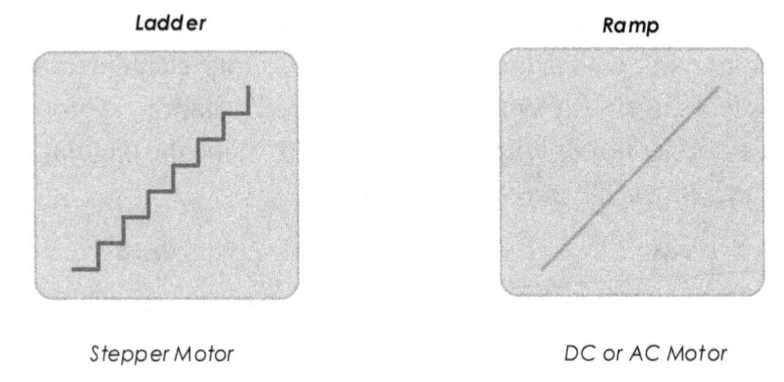

Figure 5.4 – Analogy between motors (stepper and DC/AC) and the ascend of a ladder or ramp.

A motor, therefore, is a transducer that converts electrical energy into mechanical energy. Most motors use the magnetic field, although there are some motors, called electrostatic motors, that use the electric field. Nowadays, electrostatic motors are used in micro-mechanical systems (MEMS). The motors that use the magnetic field produce forces by the interaction of the magnetic fields created by electrical currents or permanent magnets.

There are three types of stepper motors according to the constitution of their rotor:

- Permanent Magnet — The rotor is a permanent magnet.
- Variable Reluctance — The rotor is a toothed iron wheel.
- Hybrid — Combines the two types of rotors.

5.4.2 *Permanent Magnet Motor*

In Figure 5.5, the operation of a stepper motor with a permanent magnet rotor in which the stator is made of four electromagnets is depicted. In step 1, the rotor is in the vertical position. The upper electromagnet is energized, creating a south pole and a north pole in a way that the south pole of the rotor is attracted up. In the next step (step 2), the upper electromagnet is switched off, and the rightmost electromagnet is turned on. The south pole of the rotor will thus be attracted to the right towards the north pole of the left electromagnet. In step 3, this magnet is switched off, and the lower electromagnet is turned on so that a north pole is created in its upper part to attract the south pole of the rotor. Finally, in step 4, the left-most electromagnet is switched on, making the rotor rotate an extra 90° so that its south pole is facing the left. These four elementary steps are repeated successively to perform any desired rotation. This example illustrates a stepper motor with four steps of 90° since the angular position of the rotor changes 90° in each step.

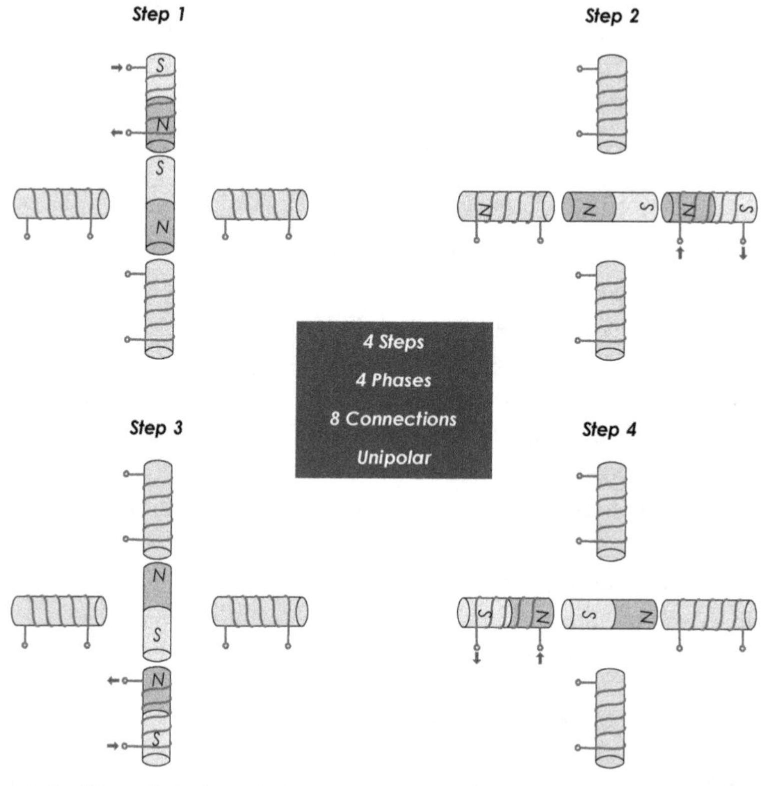

Figure 5.5 – Illustration of a unipolar stepper motor with a permanent magnet rotor with four phases and 90° steps.

There are, however, some variants that lead to increased efficiency of the stepper motor. They consist of the use of more than one electromagnet at the same time, the reversal of the direction of current in each electromagnet, and the use of different current values. These different variants are translated into advantages in the operation of the motor in terms of available torque, energy consumption, resolution, and number of connecting wires required.

Figure 5.6 shows the electric circuit that represents the stepper motor of Figure 5.5 with four phases and eight wires.

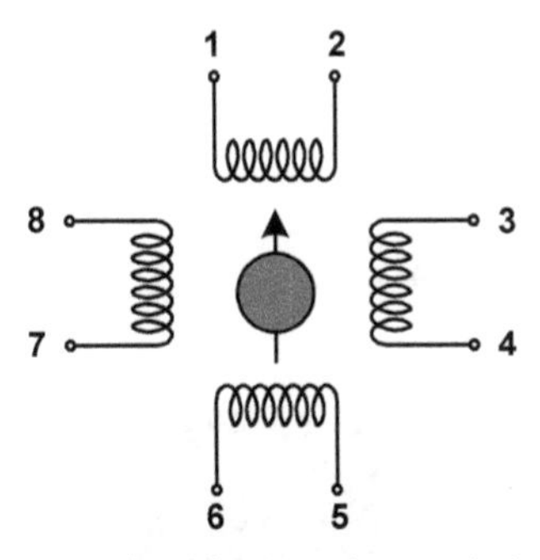

Figure 5.6 – Illustration of the number of connections and phases in the case of a unipolar stepper motor with four phases and eight wires.

The same type of motor, with 4 phases, could have only five wires if one of the terminals of each winding was connected to the same point, as illustrated in Figure 5.7.

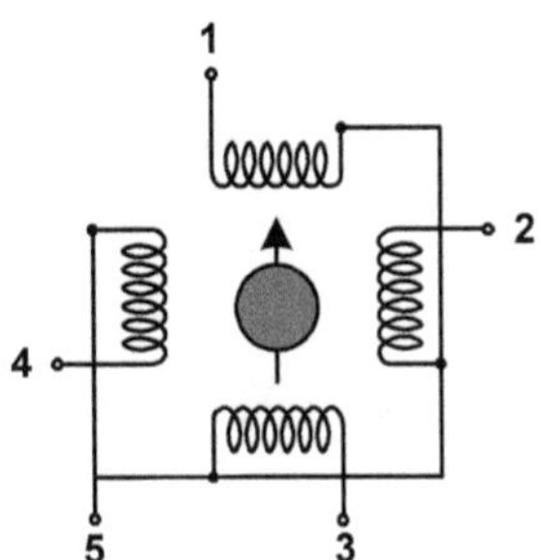

Figure 5.7 – Illustration of the number of connections and phases in the case of a unipolar stepper motor with four phases and five wires.

One of the variants that we find in stepper motors has to do with the number of phases. The phases are the number of separate windings. The motor illustrated in Figure 5.5 is a **4-phase** motor since each of the four electromagnets operates independently. Regarding the direction of the current, it is said that this engine is **unipolar** because the direction of the current in each winding is always the same. As for the number of connections, eight wires are required to connect four windings. This type of engine is the one that has less torque because only one winding exerts a force for each time.

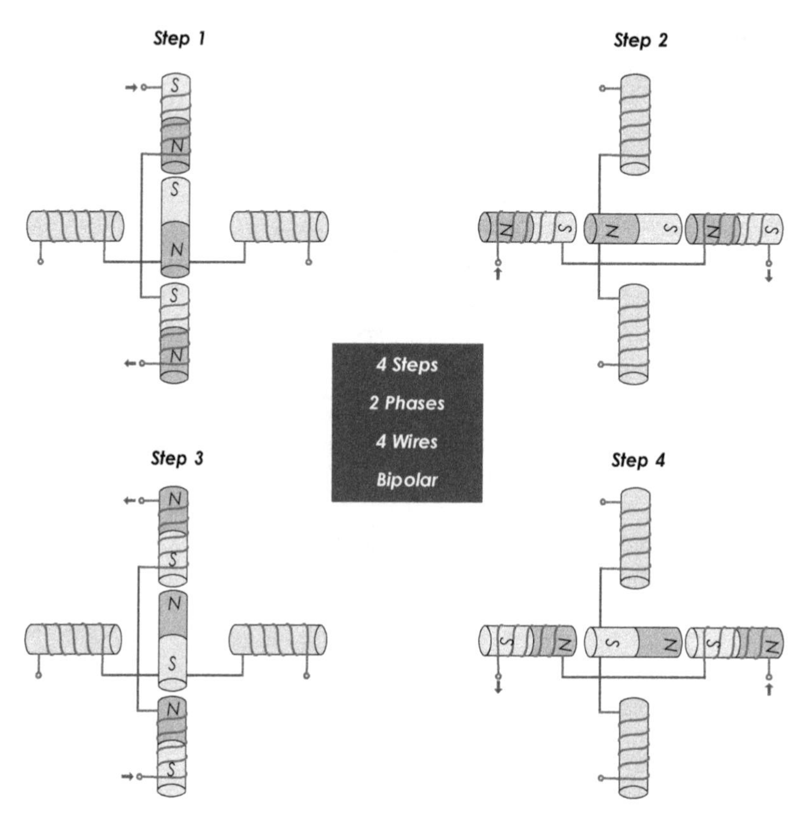

Figure 5.8 – Illustration of a bipolar stepper motor with a permanent magnet rotor with two phases and 90° steps.

Figure 5.8 illustrates the operation of a stepper motor like that of Figure 5.5 but in which two opposing windings are energized at the same time. Each pair of windings is in series and connected so that they produce a magnetic field in the same direction. Therefore, in this case, there are only two phases.

In terms of wires, only four are needed, as shown in Figure 5.9.

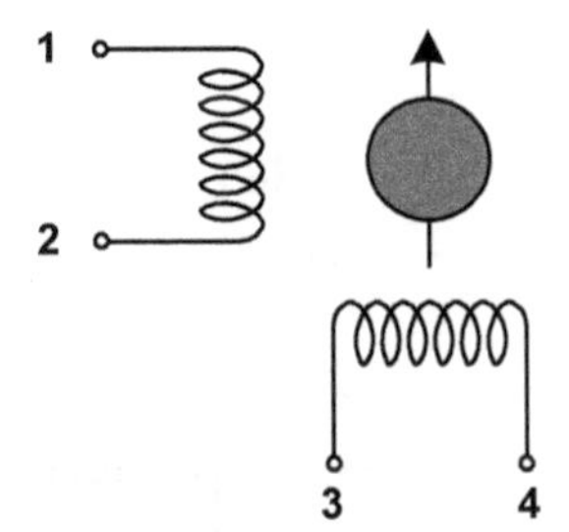

Figure 5.9 – Illustration of the number of connections and phases in the case of a bipolar stepper motor with two phases and four wires.

There are situations where the common point of two electromagnets of one phase is also available externally, as shown in Figure 5.10. This requires more connections (six instead of four) but allows the motor to operate in unipolar or bipolar mode.

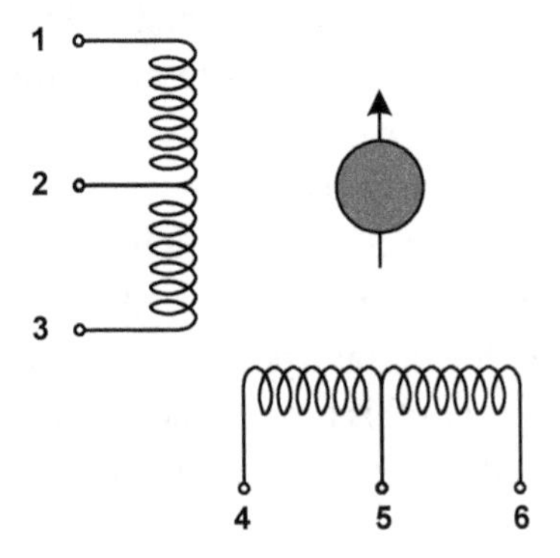

Figure 5.10 – Illustration of the number of connections and phases in the case of a bipolar stepper motor with two phases and six wires.

The use of unipolar or bipolar excitation depends on the motor controller and its capacity to reverse the direction of the current. In unipolar controllers, it is only possible to apply the current in one direction, and thus, to each phase, only one winding is connected to the current source (Figure 5.11). These are simpler controllers, but the motor will have less torque.

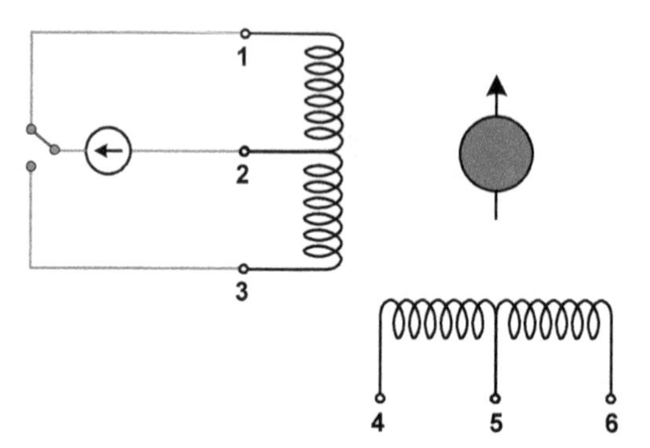

Figure 5.11 – Illustration of the connection of a current source to a bipolar stepper motor with two phases and six wires.

One way to increase the number of steps of a stepper motor is applying current in both consecutive windings so that the rotor is pointed in an intermediate direction (*half-step*), as illustrated in Figure 5.12.

Another way to reduce the step size is to gradually switch the current from one phase to the next (*micro-stepping*).

The rotor in the permanent magnet motor, in the configuration shown here, has the drawback that, for certain positions, the distance to the stator is greater, which decreases the force to which it is subjected and consequently its torque (for certain positions). The variable reluctance motor, shown below, improves this aspect since the rotor has many "teeth" so that some of them are always very close to the stator windings.

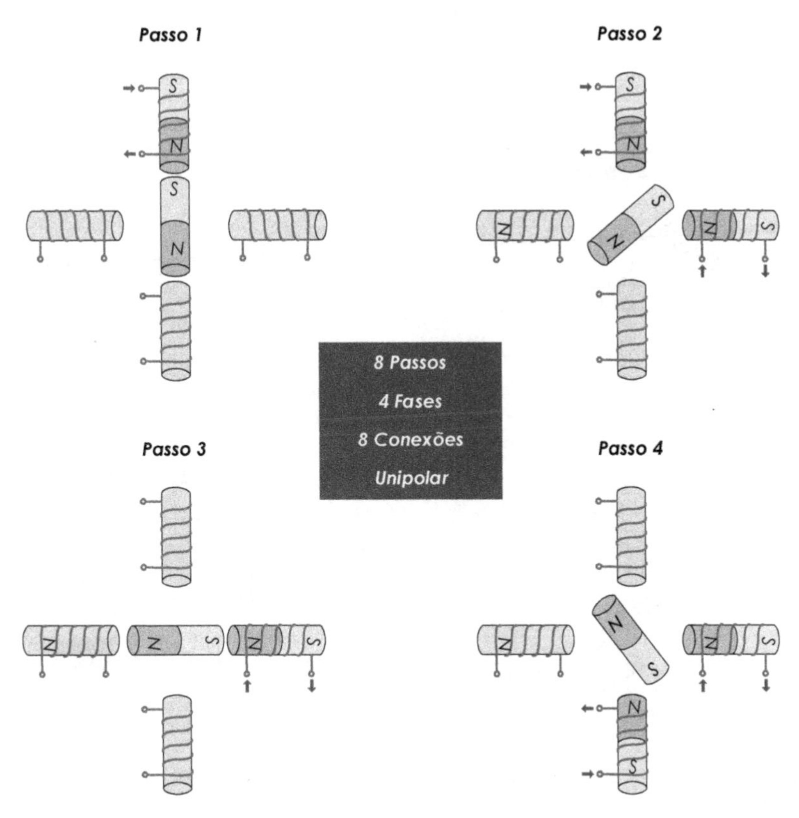

Figure 5.12 – Illustration of a unipolar stepper motor with a permanent magnet rotor with four phases and eight steps of 45° each.

5.4.3 *Variable Reluctance Motor*

Another type of stepper motor is the **Variable Reluctance** one. Figure 5.13 shows the constitution of a motor of this type. The rotor is made of a magnetizable material, typically iron, but which is not a permanent magnet. The rotor has the form of a toothed wheel. In the example of the figure, there are eight teeth. The stator is made of several electromagnets (eight in the case of the figure) connected in this case into four phases; that is, the same wire is used to make multiple winding coils (two in the case of the figure). Each phase is represented by a letter from A to D in order not to create confusion with the step number used in the description of the operation. The wire for each phase is depicted with a different color.

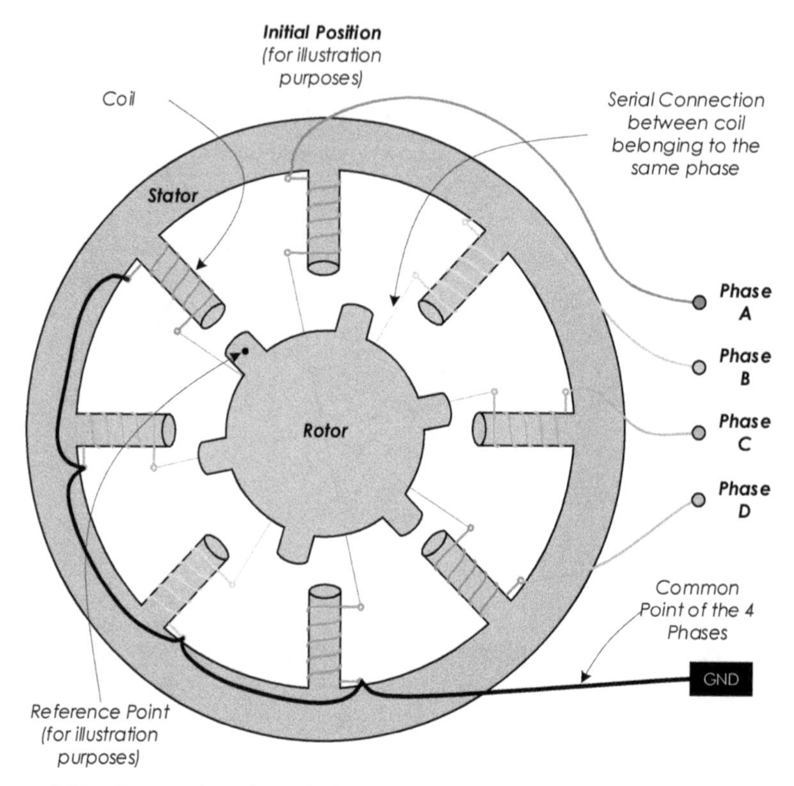

Figure 5.13 – Construction of a variable reluctance stepper motor with eight coils arranged into four phases.

There are thus eight terminals (two per phase). The phases will be energized one at a time. The number of terminals can thus be decreased from eight to five by connecting one end of each phase to a common point designated GND in the figure. There are thus four terminals corresponding to the four phases. Figure 5.14 shows the sequence of four consecutively repeated steps to make the rotor move.

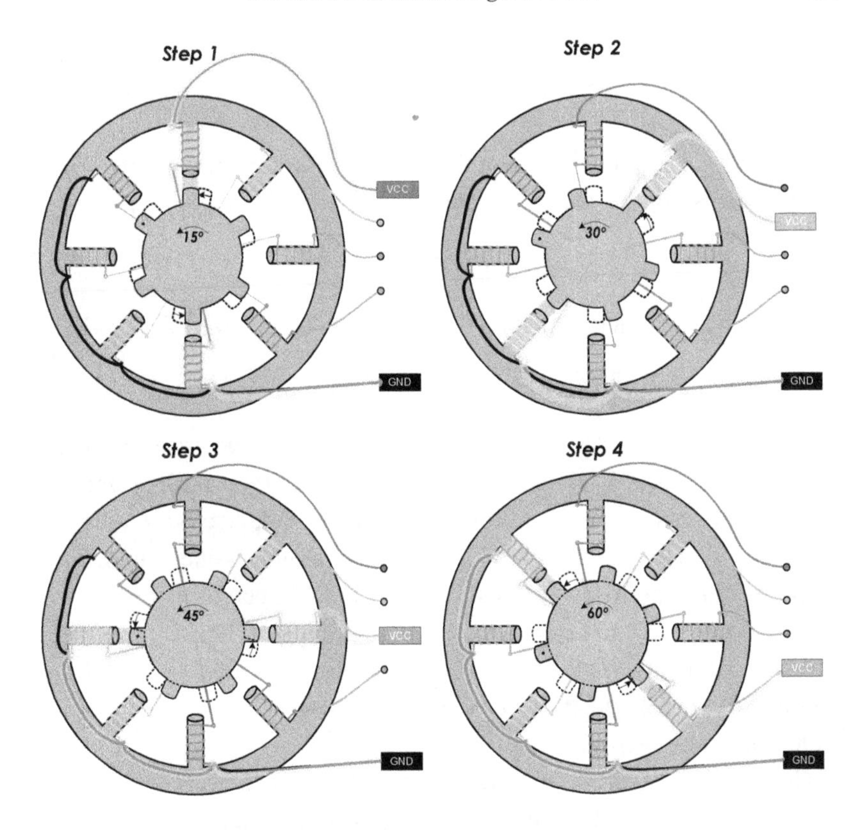

Figure 5.14 – Operation of a variable reluctance stepper motor.

From the initial position of Figure 5.13, a current is applied in phase A as shown in Step 1 of Figure 5.14. This is generally done using a DC voltage source represented in the figure by the VCC and GND labels. The magnetic field created in the two windings of phase *A*, which are positioned vertically and drawn in brown color, magnetizes the rotor attracting it so that its teeth are aligned with these windings. In the example shown, this corresponds to a rotation of 15° counterclockwise, as indicated in Figure 5.14 (*Step 1*). Note that in the other windings, there is no current applied on this step.

The current in Phase *A* is then removed and is applied to phase *B* (windings shown in yellow). The rotor attempt to align itself with the magnetic field created by these two new windings making it rotate a further 15° counterclockwise (*Step 2*). The procedure for phases *C* and *D* are repeated, and the rotor rotates an additional 15° each time. It is necessary to repeat this sequence of four steps (60°) six times to perform a complete revolution (360°).

The angular distance between the N_R teeth of the rotor and the N_S poles of the stator is not the same. The step size is thus given by

$$Step\ Size = \frac{360}{N_R} - \frac{360}{N_S}. \tag{5.3}$$

In the case depicted in Figure 5.13, one has a step size of 15°:

$$Step\ Size = \frac{360}{6} - \frac{360}{8} = 60^{\underline{o}} - 45^{\underline{o}} = 15^{\underline{o}}. \tag{5.4}$$

In this example, the engine needs to perform twenty-four steps to complete a full turn. With a different number of stator windings and rotor teeth, it is possible to build motors of variable reluctance with a different number of steps per revolution. The most common value found on the market is 200 steps, which corresponds to an angle between steps of 1.8°.

The angular resolution can be improved using micro-stepping or gearboxes. In many applications, stepper motors are used to create linear motion. Depending on how it is done, it is possible to obtain spatial resolutions of a few μm.

These motors can have smaller steps than those with a permanent magnet, which allows for a more uniform movement and a more accurate positioning.

As these motors have a rotor that is not a permanent magnet, the torque is less than in the first type of motor shown. It is possible to combine the best of these two types of motors in a third type called a hybrid motor that has a rotor with a permanent magnet and teeth. The torque is thus increased because it has a greater magnetic force due to the use of a permanent magnet and, at the same time, greater proximity between stator and rotor with a consequent increase in the magnetic force.

5.4.4 *Hybrid Motor*

The **Hybrid** Stepper Motor combines the characteristics of permanent magnet motors and variable reluctance ones. The rotor has teeth that are magnetized by a permanent magnet. The teeth are engraved on one end of the rotor shaft, on the side of the north pole (left side of the figure), and separately on the right side where the south pole of the permanent magnet is located. This makes it possible to have a rotor with more teeth — typically 50. These two sets of teeth are 180° offset; that is, the teeth on one end are aligned with the space between teeth on the other end, as shown in Figure 5.15.

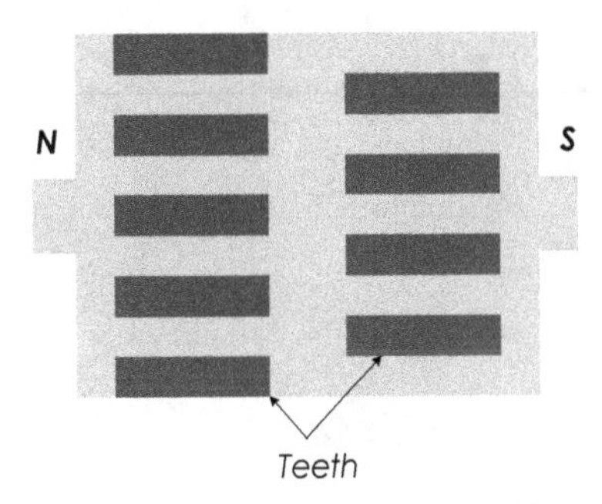

Figure 5.15 – Rotor teeth alignment in a hybrid stepper motor.

The hybrid motor stator also has teeth, in addition to several sets of windings. Figure 5.16 shows the construction of a hybrid motor with 48 teeth in the stator and 50 teeth in the rotor. In the top view shown, the appearance is the same as that of a variable reluctance motor. The difference between the two types is that the rotor has a slightly different upper and lower part (not shown in Figure 5.16). As previously mentioned, the upper part is one of the poles of a permanent magnet, and the lower part is the other pole of that permanent magnet). In addition, the teeth at the top and bottom of the rotor are misaligned, as shown in Figure 5.15.

The misalignment of the teeth of the North and South poles of the rotor makes it possible to have both a force of attraction and a repulsion between the rotor and the stator, which leads to a greater torque.

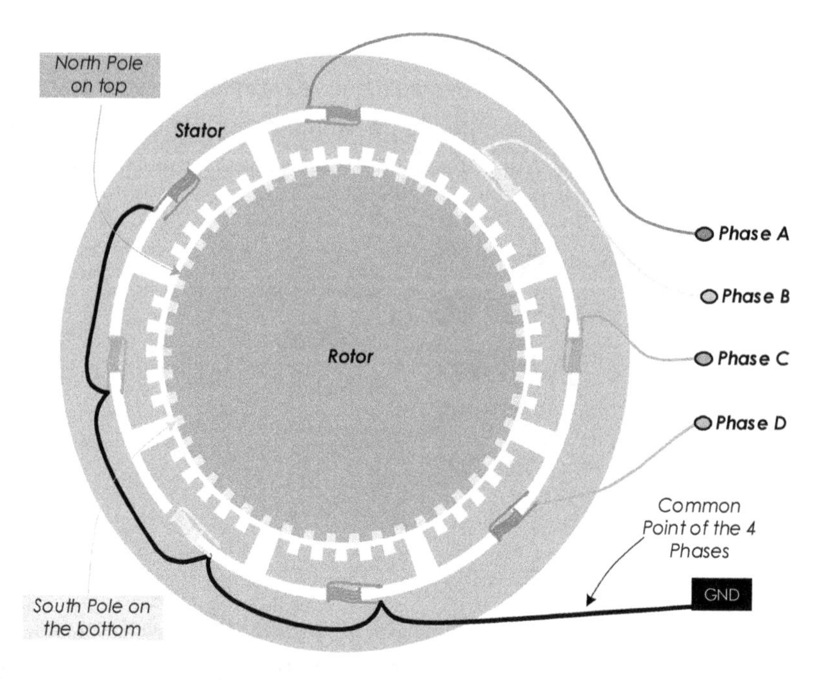

Figure 5.16 – Construction of a hybrid stepper motor with 48 teeth in the stator and 50 teeth in the rotor. The teeth of the north pole and the south pole of the rotor are out of phase.

5.4.5 *Specifications of Stepper Motors*

In stepper motors, it can happen that the rotor does not rotate the number of steps that are expected due to an elevated load torque. The motor torque depends on the angle of the rotor. The torque is created by the attraction between the stator windings and teeth (or poles) of the rotor. Only when they are **not** aligned is there an attraction force and, therefore, torque. Note that it is only the component of the attractive tangential force to the rotor that produces rotation and thus torque (Figure 5.17).

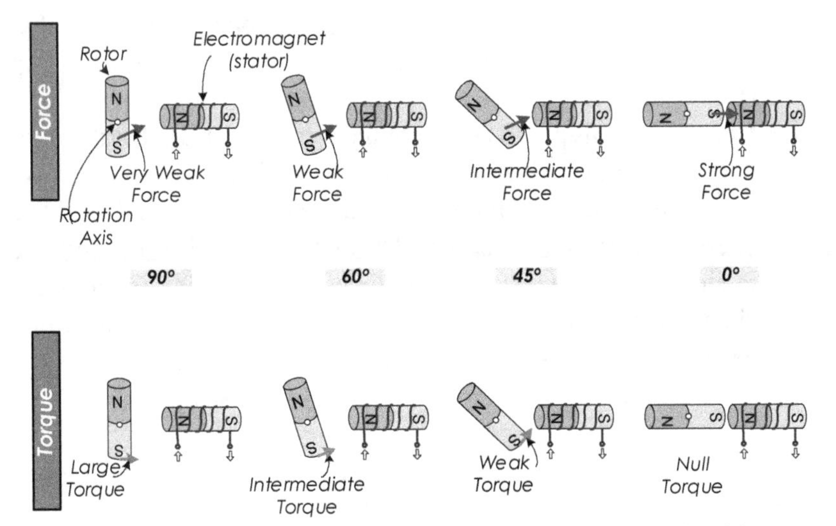

Figure 5.17 – Attractive force between a stator electromagnet and a pole of the rotor.

Another phenomenon that happens is that the closer the pole of the rotor is to the stator, the greater is the force of attraction. As shown in the figure, the force increases as the rotor approach the stator, but the torque decreases because the tangential component of the force decreases.

There are four parameters that specify the torque of a stepper motor:

- Residual torque (Detente Torque) — Torque is required to make the rotor move when no current is applied to the motor.
- Static or Binary Retention (Holding Torque) — Torque that must be applied to the rotor to move it when it is not moving, and there is current applied to the motor.
- Starter torque (Pull-in Torque) — Torque against which the rotor can accelerate from rest without missing any steps.
- Operating torque (Pull-out torque) — Torque against which the engine can function at a constant speed.

Note that the torque depends on the position as well as the velocity of the rotor.

There are two modes of operation of a stepper motor:

- **Single Step** — One step at a time. The rotor stops between the steps. If the load is heavy, it is possible that the rotor oscillates before stopping in the final position due to inertia. Each step is independent, and the rotor may change its rotation direction after each step. There is less chance that the motor will miss a step (Figure 5.18, left).

- **Continuous Step** — The rate of the steps is so high that the rotor does not stops between steps (Figure 5.18, right). The rotor always has torque, and therefore the movement is smoother. The operation is more like that of a DC or AC motor. It is not possible to stop or reverse instantly the direction of rotation (there is the chance that one or two extra steps are made). It is, therefore, necessary to increase or decrease the speed slowly.

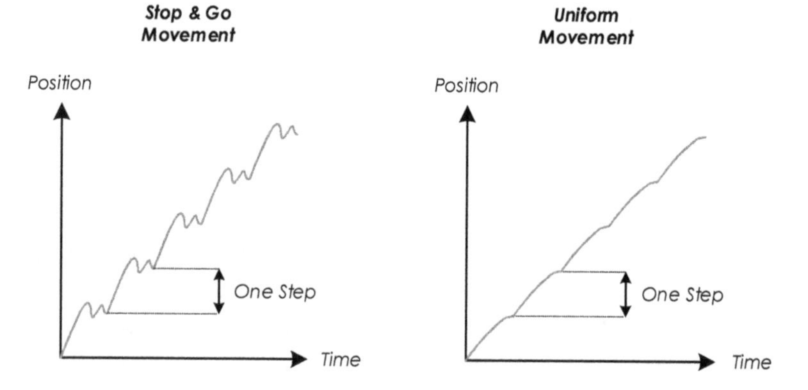

Figure 5.18 – Illustration of the evolution of rotor position in a stop&go movement (left) as well as uniform movement (right).

Stepper motors are ideal for applications where what is most important is position control because it eliminates the use of sensors to determine the current position. They are often operated in an open-loop. The steps are counted from an initial reference to know in each moment the rotor's position. The step size of a motor depends on the number of poles or teeth of the rotor and on the number of stator poles.

The use of stepper motors has several advantages over other types of motors (DC and AC):

- It is possible to operate at low speed;
- Position control without the use of external sensors;
- No brushes subject to wear (higher lifetime);

- The residual torque allows the motor to keep its position without current applied (only in the case of permanent or hybrid motor type);
- Easy to control.

This type of engine has, however, some disadvantages:

- It is difficult to produce a smooth motion;
- It has a limited torque;
- The torque decreases with speed;
- Slips can happen, and the position information is corrupted.

One can also produce linear movements using a stepper motor and a screw, as illustrated in Figure 5.19 is an example of the positioning of the reading/writing head in a computer hard disk.

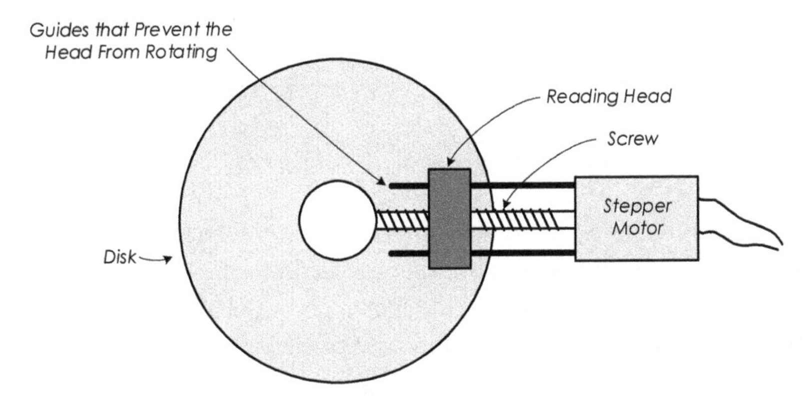

Figure 5.19 – Transformation of a rotational movement into a translational one using a screw.

Figure 5.19 presents an example of the use of a stepper motor to position a reading/writing head on a rotating disc. This is used, for example, in DVD players or computer hard drives.

5.5 Hall Effect

The Hall effect consists of the creation of an electric voltage in a conductor traveled by a current and immersed in a magnetic field. The electrons (with charge q) moving in this conductor with velocity $\vec{v}$, are subject to a force $\overrightarrow{F_B}$ (Lorentz Force) due to the external magnetic field ($\vec{B}$)

$$\overrightarrow{F_B} = q \cdot \left(\vec{v} \times \vec{B}\right) \tag{5.5}$$

which pushes them towards one side of the conductor, as shown in Figure 5.20 (note that the charge of the electron has a negative value).

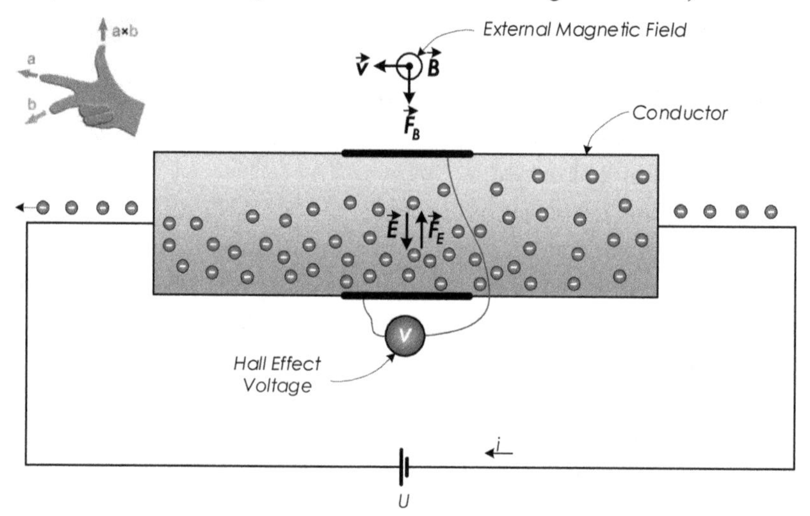

Figure 5.20 – Illustration of the Hall Effect.

This leads to more electrons on one side of the conductor than the other. This non-uniform distribution of charges leads to the emergence of an electric field ($\vec{E}$) which opposes the movement of charges which is due, again, to the Lorentz force $\overrightarrow{F_E}$ (in this case of electrical origin):

$$\overrightarrow{F_E} = q \cdot \vec{E} \,. \tag{5.6}$$

This establishes a balance between the effect of the external magnetic field and the internal electric field (perpendicular to the current).

$$\overrightarrow{F_E} = \overrightarrow{F_B} \,. \tag{5.7}$$

The value of the internal electric field is obtained from (5.5) and (5.6)

$$q \cdot v \cdot B = q \cdot E. \tag{5.8}$$

The value of the internal electric field is therefore

$$E = v \cdot B. \tag{5.9}$$

This unbalance leads to a potential difference (V) between the conductor faces spaced by l given by

$$E = \frac{V}{l} = v \cdot B. \qquad (5.10)$$

The potential difference between the two sides is therefore

$$V = l \cdot v \cdot B. \qquad (5.11)$$

If one wants to express this voltage as a function of the current flowing through the conductor, we can use the following relation between the average electron vectors and the electric current value (I):

$$v = \frac{I}{q \cdot n \cdot A} \qquad (5.12)$$

where A is the cross-sectional area of the conductor and n is the free electron density. In the case of copper, for example,

$$n = 8.5 \times 10^{28} \frac{electrons}{m^3}. \qquad (5.13)$$

Therefore, the potential difference can be determined with

$$V = I \cdot B \cdot \frac{l}{A} \cdot \frac{1}{q \cdot n}. \qquad (5.14)$$

Using the value of the free electron density in copper (n) and the charge value of the electron ($q = -1.6 \times 10^{-19}\ C$), we can write

$$V = -I \cdot B \cdot \frac{l}{A} \cdot 73{,}53 \cdot 10^6. \qquad (5.15)$$

The Hall effect allows the creation of sensors to measure current, magnetic field, position, velocity, and flow, among others.

5.6 Displacement Sensor using the Hall Effect

A sensor based on the Hall effect intrinsically measures the magnetic field or the electric current. In order to measure the position of an object, it is common to use a magnetic field source (permanent magnet) and to build an assembly in which the object to be measured interferes with the field created and which is measured by the Hall-effect sensor. This idea can be used to measure the displacement because the magnetic field intensity decreases with the distance to its source, as shown in Figure 5.21.

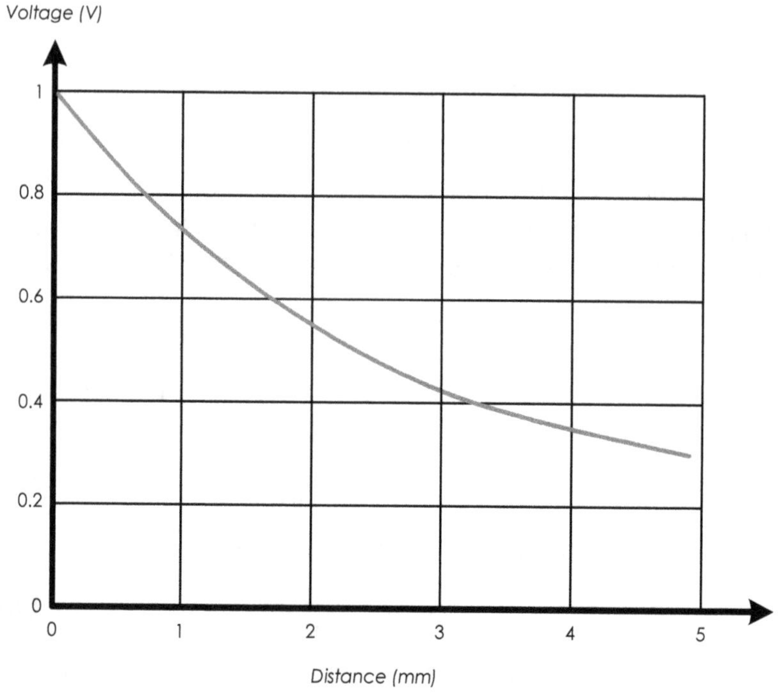

Figure 5.21 – Dependence of the voltage of a Hall Effect sensor on the distance to a magnet.

It is thus possible to create a displacement sensor using the Hall effect together with some signal conditioning electronics. The relationship between the magnetic field and the output voltage is usually quite linear.

The Hall effect sensor can also be used as a switch, as illustrated in Figure 5.22. A permanent magnet creates a magnetic field perpendicular to the Hall effect sensor. When a ferromagnetic piece is interposed between the magnet and the sensor, the measured field decreases significantly, indicating the presence of this piece.

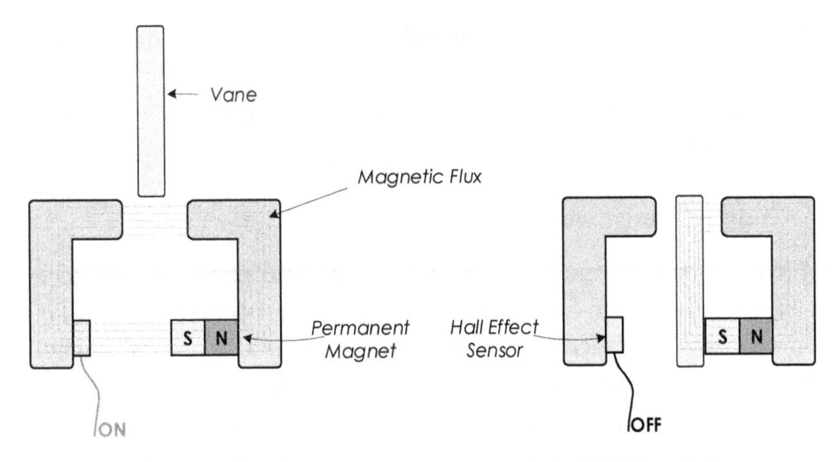

Figure 5.22 – Illustration of the operation of a Hall Effect switch.

Notice that there is no direct contact between the moving and fixed part of the sensor, which leads to less wear, a longer lifetime, and less noise.

Another example of the use of Hall effect sensors is in the measure of the rotational speed of a wheel. In Figure 5.23, an example is seen where a permanent magnet is used to create a magnetic field that is disturbed by the passage in the vicinity of the teeth of the wheel. The placement of a Hall Effect sensor between the wheel and the magnet makes it possible to detect the small variations of the magnetic field created by the rotation of the wheel.

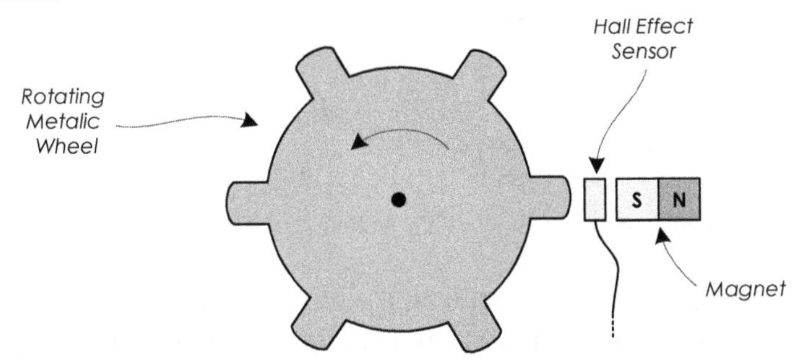

Figure 5.23 – Illustration of how a Hall Effect sensor can be used to measure rotational velocity.

This kind of sensor is often used in automobile engines to control the ignition of the sparks as the motor turns.

Hall Effect sensors have, in general, a high resolution and robustness, but they are susceptible to temperature. They also present a residual voltage due to the presence of external magnetic fields (like the Earth's magnetic field).

5.7 Magnetoresistance

Magnetoresistance is the variation of the electrical resistance due to the presence of a magnetic field. The giant magnetoresistive effect (GMR — Giant Magnetoresistance) is obtained using multiple layers of magnetic and non-magnetic material (Figure 5.24). The layers of magnetic material have opposing polarities in the absence of an external magnetic field. This increases the difficulty with which the electrons can flow through the material since each time they penetrate a magnetic layer; they have to orient their spin according to the direction of the local magnetic field. Since this orientation changes between the magnetic layers, the electrons keep changing their orientation which means that their kinetic energy is reduced.

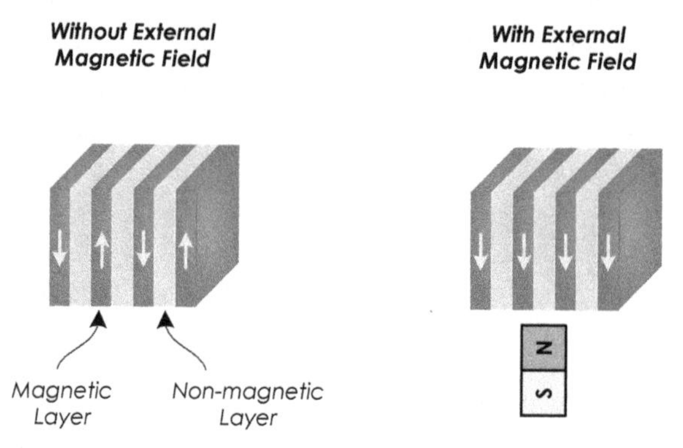

Figure 5.24 – Illustration of the operation of a GMR sensor.

When an external field is applied, as illustrated in Figure 5.24(b), all the magnetic layers get the polarization aligned which facilitates the flow of current.

This type of sensor is typically connected to a Wheatstone bridge to produce a voltage proportional to the field. However, as shown in Figure 5.25, the transfer characteristic is not linear for the entire range of values of the magnetic field. There is a discontinuity in the slope of the curve around $H = 0$.

What we do in practice is to polarize the sensor with a value of magnetic field so that the operating point is in the middle of the curve to have a more sensitive and linear characteristic. This type of linearization is called **Linearization in the Source by Polarization of the Sensor**. It is used when what is important is the variation of the quantity to be measured and not its absolute value.

One of the applications of this type of sensor is the measurement of the magnetic field in computer hard drives due to its high sensitivity and ability to detect small magnetic fields. This has permitted the increase of the number of bits stored in the surfaces of magnetic disks.

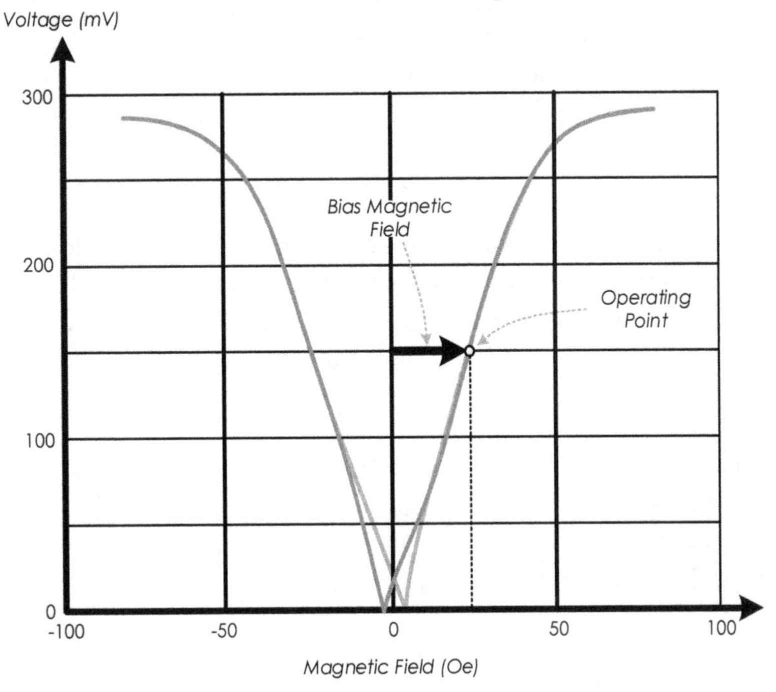

Figure 5.25 – Output voltage of a GMR sensor as a function of the magnetic field.

5.8 Magnetostriction

Another effect of magnetic origin is **Magnetostriction**. This effect is the deformation of crystalline structures due to the application of magnetic fields in ferromagnetic materials (Figure 5.26).

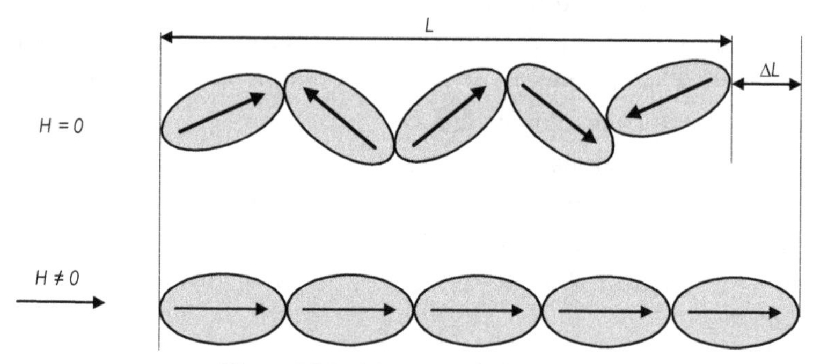

Figure 5.26 – Magnetostriction Effect.

The strain,

$$\epsilon = \frac{\Delta L}{L}, \tag{5.16}$$

is, in general, very small and depends on the intensity and direction of the magnetic field applied. The relationship between deformation and the field intensity is not linear and depends on the type of material. In the case of iron, for example, the length increases for small fields but decreases for large fields. In the case of Nickel, the length always decreases whatever the value of the applied field.

Typical values of maximum strain in materials that occur in nature are in the order of 10 ppm. An artificial material in which the effect of magnetostriction is particularly pronounced is an alloy formed by Terbium, Iron, and Dysprosium, with a maximum strain on the order of 1000 ppm.

This effect is responsible for the sounds that are heard near transformers and ballasts for fluorescent lamps and can be used, for example, in the creation of ultrasound.

The reverse is also true, i.e., the deformation of a ferromagnetic material can cause the appearance of a magnetic field. This effect, called the **Villari Effect,** can be used to make torque, pressure, and ultrasonic sensors, for example.

Two other effects related to the magnetostriction effect are the Wiedemann and Matteucci effect. The **Wiedemann Effect** is the torsion of a material in the presence of a helicoidal field. The **Matteucci Effect** is the opposite, i.e., the appearance of a magnetic field when a material suffers torsion. This effect can be used to create, for example, torque sensors.

5.9 Magnetostrictive Torque Sensor

Magnetostriction can be used to create torque sensors. The change in the magnetic characteristics of a material, when submitted to deformation, can be used to measure a torque that causes the deformation.

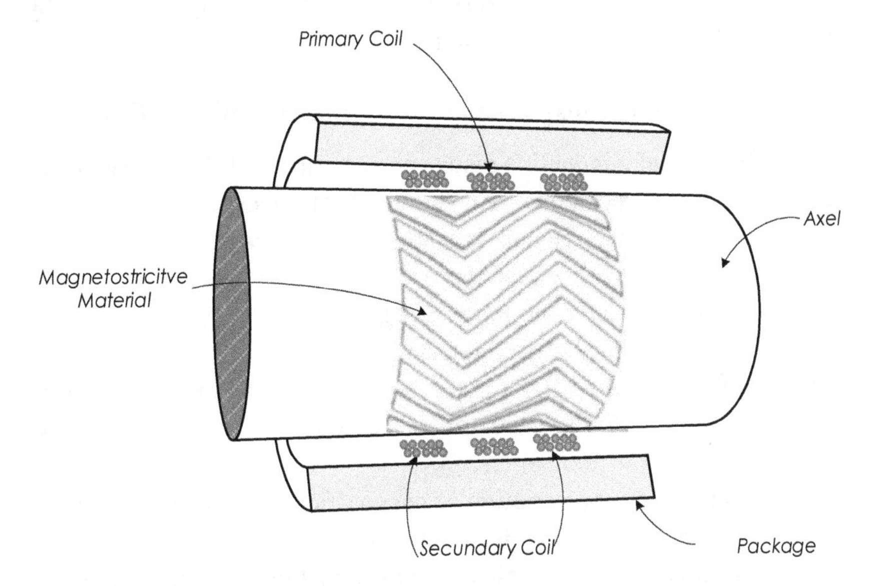

Figure 5.27 – Illustration of the coils used in the signal conditioning of a magnetostrictive torque sensor.

This sensor is made of a shaft that has part of its length covered by a magnetostrictive material. The torsion of the shaft changes the magnetic properties of the material, including its permeability (μ). By placing this material in an electrical circuit, it is possible to produce a voltage proportional to the torque. Figure 5.27 shows the arrangement of the three coils placed around the shaft. The magnetostrictive material is placed in the shaft around the area of each coil so that torsion has opposite effects in each coil.

This type of non-contact torque sensor has extensive applications, especially in the automotive industry. It can be used to measure the engine torque and to obtain the torque/rotational speed curve in real-time. It can also measure the torque caused by each individual cylinder of the engine. Another application is in a car's drive train allowing one to monitor the operation of the gearbox to optimize its operation and alert of any anomalies due, for instance, to wear. Also, in the differential, it is possible to improve the traction by preventing slipping of the wheel by measuring the torque immediately before the slipping occurs. Also, in cars, it is possible to create assisted steering systems and adaptive ones by combining a torque sensor with a motor.

5.10 Inductive linear displacement Sensor of variable Air Gap

It is possible to make inductive displacement sensors if the displacement changes the flow connected to a given circuit. This can be done by:

- Variation of the reluctance: variation of the coefficient of mutual induction or between circuits:
  - $L\ (l)$ - variable air gap sensors;
  - $L\ (\mu)$ - movable core sensors;
- Induced electromotive forces: variation of electromotive force in a circuit due to the variation of the area illuminated by the flow provided from another circuit.
- Induced currents: variation of the coefficient of induction (L) due to the presence of a body conductor.

The definition of the coefficient of self-induction is the ratio between the flow attached to the surface defined by a coil conductor and the current that flows in that conductor,

$$L = \frac{\Psi}{i}. \tag{5.17}$$

This means that a 1 A current through a coil with an induction coefficient of 10 mH, for example, leads to a magnetic flux of 10 mWb.

Figure 5.28 illustrates the construction of an inductive displacement sensor consisting of a coil wound around a piece of iron into a horseshoe shape. The magnetic circuit (iron piece) is closed with a bar, also of iron, which is attached with the body whose displacement is to be measured.

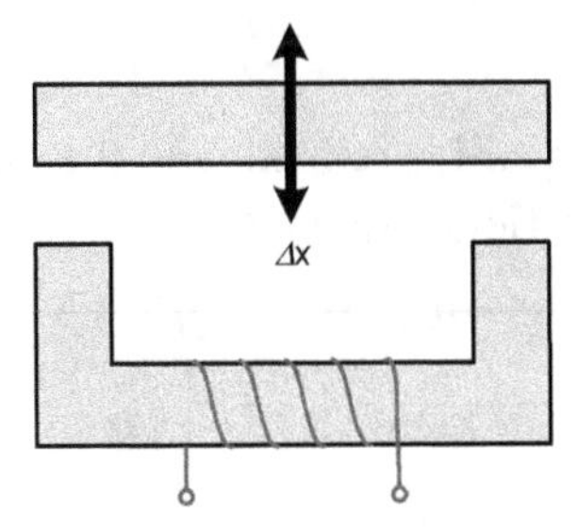

Figure 5.28 – Illustration of the construction of an inductive displacement sensor with a variable air gap.

Using the Magnetic Circuit Law,

$$\oint_{\gamma} \vec{H} \cdot \overrightarrow{d\gamma} = i, \tag{5.18}$$

where H is the magnetic field, γ is the path and i is the current that crosses the surface supported on the path, on the circuit of Figure 5.28, one gets

$$H_{fe}l_{fe} + H_0 l_0 = Ni, \tag{5.19}$$

where l_{fe} is the path length through the iron and l_0 is the path length through the air gaps. It is assumed that the magnetic field is uniform in each piece.

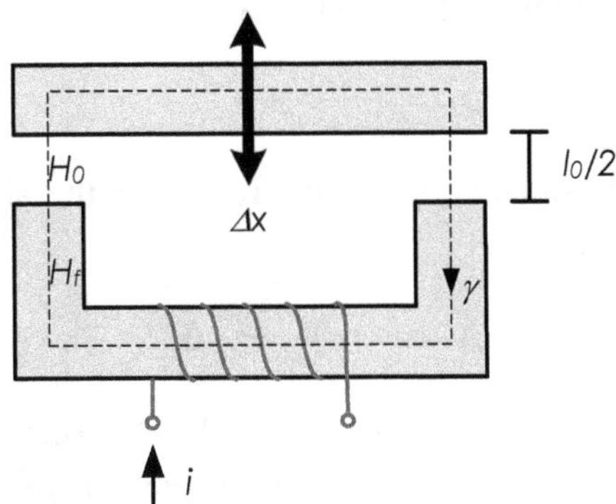

Figure 5.29 – Illustration of the employment of the Magnetic Circuit Law.

Making $B = \mu_{fe}H_{fe} = \mu_0 H_0$ one gets

$$B = \frac{N}{\frac{l_{fe}}{\mu_{fe}} + \frac{l_0}{\mu_0}} \times i. \tag{5.20}$$

The magnetic flux of a coil is related with the number of turns (N), the area (A) and the magnetic field value (B),

$$\Psi = N \cdot A \cdot B. \tag{5.21}$$

Inserting (5.20) and (5.21) into (5.17) leads to

$$L = \frac{N \cdot A \cdot B}{\frac{B}{N}\left(\frac{l_{fe}}{\mu_{fe}} + \frac{l_0}{\mu_0}\right)} = \frac{N^2 A}{\frac{l_{fe}}{\mu_{fe}} + \frac{l_0}{\mu_0}}. \tag{5.22}$$

Considering the magnetic permeability of the iron much greater than that of air, one can approximate (5.22) with

$$L = \frac{\mu_0 N^2 A}{l_0}. \tag{5.23}$$

A displacement of the movable iron piece by Δx leads to a change to the air-gap path length, l_0, of $2\Delta x$. Therefore, the change in mutual inductance coefficient is

$$\Delta L = \frac{\mu_0 N^2 A}{l_0 + 2\Delta x} - \frac{\mu_0 N^2 A}{l_0}, \tag{5.24}$$

which, after some simplification, leads to

$$\Delta L = \frac{-2\mu_0 N^2 A}{(l_0 + 2\Delta x) l_0} \cdot \Delta x. \tag{5.25}$$

For small displacements, in comparison with the airgap, $\Delta x \ll l_0/2$, (5.25) can be approximated by

$$\Delta L \approx -\frac{2\mu_0 N^2 A}{l_0^2} \cdot \Delta x. \tag{5.26}$$

This expression shows a linear relationship between displacement and self-inductance coefficient for typical displacements of a few millimeters.

To produce a voltage proportional to the displacement, one can use the bridge circuit with two coils and two resistors, as shown in Figure 5.30. One of the coils is the one wrapped around the iron piece (Figure 5.28), and whose self-induction coefficient changes with the displacement, and the other coil has a fixed value with the same self-induction coefficient as the other coil in the case of no displacement (L).

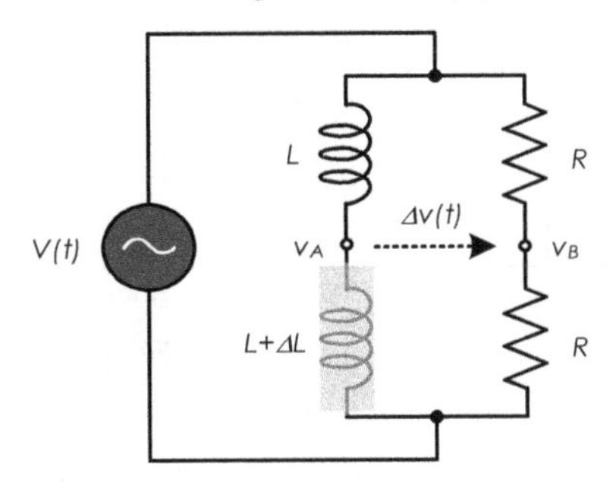

Figure 5.30 – Signal conditioning for an inductive displacement sensor.

If the bridge is powered by a sinewave,

$$v(t) = V \cdot \cos(\omega \cdot t), \tag{5.27}$$

the complex amplitude of the voltage at A is

$$\overline{V}_A = \frac{j\omega(L+\Delta L)}{j\omega + j\omega(L+\Delta L)}\overline{V} = \frac{L+\Delta L}{2L+\Delta L}\overline{V}, \tag{5.28}$$

and at B is

$$\overline{V}_B = \frac{R}{R+R}\overline{V} = \frac{1}{2}\overline{V}. \tag{5.29}$$

The differential bridge voltage is then

$$\overline{\Delta V} = \overline{V}_A - \overline{V}_B = \frac{\Delta L}{2L+\Delta L} \cdot \frac{\overline{V}}{2}. \tag{5.30}$$

Introducing (5.23) and (5.26) leads to

$$\overline{\Delta V} \approx \frac{\overline{V}}{2l_0}\Delta x. \tag{5.31}$$

The unbalance voltage is thus sinusoidal with amplitude proportional to the displacement (Δx). To obtain a DC voltage, one can use an RMS/DC converter or a peak detector if it is not important the information about the direction of travel. Otherwise, synchronous detection must be used.

A solution widely used in the field of sensors to increase the sensitivity and linearity is to use two elements that react contrary to the quantity that is to be measured. In the case of Inductive displacement, sensors can be used two windings in two identical pieces of Iron placed on opposite sides of the movable iron piece, as illustrated in Figure 5.31. This assembly is called "Push-Pull assembly."

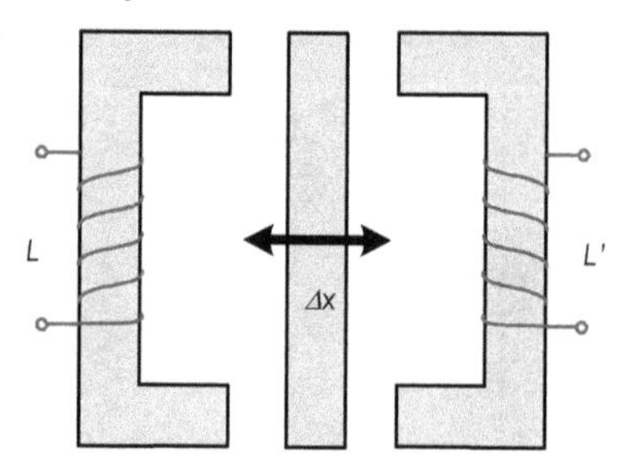

Figure 5.31 – Illustration of the construction of an inductive displacement sensor with a differentially variable air gap.

When the moving piece is displaced to the right, the self-induction coefficient L increases in absolute value, and coefficient L' decreases. According to (5.25) one has

$$\Delta L = \frac{-2\mu_0 N^2 A}{(l_0+2\Delta x)l_0}\Delta x \text{ e } \Delta L' = \frac{-2\mu_0 N^2 A}{(l_0-2\Delta x)l_0}\Delta x. \quad (5.32)$$

To carry out the signal conditioning, one usually uses the bridge circuit in Figure 5.32.

In this case, the differential voltage is

$$\overline{\Delta V} \approx \frac{\overline{V}}{l_0}\Delta x, \quad (5.33)$$

which corresponds to twice the sensitivity as can be seen by comparing it with (5.31).

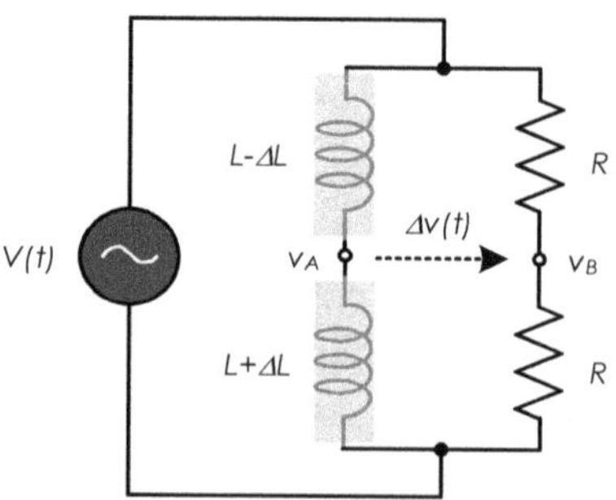

Figure 5.32 – Signal conditioning circuit for an inductive displacement sensor with moving bar in a Push-Pull assembly.

5.11 Linear Variable Differential Transformer (LVDT)

A linear variable differential transformer (LVDT) is a displacement sensor that looks like a transformer. This transformer has primary and two secondary windings with the magnetic connection between them depending on the position of the core, which is made of a ferromagnetic material. The position of the core changes with the displacement to be measured (Figure 5.33).

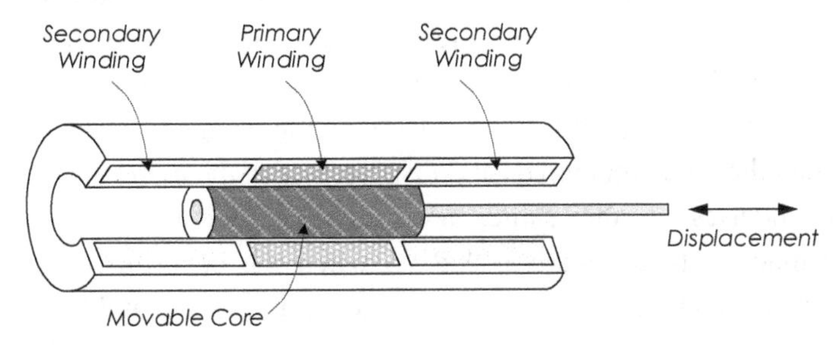

Figure 5.33 – Illustration of an LVDT displacement sensor. Image created by Honeywell International.

Figure 5.34 shows the equivalent electrical circuit of the LVDT. The resistors R_1, R_2' and R_2'' and inductances L_1, L_2' and L_2'' represent the electric load resistance and the coefficients of self-induction of the windings. These parameters do not depend, however, on the displacement to be measured. The parameter that depends on this displacement is the coefficient of mutual induction between the primary and each secondary (L_M' and L_M'').

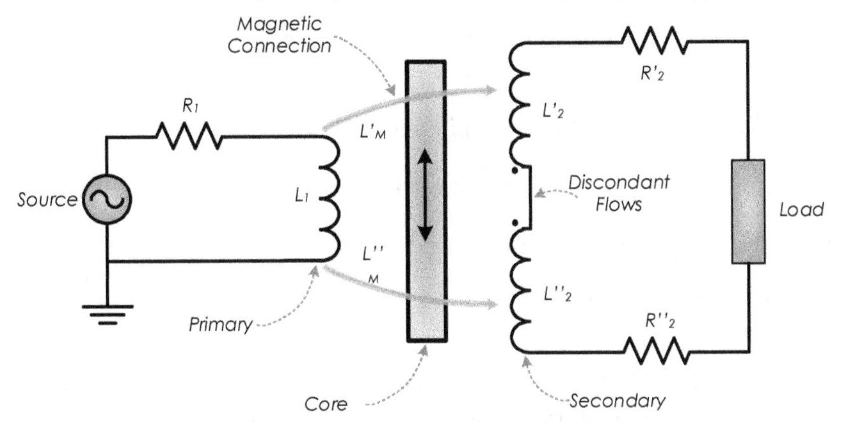

Figure 5.34 – Equivalent electric circuit of an LVDT displacement sensor.

A sinusoidal voltage is applied to the primary of the LVDT. The displacement of the transformer core leads to a change in the magnetic coupling between primary and secondary and a consequent change in the amplitude of the sinusoidal voltage induced in the secondary. In the situation of Figure 5.34, for instance, when the core moves up, the magnetic coupling between the primary and the upper secondary becomes stronger, and therefore the induced voltage is increased. At the same time, the connection between the primary and the lower secondary gets weaker, and the induced voltage in that circuit becomes smaller. The two secondary windings are connected in series but with opposite magnetic fluxes so that the voltage at its terminals is the difference of voltage in each secondary winding. Thus, when the core is centered (null displacement), the induced voltage in the two secondaries is the same, and their subtraction is zero. When the displacement is negative, that is, when the core moves downward, the induced voltage in the lower secondary is higher, and the subtraction of secondary voltages is negative. Since we are talking about sinusoidal signals, a "negative voltage" means a sinusoidal voltage that is in phase opposition to the primary voltage (Figure 5.35).

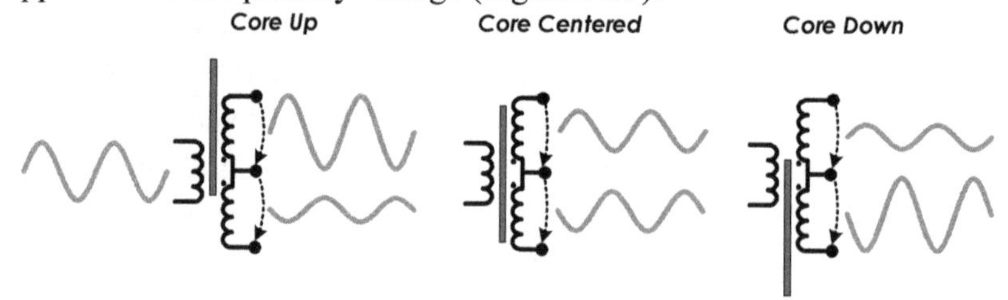

Figure 5.35 – Illustration of the waveforms in the LVDT secondary windings for different core positions.

The Magnetic Circuit Law, applied to the primary, gives

$$\overline{E_1} = (R_1 + j\omega L_1) \cdot \overline{I_1} + j\omega[L_{M'}(x) - L_{M''}(x)] \cdot \overline{I_2}, \tag{5.34}$$

and to the two secondary windings connected to a load R_c, leads to

$$0 = [R_2' + R_2'' + R_c + j\omega(L_2' - L_2'')]\overline{I_2} + j\omega[L_{M'}(x) - L_{M''}(x)] \cdot \overline{I_1}. \tag{5.35}$$

Combining the two upper equations, one gets the secondary's voltage as a function of the primary's voltage and the transformer's parameters,

$$\overline{V}_m = R_c \overline{I}_2 = \frac{j\omega R_c [L_{M''}(x) - L_{M'}(x)]}{R_1(R_2+R_c) + j\omega[L_2 R_1 + L_1(R_2+R_c)] - \omega^2\{L_1 L_2 + [L_{M'}(x) - L_{M''}(x)]\}} \overline{E}_1 \tag{5.36}$$

where

$$R_2 = R_2' + R_2'', \tag{5.37}$$

and

$$L_2 = L_2' + L_2''. \tag{5.38}$$

When the displacement is zero ($x = 0$) the mutual induction coefficients are equal, and we obtain a null voltage on the secondary.

$$x = 0 \rightarrow L_{M'}(0) = L_{M''}(0) \rightarrow \overline{V}_m = 0. \tag{5.39}$$

Usually, the load resistance, R_c, is high, which allows one to write (5.36) as

$$\overline{V}_m \approx \frac{j\omega[L_{M''}(x) - L_{M'}(x)]}{R_1 + j\omega L_1} \overline{E}_1. \tag{5.40}$$

The mutual induction coefficients depend on the core position according to

$$L_{M'}(x) = L_M(0) + C_1 \cdot x + C_2 \cdot x^2 \text{ and}$$
$$L_{M''}(x) = L_M(0) - C_1 \cdot x + C_2 \cdot x^2, \tag{5.41}$$

where C_1 and C_2 are constants that depend on the geometry of the LVDT. Introducing (5.41) into (5.40) leads to

$$\overline{V}_m = -\frac{2j\omega C_1 \overline{E}_1}{R_1 + j\omega L_1} \cdot x. \tag{5.42}$$

The rms value of the voltage on the secondary is given, in this case, by

$$V_{mef} = \frac{2\omega C_1 E_{1ef}}{\sqrt{R_1^2 + \omega^2 L_1^2}} \cdot x. \tag{5.43}$$

We conclude that the voltage on the secondary depends linearly on the displacement x.

The LVDT sensitivity depends on the operating frequency, the amplitude of the primary voltage, and the parameters of the LVDT.

The information on the absolute value of the displacement lies in the amplitude of the signal in the secondary winding. The algebraic sign, however, depends on the phase shift between the voltage on the primary and secondary. To obtain a DC voltage proportional to the displacement, we use a circuit called a **synchronous demodulator**. This circuit, shown in Figure 5.36, makes a rectification of the signal followed by a low-pass filter to extract the average value. This produces a DC voltage proportional to the amplitude of the sinusoidal voltage at the secondary terminals of the LVDT. The rectification, however, cannot be carried out using a typical circuit like, for example, a diode bridge because the result of this correction would always be a positive voltage, whatever the algebraic sign of displacement. The rectification, in this case, is made with a switch that selects the secondary voltage or its symmetric according to the value of the voltage in the primary.

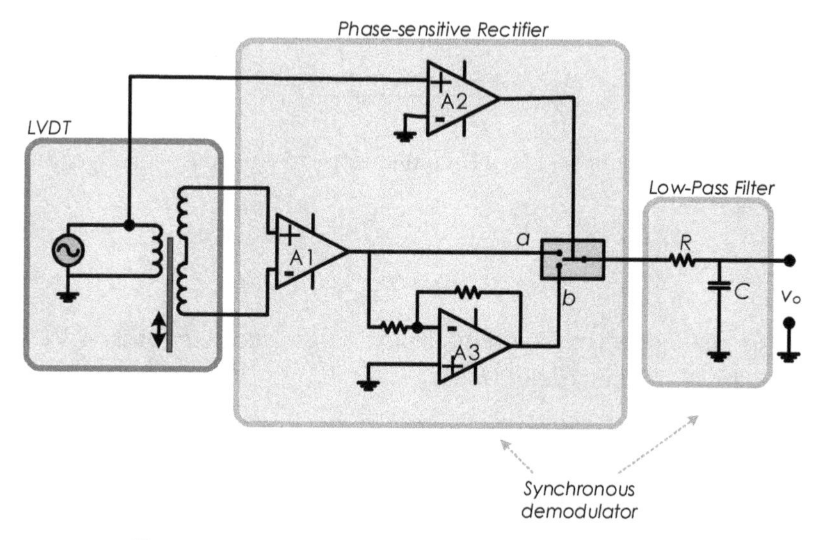

Figure 5.36 – Electric circuit of a synchronous demodulator.

Amplifier A1 converts the differential output voltage of the LVDT secondary into a single-ended voltage (referred to as the ground potential). The amplifier A2 compares the primary signal with 0 to produce a signal with two levels indicating whether the voltage at the primary is positive or negative. This signal controls the switch whose function is to select the output signal of the secondary (position a) or its symmetric (position b). This symmetry is created by an operational amplifier (A3) connected to an inverter.

In Figure 5.37, we can see synchronous demodulator output waveforms for different positions of the core.

This displacement sensor has a theoretically infinite resolution. There is no contact between the core and the windings, which means that there is no friction. This allows the lifetime is very large (200 years MTBF — mean time between failures). Another advantage of this sensor is its high linearity, sensitivity, and repeatability.

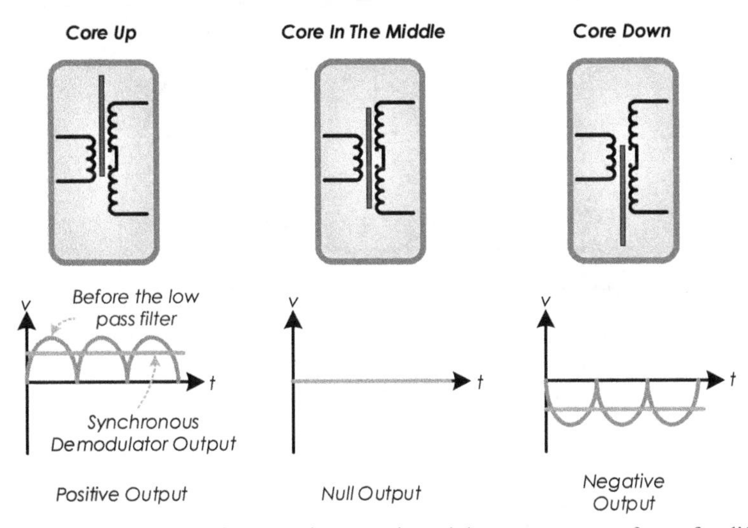

Figure 5.37 – Illustration of the synchronous demodulator output waveforms for different positions of the core.

This type of sensor has become more and more economical, and there are even cases in which the signal conditioning circuit, which is somewhat complicated, is integrated into the sensor itself, which facilitates its installation and the interface with the rest of the system. This has led to an increase in the applications where it is currently used. An example of this is the case of hydraulic systems, in which they are combined with actuators to form a retroactive system on cranes, excavators, and even on airplanes. Also, in the industry, they are used in combination with robotic arms to be able to determine their exact position. Other applications include ATMs, the manufacture of parts whose dimensions must be well controlled, and the positioning of solar panels and mirrors [15].

5.12 Angular Displacement Sensor (Microsyn)

The Microsyn is a device with a 4-pole stator and a rotor, both ferromagnetic. In each stator pole, there are two windings on the bottom, forming a transformer with a primary and a secondary. The windings in each pole are connected in series (Figure 5.38).

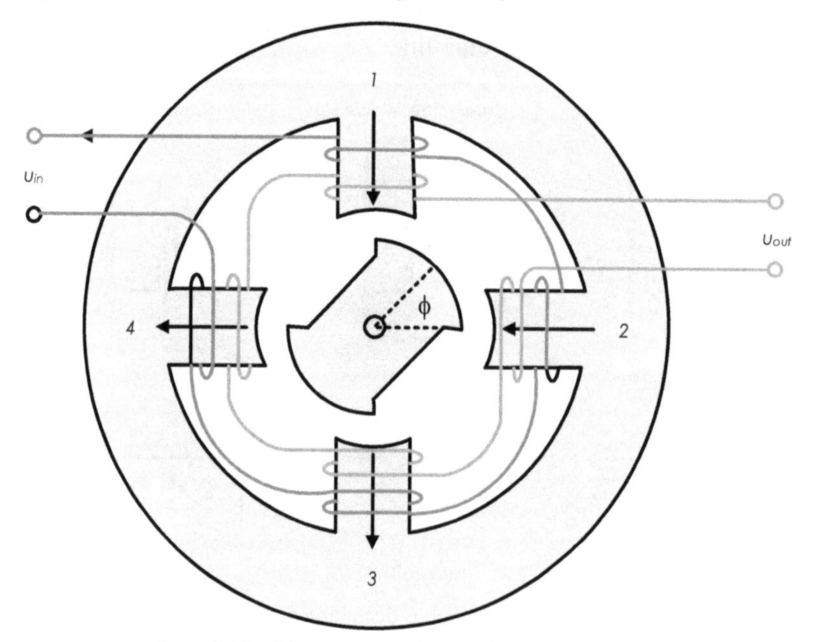

Figure 5.38 – Microsyn angular displacement sensor.

To the primary, a sinusoidal alternating voltage is applied (we assume here that it has unit amplitude),

$$u_{in}(t) = \cos(\omega t). \tag{5.44}$$

The angular position of the rotor, which is not wound, influences the magnetic connection between the windings of the primary and of the secondary.

The windings are connected so that in the neutral position, as illustrated in Figure 5.38, the voltages induced in windings 1 and 3 are symmetrical of the induced voltages in windings 2 and 4, for which output voltage is zero. When the rotor moves, it causes an increase in the magnetic reluctance of two opposite windings and a decrease in the other two.

The voltage that is in output from the secondary is therefore

$$u_{out}(t) = 2\omega(\Delta\phi_{13} - \Delta\phi_{24}) \cdot \sin(\omega t), \quad (5.45)$$

in which the variation of flux at the poles 1 and 3 is given by

$$\Delta\phi_{13} = C_a\Delta\alpha + C_b(\Delta\alpha)^2, \quad (5.46)$$

and in the poles 2 and 4 is given by

$$\Delta\phi_{24} = -C_a\Delta\alpha + C_b(\Delta\alpha)^2, \quad (5.47)$$

where $\Delta\alpha$ is the variation of the angular position and C_a and C_b constants that depend on sensor characteristics (number of turns, the primary voltage, etc.)

Introducing (5.46) and (5.47) into (5.45) results in

$$u_{out}(t) = 4C_a\omega \cdot \sin(\omega t) \cdot \Delta\alpha. \quad (5.48)$$

The amplitude of the output voltage is thus proportional to the variation of the angular position. The sign of the phase shift between the primary and secondary (0 or 180°) depends on the direction of change of the angular position.

Note that the series combination of the windings of the poles 1 and 3 with the windings of poles 2 and 4 makes it possible the increase the sensitivity of the sensor as well as correct the quadratic nonlinearity given by the constant C_b in (126) and (127). This type of non-linearity correction is called **Association of Sensors with Symmetrical Non-linearities** and is also used, for example, in metallic temperature sensors.

With this type of sensor, one can measure angular displacement typically up to 10° with a resolution of 0.01°. The sensitivity is on the order of 5 V/°, and linearity ranges from 0.5% to 1%. It is used in many of the same areas where the LVDT is used in industry and robotics.

5.13 Magnetic Resonance Imaging

5.13.1 *Introduction*

Magnetic resonance imaging (MRI) is a technique that allows the creation of 2D and 3D images of the inside of the human (and other animals) body. Figure 5.39 shows an example of the type of images you can obtain with

MRI. It is a two-dimensional image of a sagittal section of a human head where you can observe, among other structures, the brain.

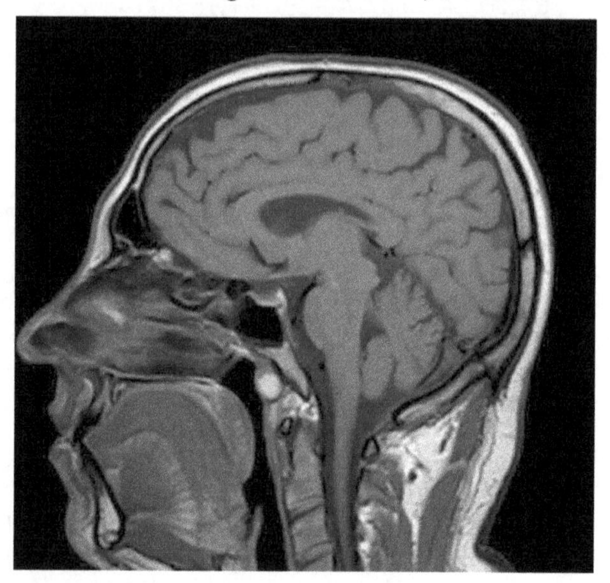

Figure 5.39 – Example of an MRI image of a human head.

This imaging technique uses an intense and constant magnetic field, both in time and in space, to orient the water molecules existing in all human tissues in a particular direction. This field has a typical intensity between 1 and 10 T and is thus several orders of magnitude higher than the terrestrial field (50 μT). Besides this field, there are other weaker magnetic fields that are also used but whose intensity varies linearly in space to create within the human body areas where the magnetic field has different values. This is what will allow the distinction between different points within the body in order to construct an image. These three magnetic field gradients are gradual changes in intensity along a perpendicular direction. Furthermore, the field gradients are switched on and off repeatedly during the process of building an image which can take from several seconds to several minutes.

There is also a third magnetic field involved in MRI. It is the magnetic field of electromagnetic waves, the molecules of the body of water produced in certain circumstances and which are detected by sensors placed around the body. It is these waves that allow us to obtain information on what is happening inside the body without even touching it. These very weak magnetic fields vary with time (with frequencies in the order of several MHz) and therefore have an associated electric field.

Magnetic resonance imaging is an indispensable means of medical diagnosis, allowing the detection and evaluation of problems that would otherwise be impossible. An example of this importance is the determination of the type of cerebrovascular accident. There are two types that require completely different treatments. Ischemic stroke is the formation of a clot (obstruction) in a cerebral vessel, while hemorrhagic stroke is the rupture of one of these vessels. The treatment of ischemic stroke consists of administering thrombolytic drugs with the aim of dissolving the clot. The administration of this medicine in the case of hemorrhagic stroke has tragic effects due to their anticoagulant effect, which will lead to greater brain damage. In order to administer the right medication, it is therefore of utmost importance to be able to distinguish between the two types of stroke.

5.13.2 *Magnetic Resonance*

To understand how the images are created in the MRI system, it is necessary to understand the physical phenomena that take place in the atoms and molecules of our body in the presence of magnetic fields.

Protons present in the nucleus of atoms that make up our body have a kind of rotation around themselves (angular momentum or spin). The Protons, as electrically charged particles, produce, due to this rotational motion, a magnetic field with lines of force, as illustrated in Figure 5.40. It is a magnetic field with the same geometry as that of Earth.

The axis of rotation of each proton existing in our body is typically oriented in a random direction. The combined effect of all the protons is a null magnetic field. This means that our body, at the macroscopic level, does not have a magnetic field. If, however, our body is placed in a strong magnetic field, each proton will tend to orient itself along the lines of force of the field, as shown in Figure 5.41.

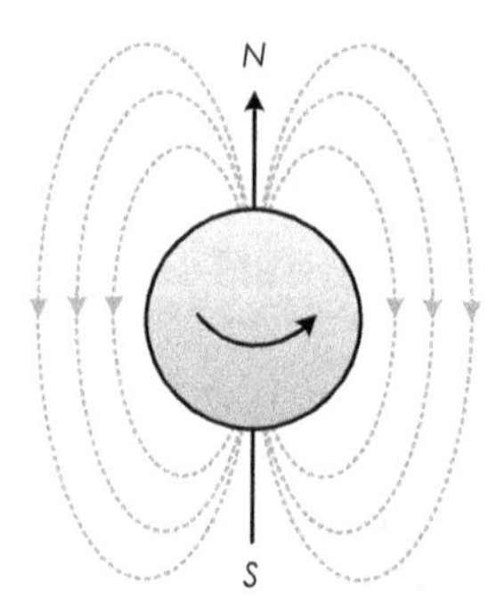

Figure 5.40 – Illustration of the magnetic field of a proton.

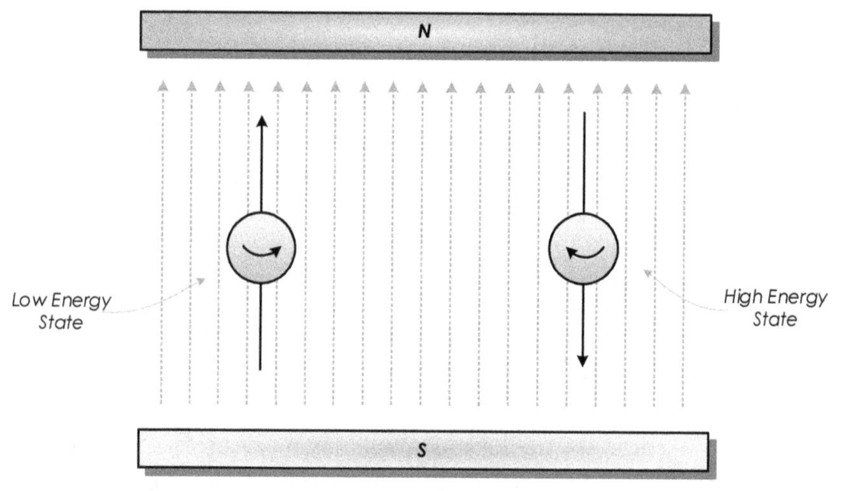

Figure 5.41 – Illustration of a proton orientation imposed by an external magnetic field.

In practice, however, things are not so simple. Due to quantum effects, protons may be oriented according to the external field, as illustrated in Figure 5.41(left) but may also be oriented in exactly the opposite direction, as illustrated in Figure 5.41(right). The first case represents a state of lower energy, while the latter case represents a state of higher energy.

At a temperature (T) of 0 K, all the protons are in the lowest energy state. At the typical human body temperature (≈ 37°C), however, there are a slightly greater number of protons in the lower energy state, N_{low}, then in the higher energy state, N_{high} (about 1 ppm). This ratio is given by the expression

$$\frac{N_{high}}{N_{low}} = e^{-\frac{\Delta E}{kT}}, \tag{5.49}$$

where k is the Boltzman constant and ΔE is the energy difference between the two states. For example, at the normal body temperature and when the external magnetic field is 0.25 T, there is an excess of protons in the low energy level of about 1 ppm.

In addition to the phenomena of each proton magnetic field and its alignment with an external magnetic field, there is another phenomenon essential for the functioning of MRI — precession of the magnetic moment. This precession is in an oscillating movement of the axis of rotation of protons around the lines of force of the external magnetic field, as illustrated in Figure 5.42.

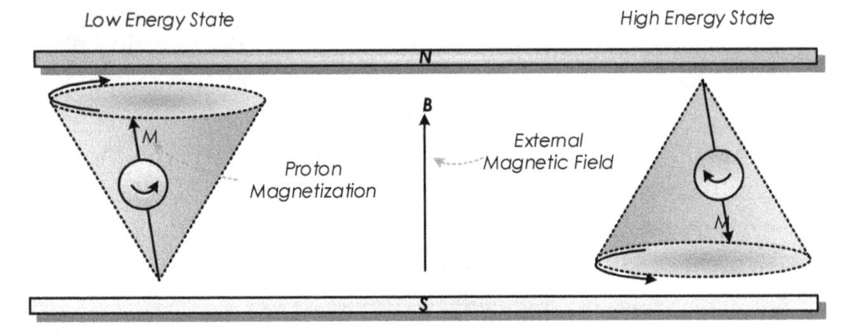

Figure 5.42 – Illustration of the precession of the magnetic moment of protons in an external magnetic field. A vector M (magnetization) is used to represent the proton's angular momentum.

We can represent the angular moment of the proton by a vector *M as shown in* Figure 5.42.

Notice that the protons continue to rotate about themselves (spin) while the rotation axis describes a conical shape around the lines of force of the external magnetic field (precession axis). The precession frequency linearly depends on the value of the magnetic field according to the Larmor equation,

$$f = \gamma \cdot B, \tag{5.50}$$

where γ is 42.58 MHz/T for the hydrogen atoms. For a magnetic field of 1 T, for example, the precession frequency is 42.58 MHz.

So far, we analyzed the rotational motion of the protons around themselves (spin) and precession around the lines of force of the external magnetic field. As will be seen later, you can change these movements by focusing an electromagnetic wave on a given sample, disturbing it momentarily. When this disruption ends, it comes back to the initial state and, in the process, produces electromagnetic radiation that is detected outside the body and used to build a picture of the structures inside it. The radiation detected is, however, the result of many protons. Therefore, it is common to talk about the result of the macroscopic magnetization vector result of the sum of the magnetization of each individual proton in a given space area instead of the magnetization of each proton. The macroscopic magnetization in the case of the system at equilibrium (unperturbed by external electromagnetic waves) is parallel to the external magnetic field.

By convention, the external magnetic field has the direction of the z-axis. The macroscopic magnetization, therefore, only has a component along z ($M_x = 0$, $M_y = 0$). This happens because each proton has a precessional motion around the same axis (direction B) with the same frequency (Larmor frequency) but with different "phases." The x and y components in the individual magnetizations are uniformly distributed, and thus their sum is 0. But the component z has two possible values: along B or opposite to B depending on the energy level of each proton. As seen earlier, at the typical temperature of the human body, there is a slight excess of protons with the magnetization according to B than in the opposite direction. The vector sum of all magnetizations thus leads to a z component different from 0 in macroscopic magnetization (and in the direction of B). Figure 5.43 illustrates the difference between single magnetizations and macroscopic magnetization.

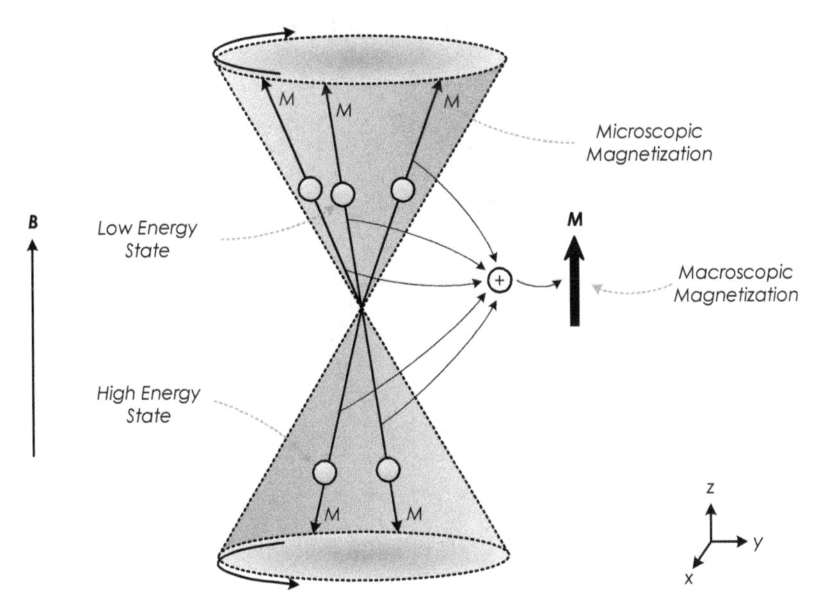

Figure 5.43 – Illustration of the vector sum of the magnetization of many protons.

5.13.3 *Excitation and Relaxation*

The spin and the movement of the precession of protons are likely to be altered by the action of an electromagnetic wave with a frequency equal to the frequency of precession.

At the typical temperature of the human body, there are more protons in the lower energy level than the high energy level (diagram on the left of Figure 5.44). When the sample is irradiated by an electromagnetic wave with a certain frequency (Larmor frequency), some protons go from the low energy level to high energy level (the second diagram of Figure 5.44), which corresponds to a spin in the opposite direction of the external magnetic field (Figure 5.43).

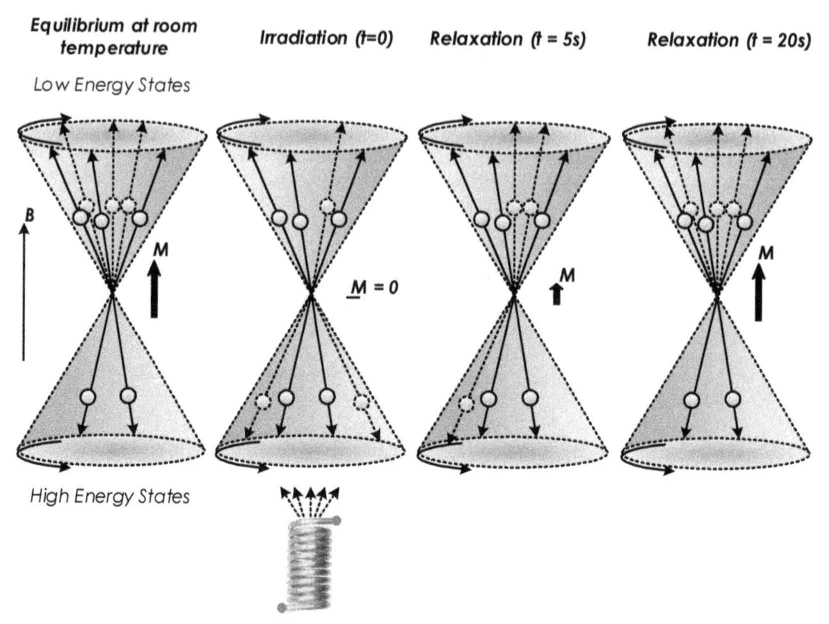

Figure 5.44 – Illustration of the effect of an external electromagnetic wave on the energy level of the protons in the sample.

If this radiation occurs for a sufficiently long period of time, the same number of protons will be found with each of the two possible energy levels. Therefore, the component according to z (parallel to the external magnetic field) of the macroscopic magnetization is zero; that is, the z components of the magnetization of each individual proton cancel each other. At the same time, the movement of the precession of protons becomes phase-locked due to the external electromagnetic wave. Thus, the vectorial sum of the transverse components of individual magnetization, M_{xy} (projection M in the xy plane), ceases to be null as it was in equilibrium (due to its uniform distribution of phases). The macroscopic magnetization thus acquires a transverse component.

In macroscopic terms, the effect of external irradiation by an electromagnetic wave can be described as causing a rotation of the macroscopic magnetization vector of the z axis in the xy plane, as illustrated in Figure 5.45.

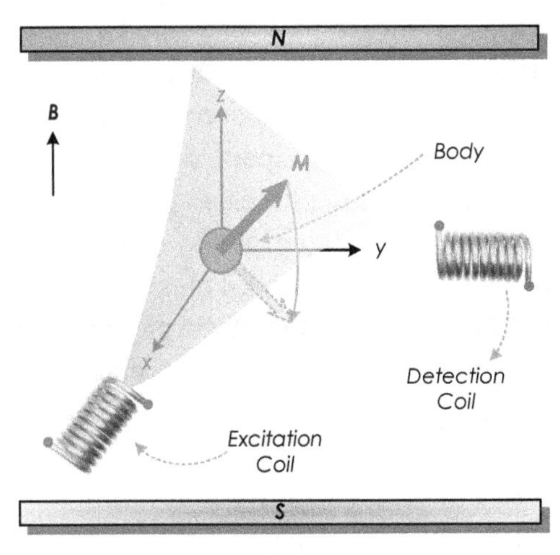

Figure 5.45 – Illustration of the effect an electromagnetic wave has on the magnetization of the protons.

Note that there remains a rotation around the precession axis, which remains unchanged, but such rotation takes place entirely in the horizontal plane (plane perpendicular to the axis of precession and where the excitation coil is), as shown in Figure 5.46.

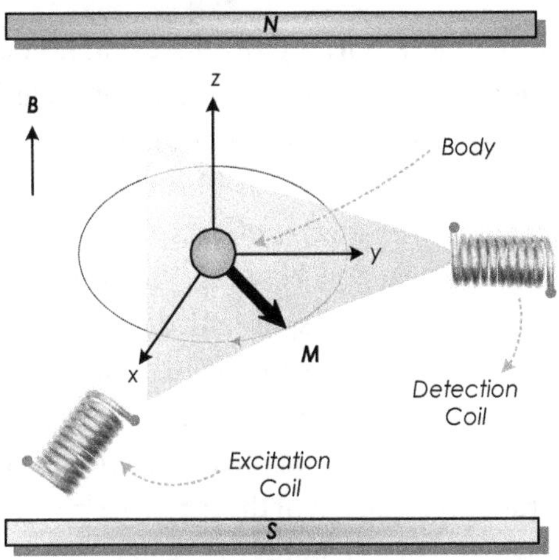

Figure 5.46 – Illustration of the precession movement of the protons and the detection of the electromagnetic wave produced by it.

We should not forget, however, that there are two physical effects that occur simultaneously during irradiation by an external wave: changing the direction of the spin of the proton and coherence of the precessional motion of the protons. This will be important to understand what happens to the magnetization when the sample ceases to be irradiated.

The movement of the precession of the protons produces itself an electromagnetic wave that can be detected outside the human body with a coil. The rotation of the magnetization of the protons causes the intensity of the signal obtained in the detection coil to change sinusoidally with time, as shown in Figure 5.47.

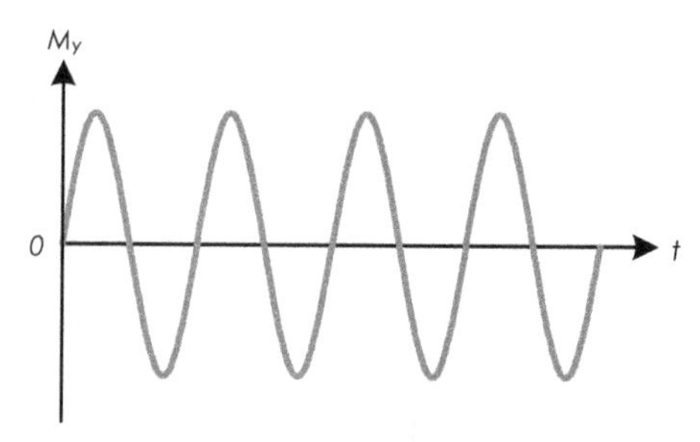

Figure 5.47 – Signal picked up by the sensing coil in an ideal situation (no relaxation).

In practice, however, after turning off the excitation coil that changed the orientation of the magnetization of protons into one in the horizontal plane, this magnetization will return to its original value (before excitation by the external electromagnetic wave) as illustrated in Figure 5.48. This is called *relaxation* and is due to several effects.

One of them is the interaction between the water molecules and the surrounding tissues. It is named the spin-lattice interaction. This interaction causes some of the protons that were in a state of high energy to return to a lower energy state. So, there is again an excess of protons in the state of lowest energy, that is, with a spin parallel to the external magnetic field, which makes the parallel component of the macroscopic magnetization (M_z) to go back to its original value (equilibrium value). Figure 5.49 illustrates this for the case of pure water.

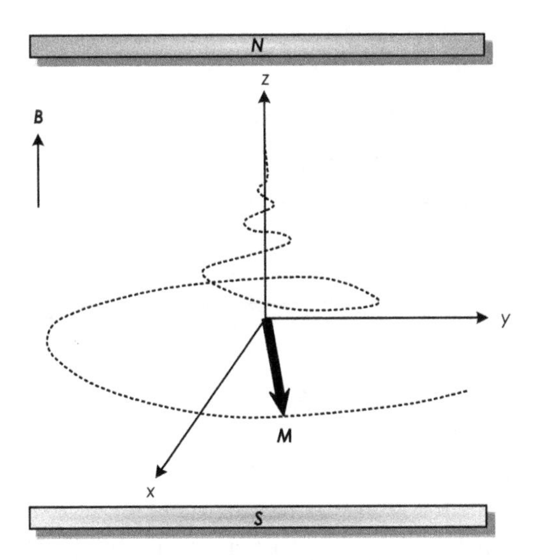

Figure 5.48 – Representation of the path described by affix the magnetization vector after disconnecting the excitation coil.

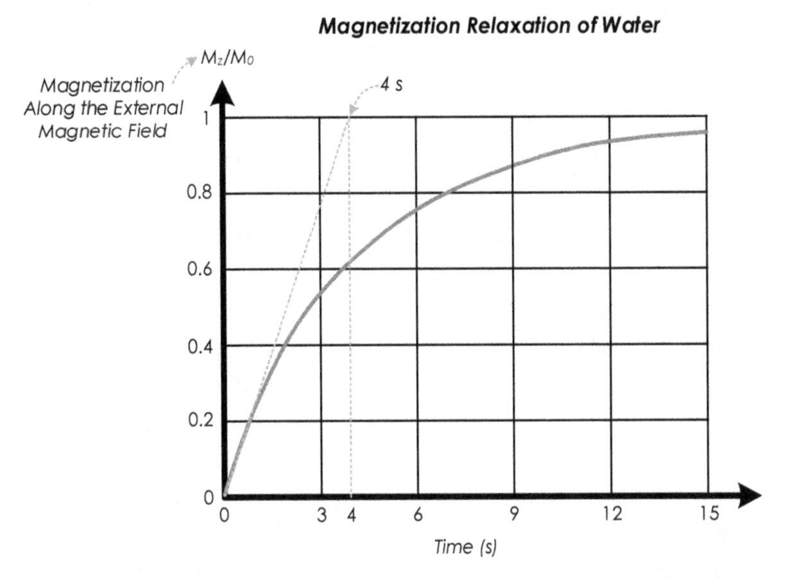

Figure 5.49 – T1 relaxation of the M_z component of the magnetization after removal of the electromagnetic wave excitation.

This variation is exponential according to the equation

$$M_z(t) = M_0(1 - e^{-\frac{t}{T_1}}), \tag{5.51}$$

where M_0 is the value of the macroscopic magnetization in equilibrium and T_1 is the time it takes magnetization in z-direction back to 66.6% of its original value.

As for the x and y components, one would expect them to return to their equilibrium value at the same rate as the component along z. This idea is, however, wrong and comes from considering the macroscopic magnetization as a vector whose magnitude is conserved — if the z component increases, the x, and y must decrease accordingly. In fact, macroscopic magnetization is the result of the vector sum of the magnetization of each individual proton. That individual magnetization always has the same amplitude (but varying orientation). Moreover, the physical phenomenon that makes the longitudinal component of the macroscopic magnetization return to baseline (M_0) is not the same as those that lead the perpendicular components to return to their original value (a null value). It is for this reason that the time it takes the macroscopic components M_z and M_{xy} to return to baseline is different.

The effect that contributes to the relaxation of the transverse component of the macroscopic magnetization is the interaction between different water molecules. This leads to an exponential decay of the M_{xy} component with a time constant T_2. This constant, which is less than T_1, is called the constant of transverse relaxation or spin-spin relaxation.

There is, however, another important effect that has nothing to do with the type of substance the sample itself is made but with the magnetic field present, particularly its inhomogeneity, which may be due to local variations in magnetic susceptibility or imperfections in the external magnetic field. These small variations cause the precession frequency given by the Larmor equation (130) to have slightly different values when the external electromagnetic wave is turned off. Before that, all protons are running their precessional motion in phase and at the same pace. The component of the macroscopic magnetization in the xy plane has significant value because it results from the constructive sum of the magnetizations of many protons.

After turning off the external electromagnetic wave, the individual magnetizations start shifting due to slightly different rates of precession so that their vector sum leads to components on the xy plane that get smaller and smaller (Figure 5.50).

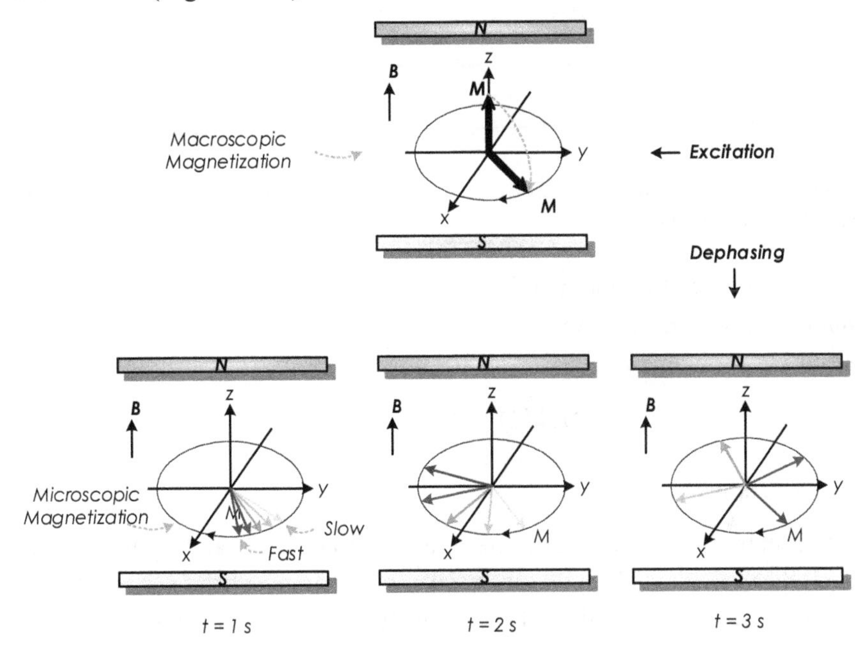

Figure 5.50 – Illustration of the "lack of coherence" that arises in the precessional movements of the protons after the external electromagnetic wave is turned off.

This mismatch effect leads to a more rapid decay of the transverse component of transverse magnetization. The effective time constant of relaxation, represented by T_2^* is less than T_2 and is given by

$$\frac{1}{T_2^*} = \frac{1}{T_2} + \frac{1}{T_{in}} \tag{5.52}$$

where T_{in} is the time constant due only to the phase mismatch caused by the magnetic field inhomogeneity.

The signal obtained from the detector coil is a damped sinusoid with an envelope that decays exponentially with a time constant T_2^*, as illustrated in Figure 5.51.

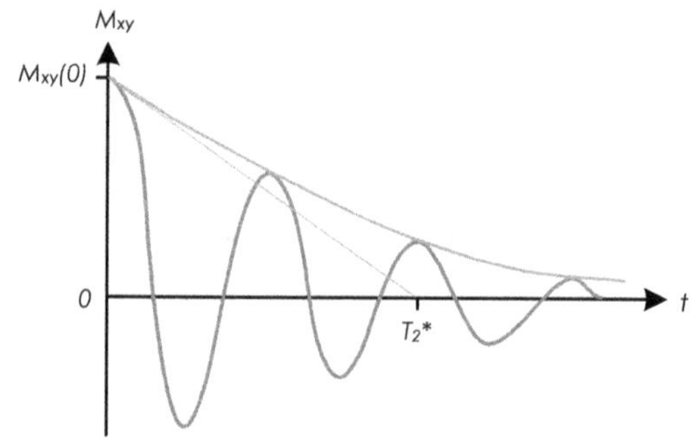

Figure 5.51 – Evolution of the transversal component of the magnetization relaxing to the equilibrium condition (blue).

By representing the amplitude M_{xy} component of the transverse magnetization (projection of magnetization in the xy plane), its time evolution is given by

$$M_{xy}(t) = M_{xy}(0)e^{-\frac{t}{T_2^*}}. \tag{5.53}$$

5.13.4 *Detection*

Different body tissues have different relaxation times T_1 and T_2. In modern MRI, these are the values that are used to create images of the human body and not the absolute value of the magnetization. Table 5.1 presents the typical relaxation times of some organ's values.

Table 5.1 – Values of relaxation times T_1 and T_2 for some tissues and the value for pure water compared to a homogeneous magnetic field.

TISSUE	T_1 (ms)	T_2 (ms)
Fat	220	90
Liver	440	50
Spleen	460	80
Muscle	600	40
White Matter	700	90
Grey Matter	820	100
Blood	800	180
Water	2500	2500

Recall that the relaxation time T_2^* has a significant contribution to the decay of the transverse magnetization but does not depend directly on the type of tissue. What matters is creating an image of the entire human body. It depends on the homogeneity of the magnetic field. For this reason, techniques have been developed for the specific purpose of measuring time T_2. The main one is called *Spin Echo.*

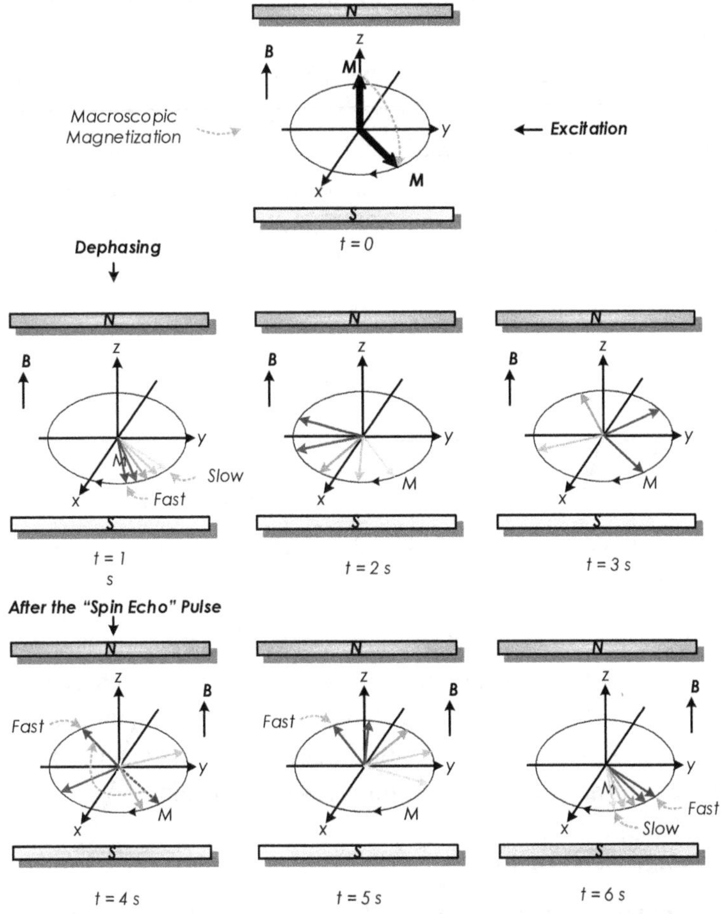

Figure 5.52 – Illustration of re-focusing of individual proton magnetization process.

This technique tries to overcome the change of phase caused by different rates of the precession of a set of protons using irradiation with an electromagnetic wave that causes a 180° rotation around the x-axis of magnetization of each proton. This delays the protons that were in advance in relation to others and vice versa. This, however, does not affect the different rates of precession, which causes those protons which precession is behind the others to eventually catch up and be in phase again. Similarly, the protons whose precession is in advance, as they precess at a slower pace than the other protons, will see their precession delayed relative to the rest of the protons until they are in phase with them. Figure 5.52 illustrates this process.

In summary, after the sample has been irradiated by an electromagnetic wave that made the magnetization rotate 90°, it is again, and repeatedly, at predefined time intervals, irradiated for time intervals that cause a 180° rotation in polarization (around the x axis). This causes the individual magnetizations to periodically be in phase giving rise to a signal in the detection coil stronger, as illustrated in Figure 5.53.

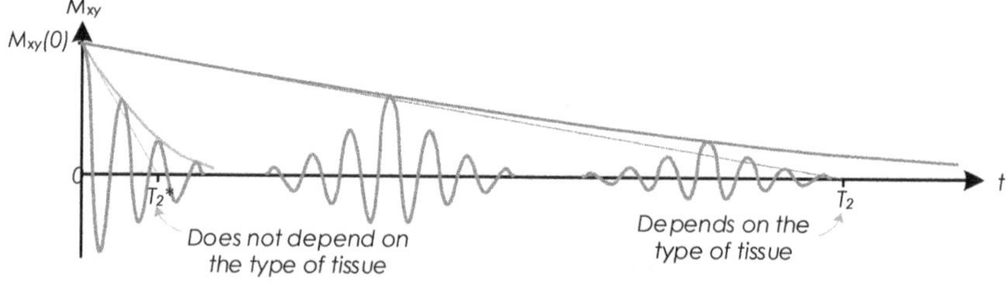

Figure 5.53 – Time evolution of the signal in the detection coil when using the spin-echo technique to measure T_2.

However, the effect of interactions between molecules that are the source of constant T_2 relaxation is not affected. Thus, each echo has a lower amplitude because fewer protons are precessing in phase. The envelope defined by the maximum of the several echoes is an exponential with a time constant T_2 which is want we want to obtain.

5.13.5 *Image Construction*

So far, we have seen how it is possible, using an external magnetic field, irradiating with electromagnetic waves and a sensing coil, to create a macroscopic magnetization, disturb it, and detect that magnetization out of the body as it returns to equilibrium. It was also seen that the time the relaxation takes allows us to distinguish the different internal structures of the body. To create an image of these structures, however, you must associate that information obtained in the detection coil with the spatial location of the protons that originated it. The method used for this purpose is discussed below.

It is impossible nowadays, from the signal measured outside the body, which is the result of the combined action of all protons in a given area, to determine what the contribution to it each individual proton was. What we can do is determine the contribution of a set of protons located in a well-defined region. For that, the 3D space is discretized in a set of such small cube-shaped elements, which are called "voxel" by analogy with "pixel" used for the discretization of a plan (Figure 5.54). A voxel is a volumetric pixel.

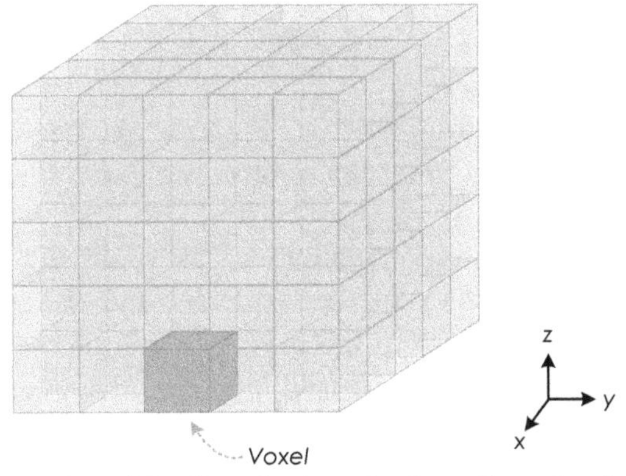

Figure 5.54 – Illustration of the discretization of a three-dimensional volume into small elements — voxel.

The creation of a 3D image is, in practice, done by creating multiple 2D images representing different slices of the body, as shown in Figure 5.55.

The first step in creating the images is to isolate the information relative to one slice of the human body. This is done using a mathematical relationship between the energy required to bring a proton from a low power state for the high-energy state (excited state) and the value of the magnetic field. The frequency of the photons that have that energy, which is the same frequency of precession of the spin of protons around the lines of force of the magnetic field, is given by (5.50). The selection of a slice is then made by applying a gradient magnetic field in a given direction, in such a way that its value varies slightly in that direction, and irradiating the body with electromagnetic radiation with a single frequency (f). Thus, only the protons that are in a zone where the magnetic field is f/γ will be excited and will have their magnetization rotated from the z-axis to the xy plane.

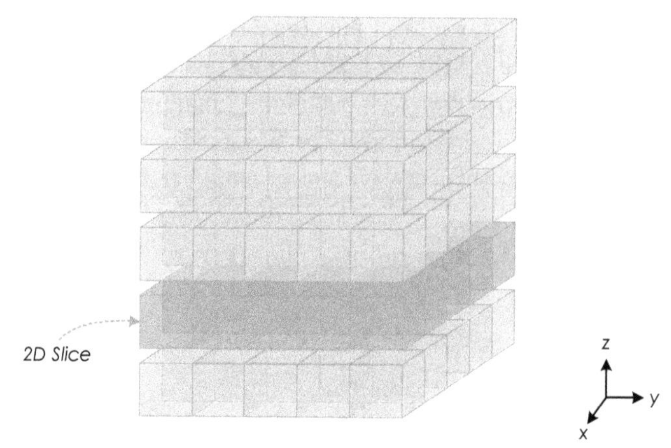

Figure 5.55 – Illustration of dividing space into a series of slices.

Figure 5.56 illustrates the relationship between the value of the magnetic field, the frequency of the radiation, and the position of the excited protons. The variation of the magnetic field happens in a single direction (e.g., z). In the other two perpendicular directions (x and y) the field has a constant value.

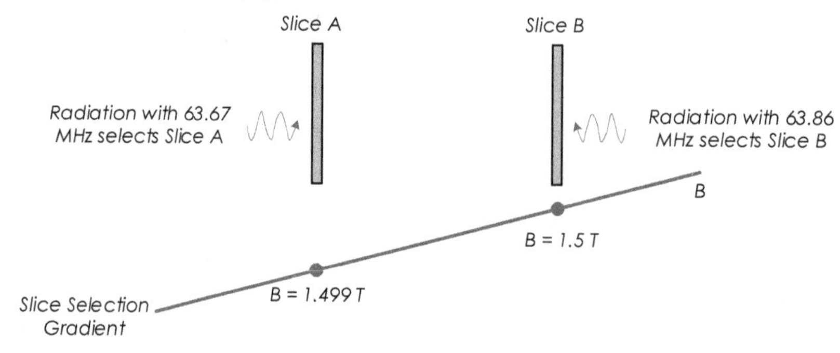

Figure 5.56 – Illustration of the relationship between radiation frequency, magnetic field value, and the position of the excited protons.

Figure 5.56 presents an example where the magnetic field varies from a value slightly smaller than 1.5 T to a value slightly greater than 1.5 T.

When the radiation has a frequency of 63.67 MHz, only the protons located in slice A are excited. All others remain on the level of energy they had. In another case where the frequency of the radiation has a different value (63.86 MHz), the excited protons are those found in slice B. Only those protons which are in that slice will produce an electromagnetic wave during relaxation and will contribute to the signal obtained in the detection coil.

Note that in practice, the radiated wave does not have a single frequency but a range of close-by frequencies. It means it has a bandwidth greater than 0. The bigger the bandwidth, the bigger is the thickness of the slice obtained. On the other hand, the more abrupt the magnetic field gradient is for the same bandwidth of the radiation, the smaller the thickness of the cut, as illustrated in Figure 5.57.

For the slice to be spatially well defined, a sinusoidal signal with a sinc envelope is used as the irradiating wave, as illustrated in Figure 5.58. Recall that the Fourier transform of a rectangle is a sinc. A sinusoidal signal with a sinc-shaped envelope is made up of a sum of sinusoids with the same frequency and amplitude within a well-defined range.

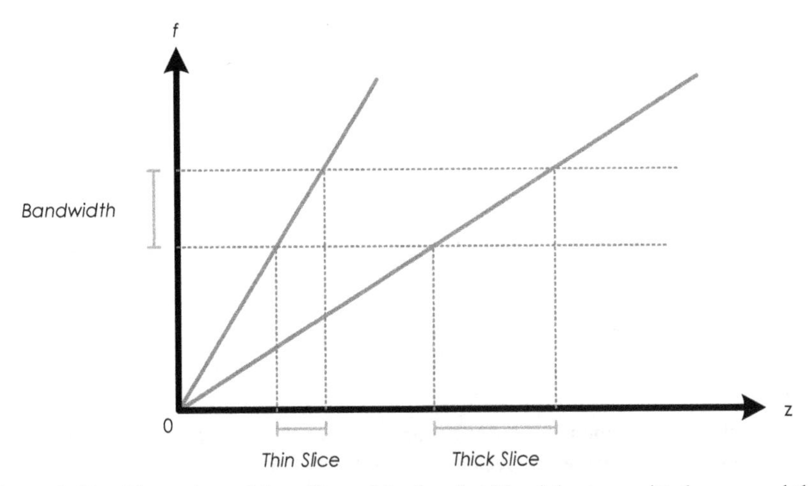

Figure 5.57 – Illustration of the effect of the bandwidth of the transmitted wave and the width of the slice obtained.

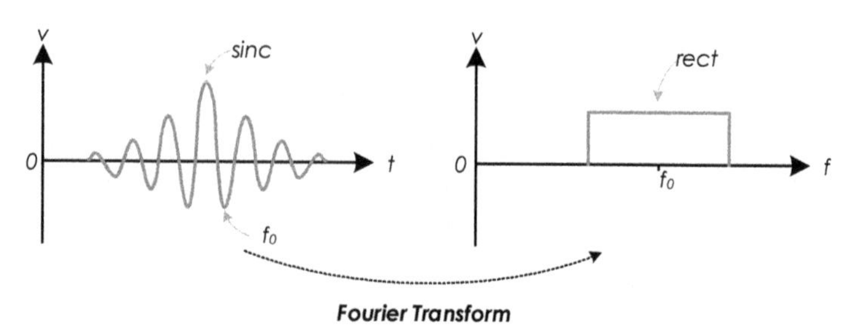

Figure 5.58 – Irradiation is done with a sinc-shaped pulse.

After selecting a given slice, we still have a huge set of protons distributed roughly along a surface, which will contribute to the detected signal.

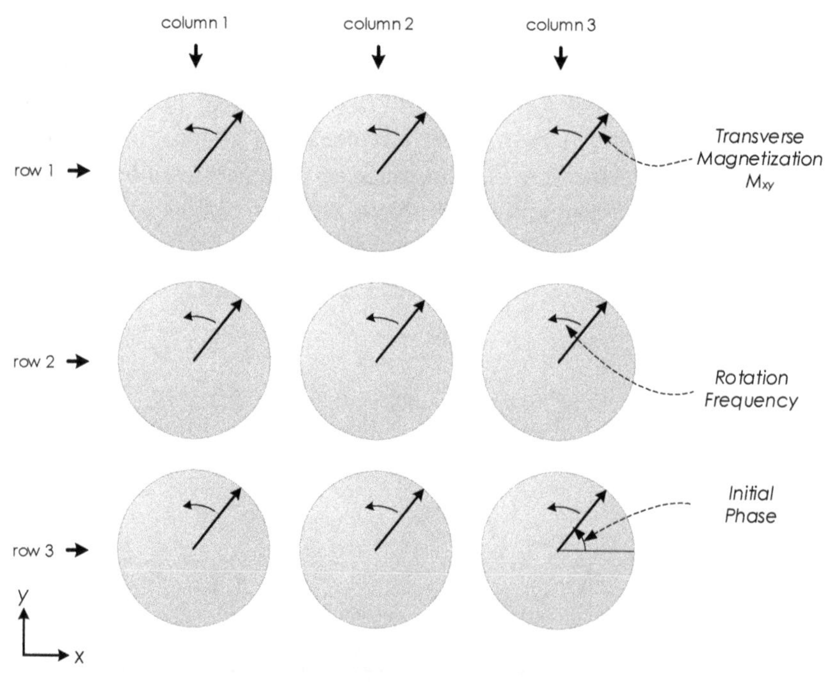

Figure 5.59 – Illustration of the rotation motion of the macroscopic transverse magnetization on nine voxels arranged in a three-by-three matrix.

All these protons have a precessional motion at the same frequency and in phase. Figure 5.59 illustrates the rotation of the macroscopic transverse magnetization in 9 voxels arranged in a 3-by-3 matrix. The frequency of rotation is represented by an arrow in an arc. The bigger the length, the higher the frequency of rotation. In the example shown, all voxels have the same magnetization rotation frequency.

If it were possible to have at every point in space a different magnetic field value, then it would be possible to excite a small number of protons at a time and thus determine the value of one voxel at a time. In practice, this is not possible. There is always a whole slice of excited protons.

Other techniques are used to separate the contribution of each group of protons according to their location inside the slice to create a 2D image, as will be seen below.

One of the ways used to distinguish protons in a slice is to make the frequency of precession different. Different precession frequencies lead to different frequencies in the electromagnetic wave radiated during relaxation. The sum of these sine waves gives a signal with a non-sinusoidal time variation and induces a non-sinusoidal voltage signal in the detector coil. From this signal, it is possible to recover individual sinusoids using Fourier transform, which is, by definition, describes any signal by a sum of sinusoids.

To make the precession frequency different, it is necessary that the magnetic field, during the signal acquisition in the detector coil, varies in space, which means that it is necessary to use a magnetic field gradient. As mentioned above, it is not possible to have a magnetic field with different values at every point in space or, in the case, in each voxel of the slice. It is only possible to vary the magnetic field in a given spatial direction. Imagine that a gradient along x is applied, as shown in Figure 5.60.

The magnetization in the first column of 3 voxels has a lower rotational frequency than the voxels in columns 2 and 3 because the value of the magnetic field is different there. The frequency of rotation of the transverse magnetization is given by

$$f_x(x, y) = f_0 + \gamma \cdot G_x \cdot x, \tag{5.54}$$

where f_0 is the frequency when there is no spatial variation of the magnetic field and G_x is the magnetic field gradient along x, which, by definition is,

$$G_x = \frac{dB}{dx}. \tag{5.55}$$

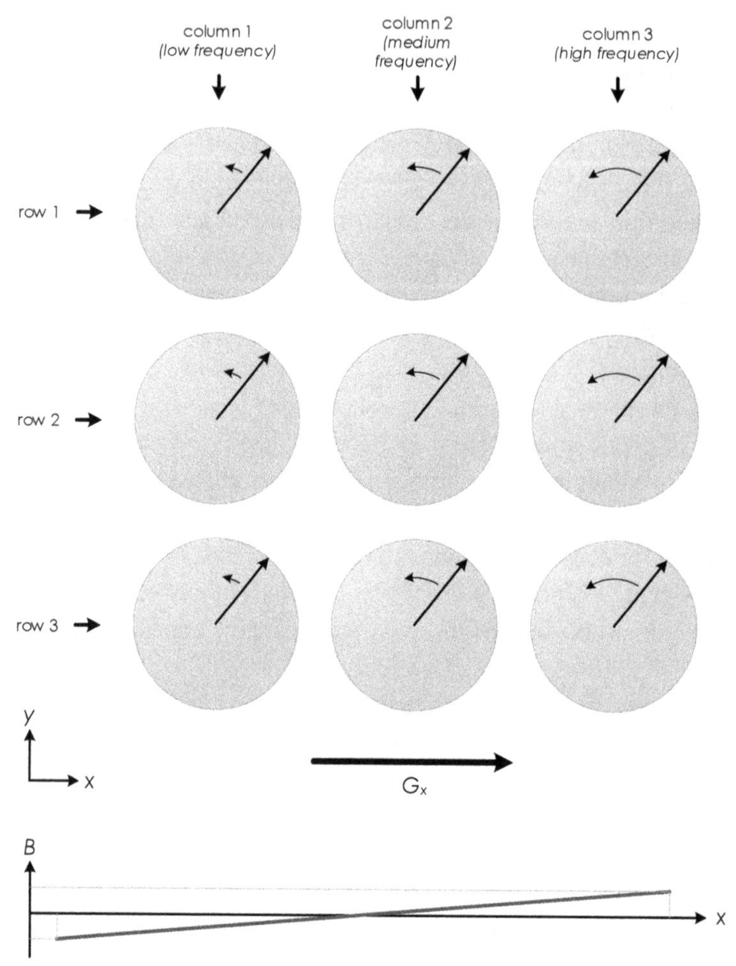

Figure 5.60 – Illustration of the movement of rotation of the transversal magnetization in 9 voxels arranged in a 3-by-3 matrix when a magnetic field gradient is applied along x.

Note in Eq. (5.54) that the frequency does not change with y, that is, all voxels of the same column have the same frequency of rotation of the magnetic field.

In the example, the detected signal is the sum of three sinusoids with different frequencies but with the same initial phase. The different amplitude of these sinusoids reflects the fact that there can be more or fewer protons and also that the transverse magnetization in each voxel decreases over time due to relaxation.

So far, it has been said that there is a detector coil, placed in the xy plane, in which a voltage is induced by the transverse component of the electromagnetic wave created by the rotation of the magnetization of the protons. In fact, there are two perpendicular coils in the xy plane: one along the x-axis and one along the y-axis. This provides information about the amplitude and initial phase of the transverse component of the magnetization. These two voltages (which are in quadrature) are regarded as part of a complex number that varies with time ($\overline{v}(t)$). These two voltages are digitalized with a sampling frequency f_s yielding two vectors with N_x real elements or one vector with N_x complex elements ($\overline{v}(k_x)$).

To this complex sampled signal, it is then applied a discrete Fourier transform (DFT) giving rise to a spectrum of this signal.

$$\overline{V}[i_x] = DFT\{\overline{v}[k_x]\} = \frac{1}{N_x}\sum_{k_x=1}^{N_x} \overline{v}[k_x] e^{\frac{-j2\pi k_x i_x}{N_x}}, i_x = 1, \ldots, N_x. \tag{5.56}$$

The spectrum, which also has N_x complex values, contains the amplitude and initial phase of each sinusoid with frequency

$$f[i_x] = \frac{i_x}{N_x} \times \frac{f_s}{2}. \tag{5.57}$$

The amplitude is the module of $\overline{V}$ and the initial phase is the argument of $\overline{V}$. The amplitude of each sinusoid indicates the intensity of the transverse magnetization at a point in space located in

$$x = \frac{f[i_x] - f_0}{\gamma \cdot G_x}. \tag{5.58}$$

The use of the gradient magnetic field along x and the computation of the DFT on the signal measured in the detection coils, sampled in complex form, allows the determination of the intensity of magnetization in a given column of the image matrix (matrix of voxels), i.e., the column corresponding to the x coordinate given by (5.58).

It is also necessary to separate the components of intensity according to the row where they come to obtain the intensity in each individual voxel. This could be done using the same technique but now using a magnetic field gradient along y rather than x. The problem is that you cannot use this technique for the two directions at the same time – you can use one or the other. The solution is to use this technique for the x-direction, as described above, and use another that will allow the discrimination in *y*. This other technique is called *phase encoding*.

Phase encoding uses the initial phase of the sinusoids to associate the transverse magnetization for a given row. One cannot, however, simply use a different initial phase value for each row because the detection coils measure the sum of sinusoids corresponding to each voxel. When adding up sinusoids with the same frequency and different initial phases (and amplitudes), one obtains a sinusoid of the same frequency but with an amplitude and initial phase that is a combination of the amplitudes and initial phases of these sinusoids, which however cannot be separated. The solution adopted is to use the change in the absolute value of the initial phase over several repetitions of the measurement.

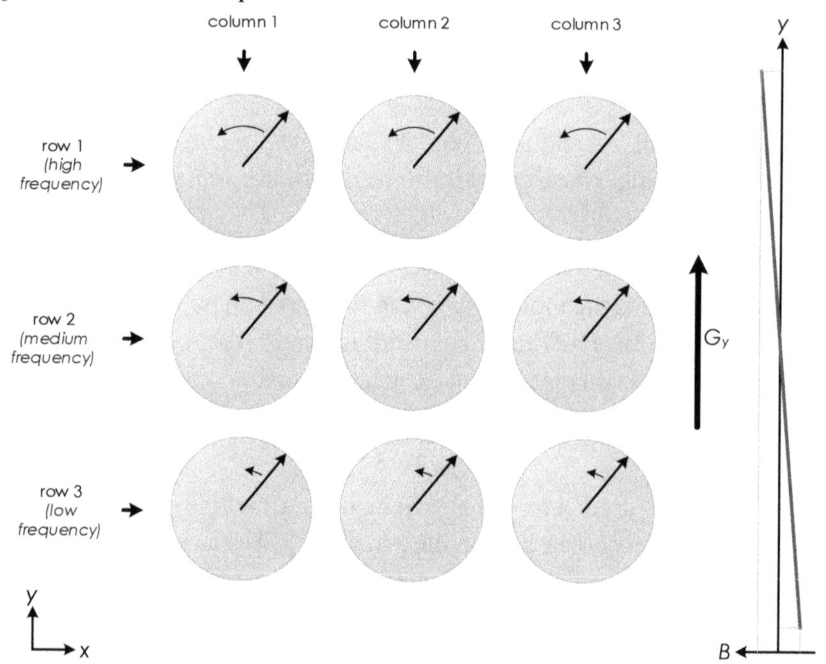

Figure 5.61 – Illustration of the rotation of the macroscopic transverse magnetization in 9 voxels arranged in a 3-by-3 matrix when a magnetic field gradient is applied along the y-axis.

The adjustment of the initial phase of the transverse magnetization is performed before the application of the gradient along *x* and the measurement of the signal in the sensing coil. This adjustment is performed with the aid of a magnetic field gradient in y. This gradient causes the rotation frequency to be different in each row of the voxel image, as illustrated in Figure 5.61.

After being under the influence of a gradient G_y during an interval t_1.

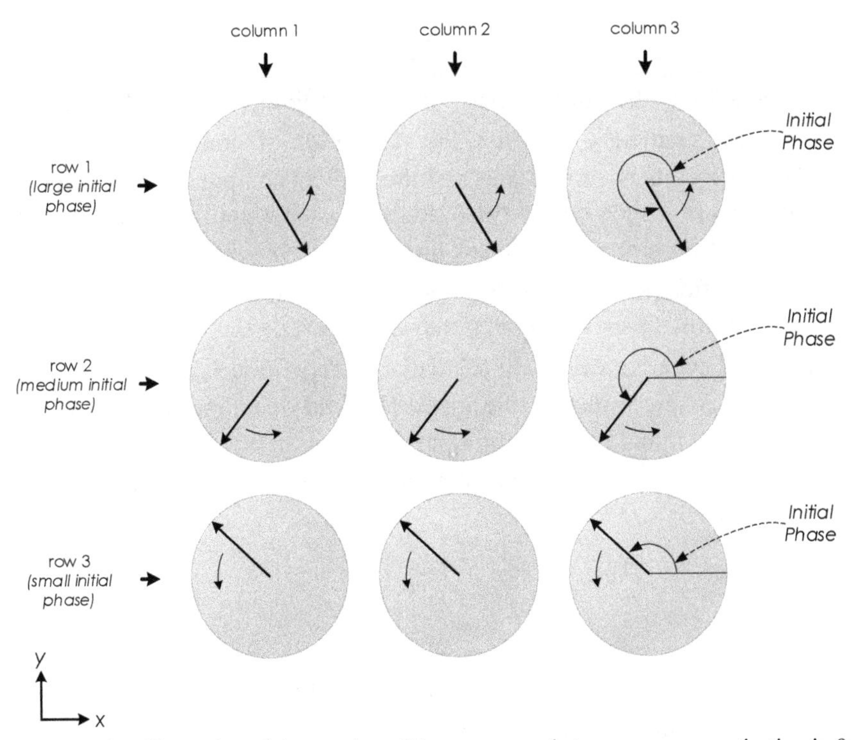

Figure 5.62 – Illustration of the rotation of the macroscopic transverse magnetization in 9 voxels arranged in a 3-by-3 matrix after the use of a magnetic field gradient along y during an interval t_1.

This gradient is applied for a certain time interval t_1. During this time, the transverse magnetization had the opportunity to rotate by an angle proportional to the rotation frequency: the higher the frequency, the greater the angle of rotation. After the time interval t_1 the transverse magnetization of the first row of voxels has a larger angle than that of rows 2 and 3 (and in row 2 greater than in row 3), as illustrated in Figure 5.62.

The initial phase of the transverse magnetization in location (x, y) is given by

$$\phi(x, y) = 2\pi f_y t_1, \tag{5.59}$$

where

$$f_y(x, y) = f_0 + \gamma \cdot G_y \cdot y, \tag{5.60}$$

and

$$G_y = \frac{dB}{dy}. \tag{5.61}$$

After setting the initial phase for the different rows with the G_y gradient, the gradient G_x is used, the acquisition of signal samples from the detecting coils is carried out, and the DFT is computed as explained. This procedure is then repeated N_y times. In each repetition, the value of G_y used is different, which causes the value of the initial phase of each row to be different. Figure 5.63 shows an example with seven rows and five repetitions. Over the repetitions, the gradient G_y varies, assuming positive and negative values. In particular, in repetition 3, in this example, the gradient is zero; that is, the magnetic field does not change, which means that the initial phase is the same in all rows.

At the end of the procedure, there are N_y spectrums computed. They are given by

$$\overline{V}\left[i_x, k_y\right] = \frac{1}{N_x} \sum_{k_x=1}^{N_x} \overline{v}\left[k_x, k_y\right] e^{\frac{-j2\pi k_x i_x}{N_x}}, \quad i_x = 1, \ldots, N_x \text{ and } k_y = 1, \ldots, N_y \tag{5.62}$$

where k_y represents the index in each repetition (each repetition was obtained with a different initial phase).

In each column of the voxel matrix, corresponding to a frequency k_x, the amplitude of the sinusoid obtained is the same in each repetition k_y (assuming no change in the sample). The initial phase of the sinusoid, however, is different. Thus, both the real part and the imaginary part of $\overline{V}\left[i_x, k_y\right]$ will vary with k_y as illustrated in Figure 5.64 for the real part.

We can thus distinguish the different rows through the frequency of the periodic function that is the real component (and also imaginary) of the acquired signal spectrum as a function of the index of repetition (i_y).

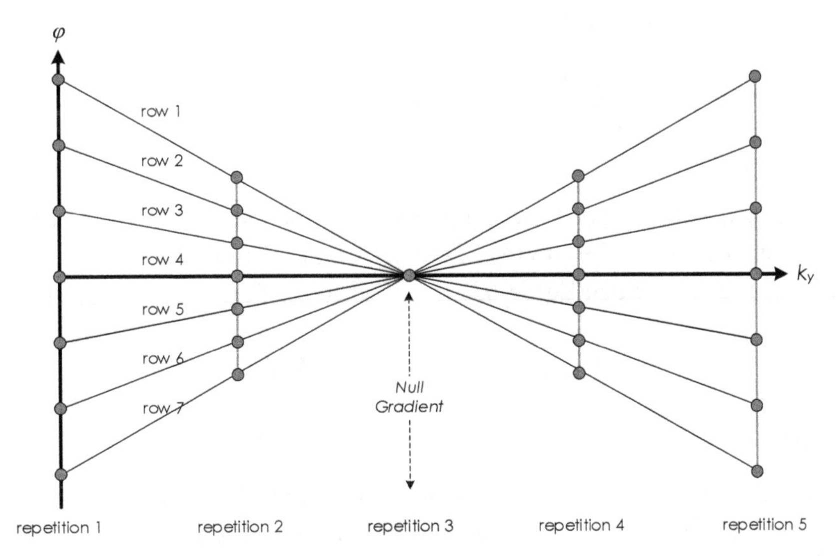

Figure 5.63 – Comparison of different initial phases of the transverse magnetization of a row of voxels for each repetition of the phase encoding procedure.

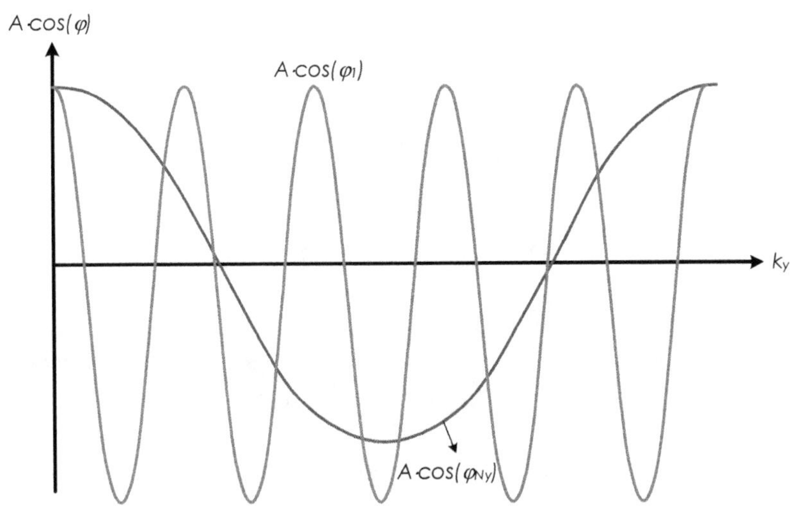

Figure 5.64 – Variation of the real component of the result of DFT according to the index of the repetition for different rows.

Using again the DFT, now on complex amplitudes given by (5.62) in relation to i_y results in

$$\overline{V}[i_x, i_y] = DFT\{V[i_x, k_y]\}$$
$$= \frac{1}{N_y} \sum_{k_y=1}^{N_y} \left\{ \frac{1}{N_x} \sum_{k_x=1}^{N_x} \overline{v}\left[k_x, k_y\right] e^{\frac{-2\pi k_x i_x}{N_x}} \right\} e^{\frac{-2\pi k_y i_y}{N_y}}, \tag{5.63}$$
$$i_x = 1, \ldots, N_x \text{ and } i_y = 1, \ldots, N_y,$$

where i_x is the index of the column, and i_y is the index of the row of the voxel matrix that makes up the 2D image obtained from the inside of the human body. Note that Eq. (5.63) is a 2D discrete Fourier transform. One has

$$\overline{V}[i_x, i_y] = DFT_{2D}\{\overline{v}[k_x, k_y]\}. \tag{5.64}$$

5.13.6 *System*

The MRI system contains a superconductor coil used to generate a strong external magnetic field that aligns all the protons on a given part of the human body. It also contains three sets of coils to generate the magnetic field gradients along three orthogonal axes (Figure 5.65).

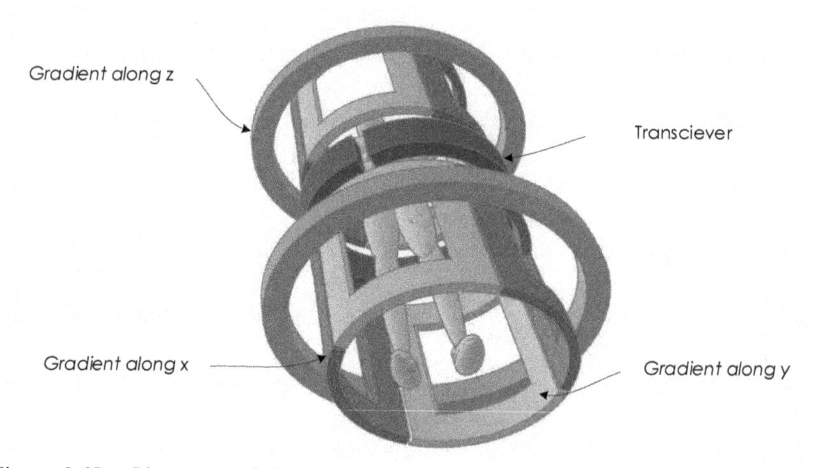

Figure 5.65 – Placement of the coils used to create gradients of the magnetic field in perpendicular directions.

It also contains a coil for generating electromagnetic waves to irradiate the body and two coils to measure the radiation produced by the relaxation of the transverse magnetization of protons. Finally, it has a data acquisition module, a computer to control the current flowing through the gradient coils and the generation of electromagnetic waves, as well as to perform the necessary calculations to build the image of the human body and present it to the user.

5.14 Questions

1. Consider a stepper motor used to control the position of the reading head of a computer hard drive, as illustrated in the figure. The motor runs 30° at each step. The screw has 20 turns per cm. The disc tracks have a spacing of 0.25 mm. How many steps does it take to turn the engine so that the reading head moves from one track to the next?

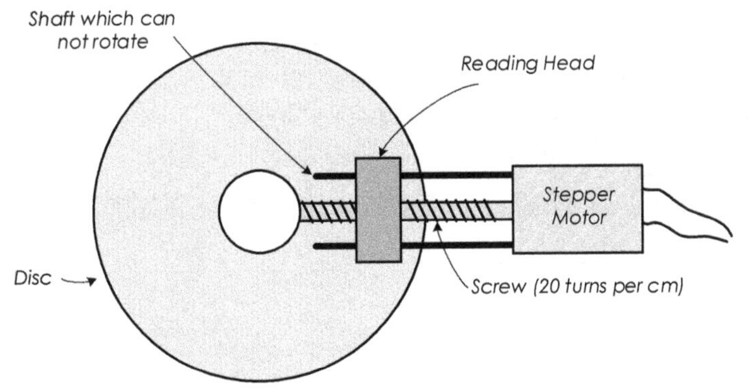

Figure 5.66 – Illustration of a computer hard drive.

2. How many steps do I need to take to complete a full turn in a stepper motor with eight poles and six rotor teeth?
3. Explain how the Hall effect sensor works.
4. Explain how a Hall effect sensor could be used to detect the opening of a door.
5. Explain how you would use a Hall effect sensor to measure the rotation speed of a shaft.
6. Explain how a magnetic field sensor based on the giant magnetoresistive effect (GMR) works.
7. What does magnetostriction consist of?
8. How does a stepper motor with a permanent magnet work?

9. What is the residual torque of a stepper motor?
10. What is the difference between the full-step and half-step operating modes of a stepper motor?
11. What is the main advantage of a stepper motor over other types of motors?
12. What is the main disadvantage of a stepper motor compared to other types of motors?
13. In a Hall effect sensor, what is the quantity that is directly measured?
    a. The electrical voltage.
    b. The electric current or electric voltage.
    c. The magnetic field or the electric current.
    d. The magnetic field or electrical voltage.
    e. None of the above.
14. What is the analytical expression of the Lorentz Force?
    a. $\vec{F} = q(\vec{E} + \vec{v} \times \vec{B})$.
    b. $\vec{F} = q\vec{E}$.
    c. $\vec{F} = q(\vec{v} \times \vec{B})$.
    d. $\vec{F} = q(\vec{E} \times \vec{v} \times \vec{B})$.
    e. None of the above.
15. What torque specification indicates the value of the torque required to apply to the rotor to move it when it is stopped, and current is applied to the motor?
    a. Residual torque.
    b. Static or holding torque.
    c. Pull-in torque.
    d. Pull-out torque.
    e. None of the above.

16. Regarding the stepper motor represented in the figure, which of the statements is true?

    a. The motor has eight windings and eight phases.
    b. The motor has four windings and eight phases.
    c. The motor has four windings and four phases.
    d. The motor has eight windings and four phases.
    e. None of the above.

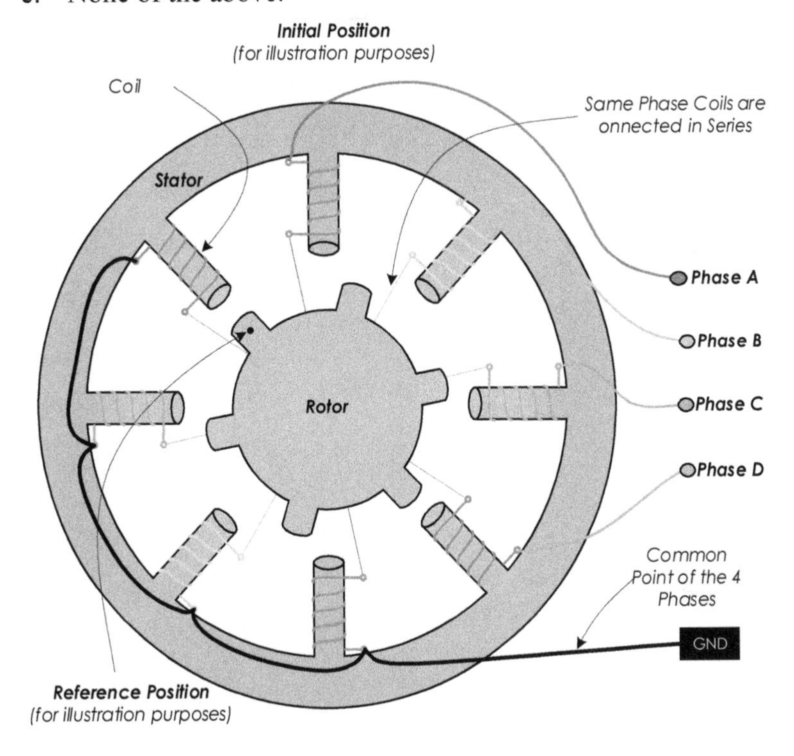

Chapter 6

Devices Based on Mechanical Phenomena

6.1 Piezoelectric effect

The piezoelectric effect consists of the appearance of a voltage in a material subject to deformation. This is due to the change in the internal distribution of the electric charge of an asymmetric crystal structure. This redistribution of electric charge gives rise to the appearance of an excess of positive and negative electrical charges on opposite faces of the material.

This effect exists in natural crystals such as quartz, certain ceramic materials, and polymers produced by Man. The ceramic materials such as lead zirconate titanate (PZT) have a piezoelectric coefficient of about two orders of magnitude greater than the magnitude of single crystals and can be produced economically. They have, however, poor temporal stability.

As an example, look at quartz which is a crystal made of silica (silicon dioxide, SiO_2) with three molecules of silica in each cell. Oxygen has six electrons in its valence shell, and silicon has four electrons in its valence shell. The two oxygen atoms and the silicon atom form a covalent bond. Two of the silicon valence electrons are shared with the oxygen atoms, and the other two electrons with another oxygen atom to complete the valence shell (eight electrons). Each oxygen atom is bonded to two silicon atoms, and each silicon atom is bonded to four oxygen atoms forming a tetrahedron, as shown in Figure 6.1.

In practice, silicon is left with a positive charge (minus four electrons) and oxygen with a negative charge (plus two electrons). Each cell has, however, a neutral charge distribution (Figure 6.2, left).

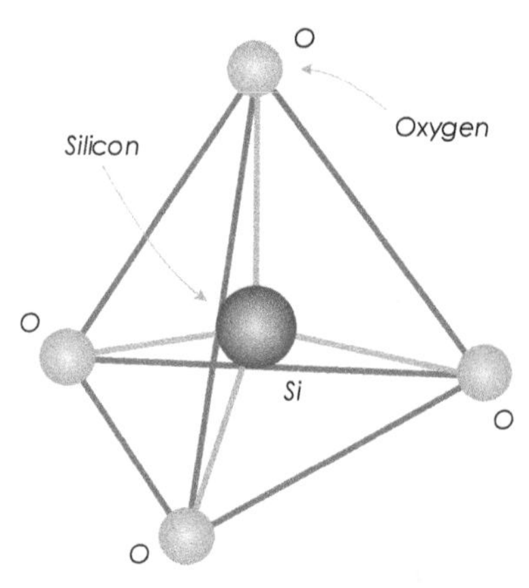

Figure 6.1 – Illustration of the constitution of a silica molecule (SiO_2).

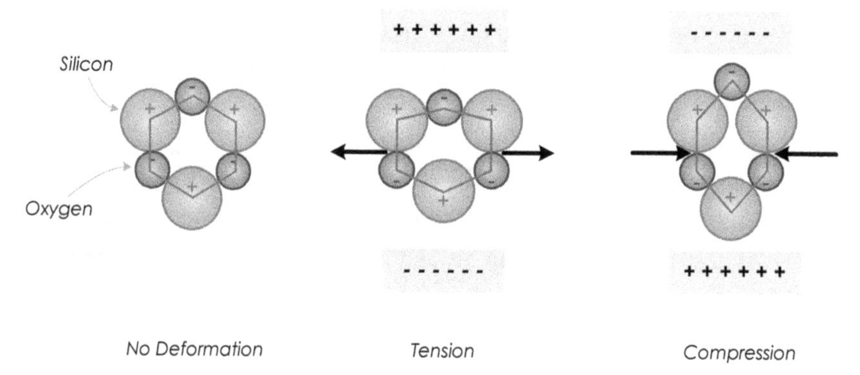

Figure 6.2 – Illustration of the piezoelectric effect in a quartz crystal.

When the crystal structure is compressed, oxygen and silicon atoms change position in such a way that each cell of the crystalline network ceases to have asymmetry of charge to have an excess of positive charge on one side and negative on the other (Figure 6.2, middle). This distribution is reversed when the structure is stretched (Figure 6.2, right).

In macroscopic terms, one of the crystal's faces will have a positive charge and the other a negative one, like with a capacitor to which a voltage is applied.

The piezoelectric effect is reversible; that is, if a voltage is applied between the faces of the crystal, it will undergo deformation.

Table 6.1 presents a comparison of the sensitivity, resolution, and dynamic range of strain sensors based on different physical principles. It is observed that the piezoelectric effect is the one that gives rise to an increased sensitivity and allows for better resolution.

Table 6.1 – Comparison of sensitivity to deformation, resolution, and dynamic range of different physical effects.

PHENOMENA	SENSITIVITY [V/ε]	RESOLUTION [ε]	RANGE
Piezoelectric	5.0	0.00001	100 000 000
Piezoresistive	0.0001	0.0001	2 500 000
Inductive	0.001	0.0005	2 000 000
Capacitive	0.005	0.0001	750

A disadvantage of piezoelectric sensors is that they are not suitable for static measurements. A static force results in a fixed number of charges on the surface of the piezoelectric material. Using conventional circuits for signal conditioning and imperfect insulating materials causes this charge to be drained over time.

Depending on how the piezoelectric material is cut to make a sensor or actuator, there are three possible modes of operation (Figure 6.3):

Transversal — Force is applied according to one of the axes (y, for example), and charges are generated according to a perpendicular direction (x-axis, for example). The amount of charge depends on the material's dimensions and of the piezoelectric coefficient (*d*): $C_x = -d \times F_y \times b/a$;

Longitudinal — Force is applied according to one of the axes (x-axis, for example), and charges are generated according to the same axis. The amount of charge does not depend on the material's dimensions. Several elements (n) can be connected mechanically in series and electrically in parallel: $C_x = d \times F_x \times n$;

Tangential — Force is applied according to one of the axes (y-axis, for example) but at different points. The charge imbalance occurs in a direction perpendicular (x-axis for example). The amount of charge does not depend on the material's dimensions. Several elements can be connected mechanically in series and electrically in parallel: $C_x = 2 \times d \times F_y \times n$.

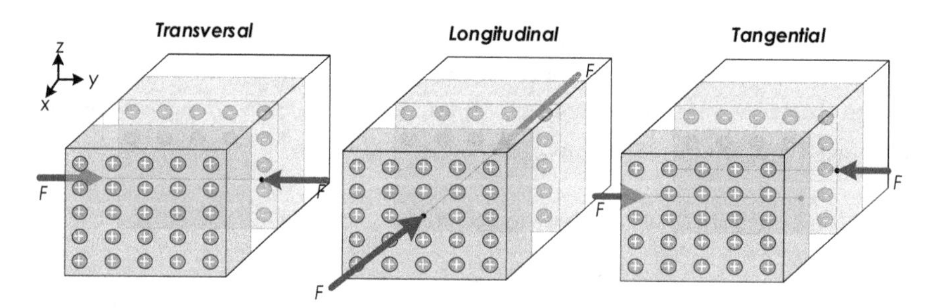

Figure 6.3 – Illustration of the different modes of operation of a piezoelectric material. The positive charges accumulate on the front side and negative on the back.

A piezoelectric material can be electrically modeled as a voltage source proportional to force, pressure, or deformation applied, followed by filtering (Figure 6.4).

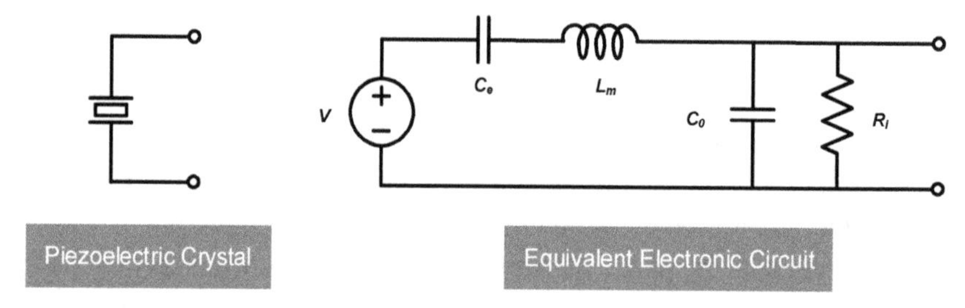

Figure 6.4 – Electrical scheme equivalent of a piezoelectric material.

This model includes effects due to the mechanical construction of the material and other non-idealities:

L_m inductance is due to the mass and the sensor's inertia.

C_e capacitance is inversely proportional to the material elasticity.

C_0 capacitance represents the static capacity of the sensor.

R_i resistor represents the resistance of the isolation.

The typical frequency response of a piezoelectric sensor is shown in Figure 6.5. There is a zone where it is flat and is used in situations where it is important that the output voltage does not vary with frequency, and there is another area on which there is a resonance that causes the output voltage to be higher. This zone is used in cases where it is important to have a high sensitivity.

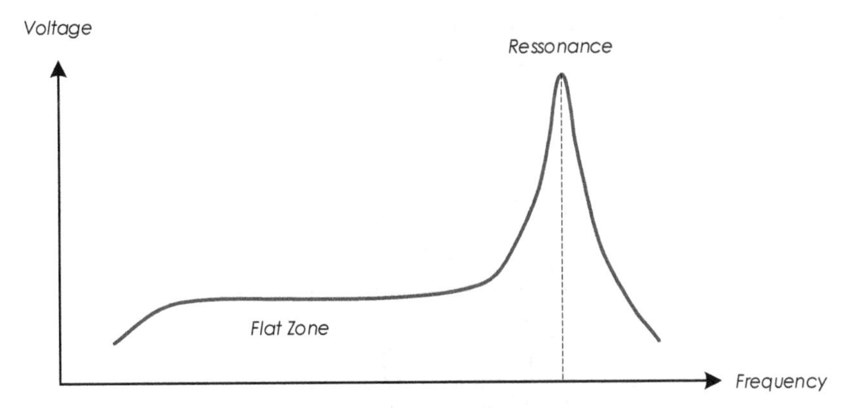

Figure 6.5 – Typical frequency response of a piezoelectric material.

6.2 Accelerometer

Piezoelectric materials can be used to create acceleration sensors. This type of sensor uses a test mass which, under acceleration, causes a force on the piezoelectric material. This force, in turn, causes a deformation that results in a voltage that is proportional to acceleration.

6.2.1 *Configurations*

There are three kinds of used configurations:

- Tangential
- Flexion
- Compression

The accelerometers which run tangentially use a piezoelectric material between a seismic mass and a fixed structure, as shown in Figure 6.6. A preload ring is used to create a rigid structure with a linear variation.

When the structure is subjected to acceleration, the seismic mass moves up and down, causing a tangential deformation on the piezoelectric material. This deformation results in a voltage proportional to the acceleration.

Due to the isolation between the piezoelectric material and the encapsulation of the sensor, this operating mode allows a good insensitivity to temperature variations and to deformations of the structure's base. Moreover, the use of the piezoelectric effect in tangential mode allows one to build small sensors, which in turn lead to high operating bandwidth and to a small mechanical load from the sensor on the device being measured. Of the three types of accelerometers specified, this is the best performer.

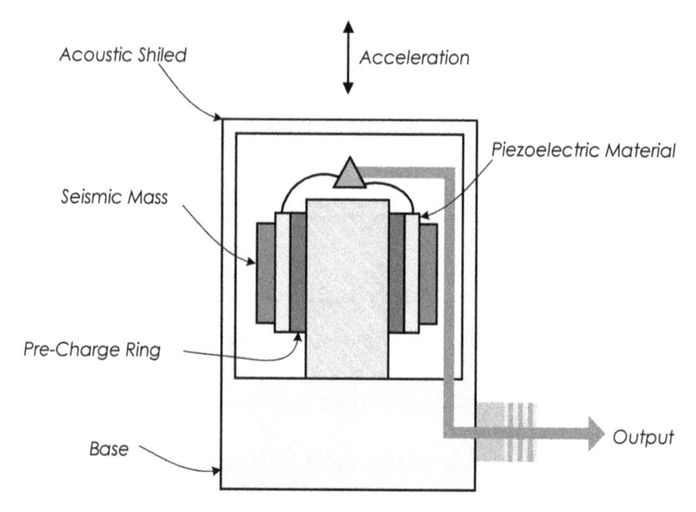

Figure 6.6 – Constitution of an accelerometer operating in tangential mode.

Figure 6.7 shows the constitution of a **flexion** accelerometer. The piezoelectric material together with a seismic mass is supported on a central point. A vertical acceleration (relative to the arrangement in the figure) causes flexion of the piezoelectric material and a consequent tension between its faces.

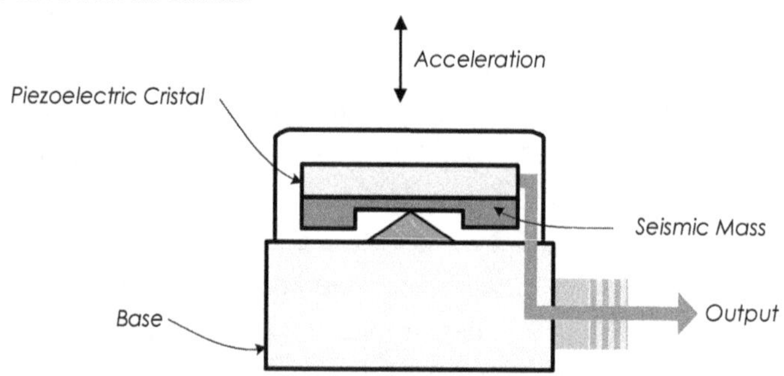

Figure 6.7 – Constitution of a flexion accelerometer.

This type of assembly is particularly immune to transverse movements and is suitable for operation at low frequency and low accelerations as found, for example, in test structures of civil construction (buildings, bridges, etc.).

The accelerometers operating in **compression** are simple structures with high rigidity. Although not widely used, it represents the traditional constitution of accelerometers (Figure 6.8).

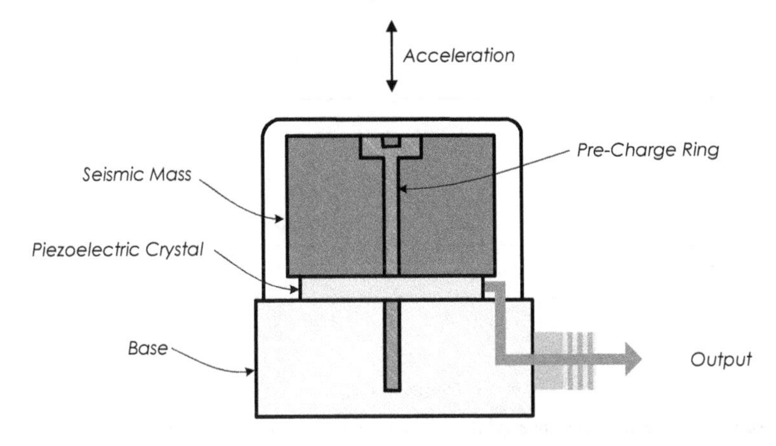

Figure 6.8 – Illustration of the constitution of an accelerometer in compression mode.

The piezoelectric material is placed between a seismic mass and the base of the sensor. A bolt applies a preload to the material. Without such pre-loading, it would not be possible to measure negative accelerations (from bottom to top in the arrangement of the figure). When an acceleration is applied, the compression force exerted by the seismic mass increases or decreases. The higher the mass, the higher the deformation exerted on the material and the higher the output signal.

This configuration is very robust and withstands high accelerations such as those found, for example, in the case of a collision. However, due to the contact between the piezoelectric material and the encapsulation of the sensor, this type of assembly tends to be more sensitive to deformations of the structure's base. For the same reason, this assembly makes the sensor more sensitive to temperature variations due to thermal expansion and contraction. These problems lead to measurement errors, especially important when the sensor is applied in thin metallic plates or at low frequencies in environments whose temperatures vary significantly, like on the outside or near air conditioning units or ventilation.

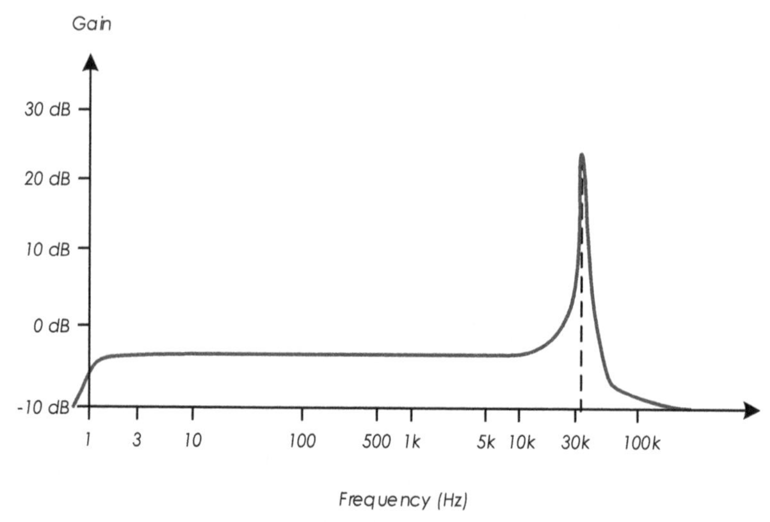

Figure 6.9 – Frequency response of the piezoelectric accelerometer from Brüel & Kjær Vibro, model ASA068.

In short, the different types of accelerometers are constructed to resolve a specific problem — the compression accelerometers have a greater rigidity to function with high accelerations, the flexion ones use a small support area of the piezoelectric material to be insensitive to transverse accelerations and accelerometers that work in tangential mode insulate the exterior environment from the piezoelectric material.

This sensor has a range of measurement of ±500 g and a sensitivity of 10 mV/g specified for 80 Hz and 25°C. The specified frequency response is shown in Figure 6.9. A range of operating frequencies between 1.5 and 13 kHz (specified at 3 dB) is shown.

6.2.2 *Signal Conditioning*

The piezoelectric sensors are sensors that have a high output impedance (low output current) that can reach 100 MΩ. This requires special care with signal conditioning circuits, in particular with the input currents of the amplifiers and the parasitic capacities of the cables.

In the case of the sensors based on the piezoelectric phenomenon, the output signal is in the form of an electric charge that is accumulated on opposite faces of the piezoelectric material. The distribution of electric charges can also be seen as a potential difference. Basically, the sensor acts as a capacitor in which the charge distribution that arises in its plates is due to the dielectric and the piezoelectric effect (and not due to an external battery that promotes the load distribution).

Contrary to the capacitive sensors, seen in Chapter 3, on which the voltage is kept constant, and a capacity change causes a change in load in the case of piezoelectric sensors, the capacity remains constant (C) and it is the change of the load (ΔQ) that gives rise to a voltage change. This change is given by

$$\Delta V = \frac{\Delta Q}{C}, \tag{6.1}$$

in the case where the transducer is not connected to any circuit.

For signal conditioning, an assembly known as a **charge amplifier** shown in Figure 6.10 is used.

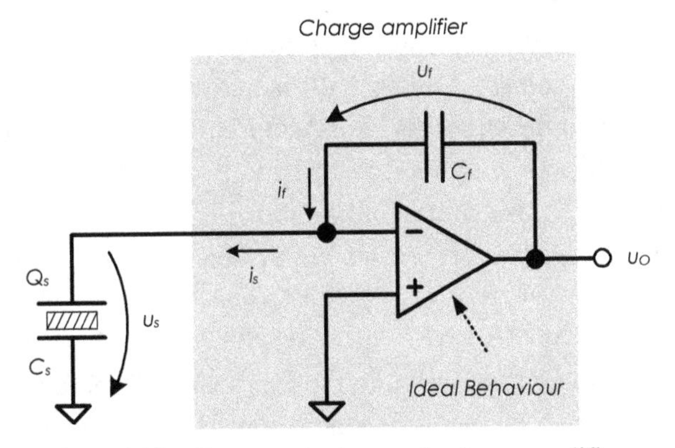

Figure 6.10 – Conceptual scheme of a charge amplifier.

Considering the currents shown in Figure 6.10 and considering an operational amplifier with an ideal behavior, in particular with an input current of zero,

$$i_s = i_f. \tag{6.2}$$

The current in the feedback capacitor is given by

$$i_f = C_f \frac{du_f}{dt}, \tag{6.3}$$

and on the sensor is given by

$$i_s = C_s \frac{du_s}{dt} = \frac{dQ_s}{dt}, \tag{6.4}$$

having used

$$Q_s = C_s u_s. \tag{6.5}$$

The output voltage is therefore

$$u_0 = u_f. \tag{6.6}$$

Considering that the amplifier is ideal, its input voltage is zero.

By inserting, inserting (6.3) and (6.4) in (6.1) and using (6.6), leads to

$$u_0 = \frac{Q_s}{C_f}. \tag{6.7}$$

The output voltage is, therefore, proportional to the load of the sensor.

In practice, however, the amplifier is not ideal, and its input current (I_p) will go through the capacitor C_f (and the sensor too) leading to an additional output voltage which will increase over time due to the integrating action of the capacitor. This would make the amplifier saturate. To reduce this effect, a high-value resistor is used in parallel with the feedback capacitor, as illustrated in Figure 6.11.

This resistance, however, creates a constant voltage deviation on the output with a value of $R_f \times I_p$. Another consequence is the fact that it introduces a time constant $R_f \times C_f$ in response to a step-change in the input. This leads to an upper limit in the frequency of operation.

Figure 6.12 shows an operational amplifier with a finite gain as well as a capacity (C_c) and insulation resistance (R_i) from the connection cable.

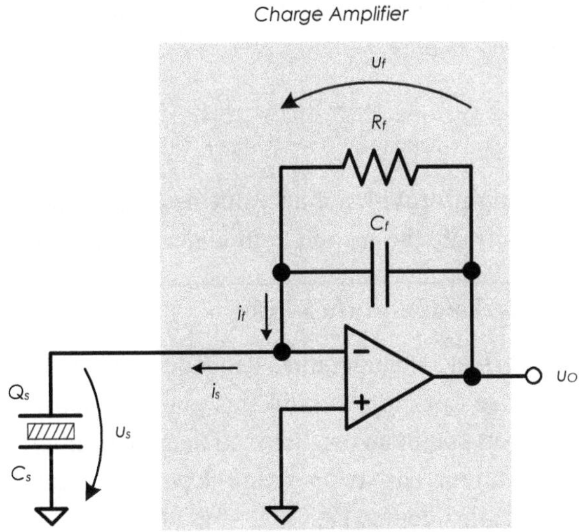

Charge Amplifier
Uf
Rf
Cf
if
is
Qs
Cs
Us
Uo

Figure 6.11 – Schematic of a charge amplifier with the feedback resistor.

Charge Amplifier
Cable
Uf
Rf
Cf
if
is
Qs
Cs
Us
Cc
Ri
Uo

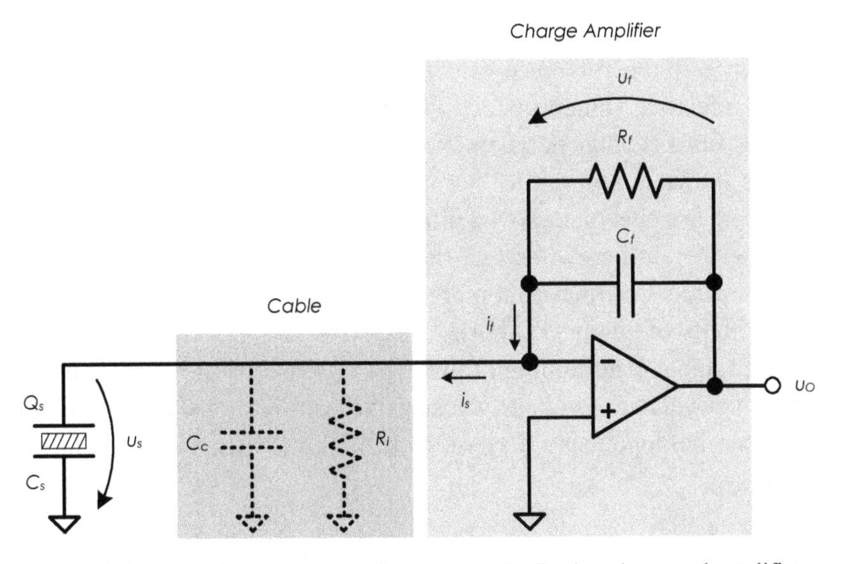

Figure 6.12 – Schematic of a charge amplifier with the feedback resistor, real amplifier, and contribution of the capacity of the cable and its insulation resistance.

In this situation, the real output voltage of the charge amplifier, without considering the resistance, is

$$u_0 = \frac{Q_s}{C_f} \times \frac{1}{1+\frac{C_f+C_c+C_s}{A \cdot C_f}}. \quad (6.8)$$

Therefore an amplifier with a gain value as large as possible must be used in order to eliminate the second term under the fraction in (6.8).

6.3 Piezoelectric Temperature Sensor

The piezoelectric effect is temperature-dependent. Thus, it is possible to create temperature sensors that exploit this phenomenon. These sensors use a piezoelectric crystal in an oscillator so that the oscillation frequency depends on temperature. This temperature dependence can be described by a polynomial of third-degree [7],

$$\frac{\Delta f}{f_0} = a_0 + a_1 \Delta T + a_2 \Delta T^2 + a_3 T^3, \quad (6.9)$$

where f_0 is the value of the reference frequency, Δf is the offset frequency in relation to this reference value, ΔT is the temperature offset in relation to the reference value to which the frequency oscillation is f_0 and a_i are the coefficients of the polynomial. These coefficients depend on the crystal cut, as shown in Figure 6.13.

There are cuts that allow a more linear relationship ($a_2 \approx 0$ and $a_3 \approx 0$), like the cut known as LC, developed by Hewlett-Packard [21], and there are cuts less linear but more sensitive (larger values of a_1) reaching a sensitivity of 90 ppm/°C [14].

The signal conditioning of this type of sensor is the conversion of the frequency value to a voltage whose value is proportional to this frequency. This can be done using a phase-locked loop (PLL), as shown in Figure 6.14.

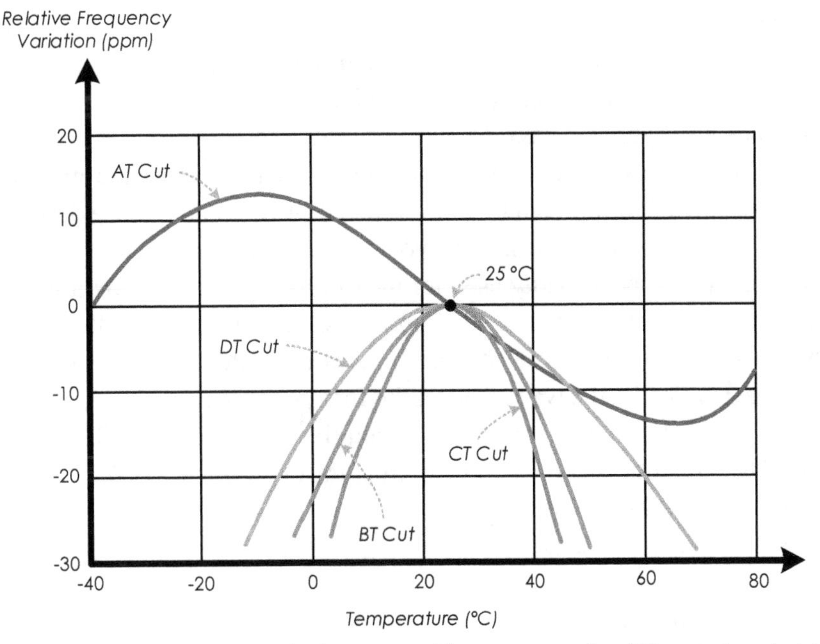

Figure 6.13 – Relative change in frequency with temperature for different piezoelectric crystal cuts.

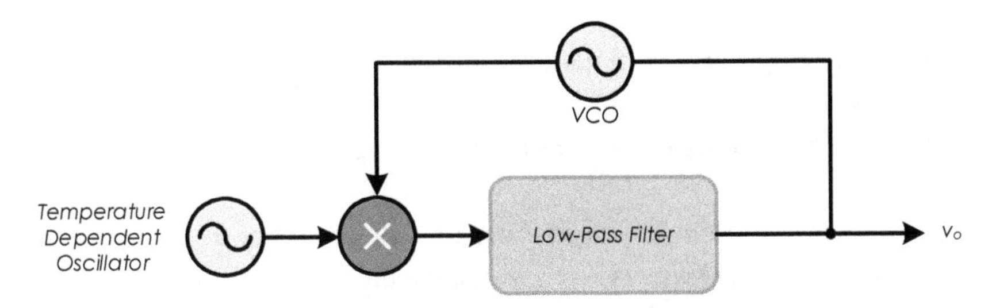

Figure 6.14 – Signal conditioning circuit for a piezoelectric temperature sensor using a PLL.

6.4 Acoustic Waves

An acoustic wave is the alternating compression and expansion of a material that can be solid, liquid, or gaseous. These waves are said to be sound waves if they have a frequency between 20 Hz and 20 kHz. Below 20 Hz, they are called infrasound, and above 20 kHz are called ultrasound.

The detection of infrasound is used in the structural analysis of buildings, earthquake prediction, and other sources of large dimensions. When infrasound has a significant magnitude, it can be felt by humans, even causing psychological effects (panic, fear, etc.).

Audible waves can be created by vibrating strings (stringed musical instruments), vibrating columns of air (wind instruments), and vibrating plates (some vibrating instruments, vocal cords, and speakers).

Ultrasounds are typically used for measuring distance or proximity. A wave is emitted at a certain point, and it takes time to return to this point after a given object reflects it; that time is measured to determine the distance traveled by the wave. One of the first practical applications of this type was used in sonar to detect submarines underwater. There are, nowadays, numerous applications where ultrasound is also used:

- Automobile parking;
- Thickness measurement;
- Temperature measurement;
- The measure of the level of liquids;
- Help for the visually impaired;
- Controlling focus lenses in camcorders and photographic cameras;
- Detecting flaws in materials;
- Creating maps;
- Detection, localization, and object recognition.

There are two commonly used types of waves (Figure 6.15):

- **Longitudinal waves** — in which the motion of the particles made according to the direction of propagation;
- **Tangential waves** — in which the motion of the particles is perpendicular to the direction of propagation.

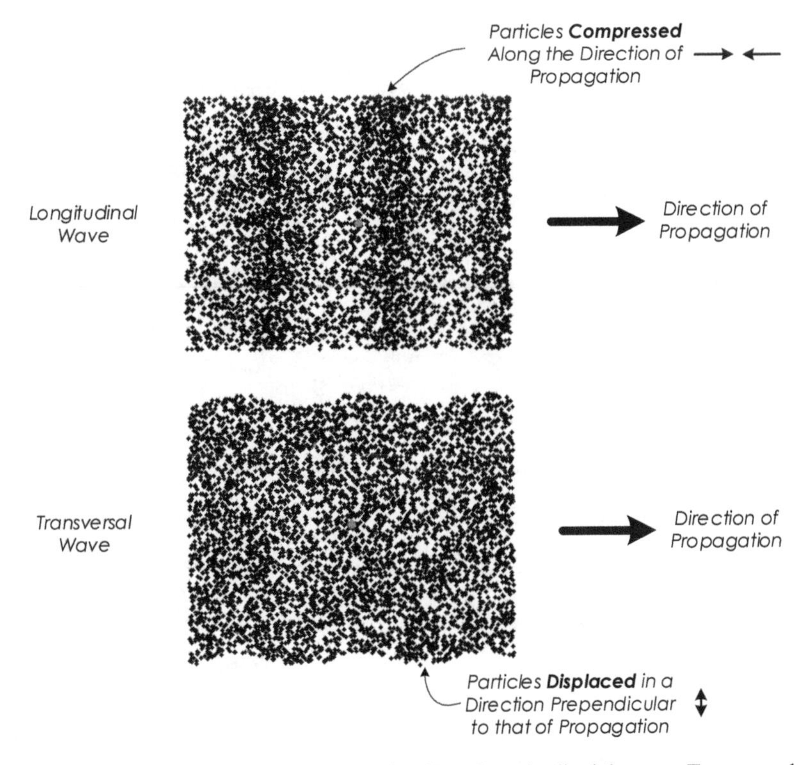

Figure 6.15 – Different propagation modes. Top: Longitudinal, bottom: Transversal.

It is necessary to know the value of the propagation velocity to determine the distance traveled by the wave from the flight time. This velocity, besides depending on the environment, also depends on temperature, relative humidity, atmospheric pressure, and carbon dioxide concentration (Figure 6.16).

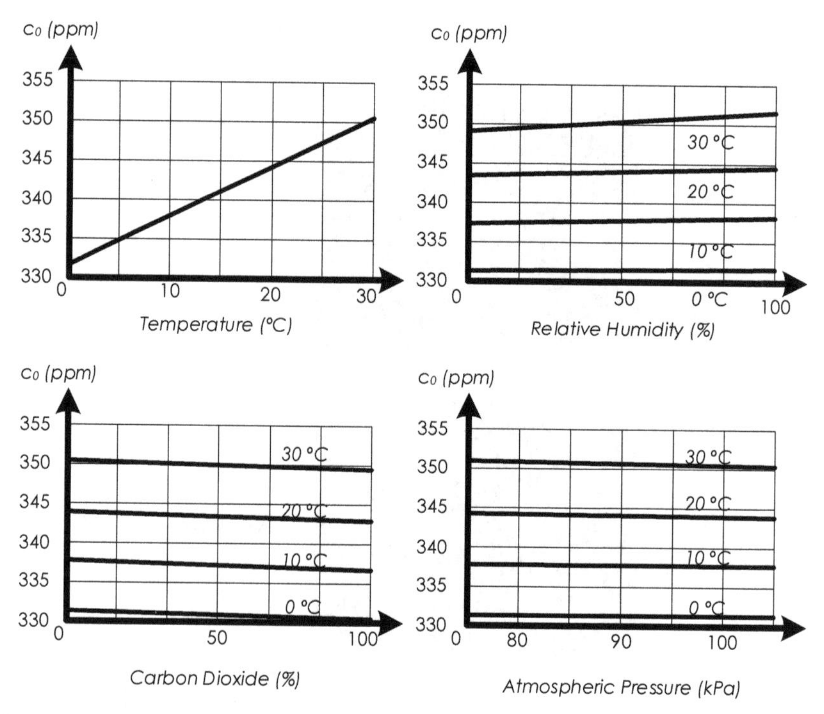

Figure 6.16 – Charts that show the change in sound velocity in the air as a function of temperature, relative humidity, carbon dioxide concentration, and atmospheric pressure.

As the wave propagates, the amplitude of the acoustic pressure becomes smaller due to friction of the molecules of the environment. This determines the attenuation range of a distance measurement system using ultrasound, for example. This attenuation increases with frequency because the speed of the particles in the environment is higher, which causes higher friction (Figure 6.17).

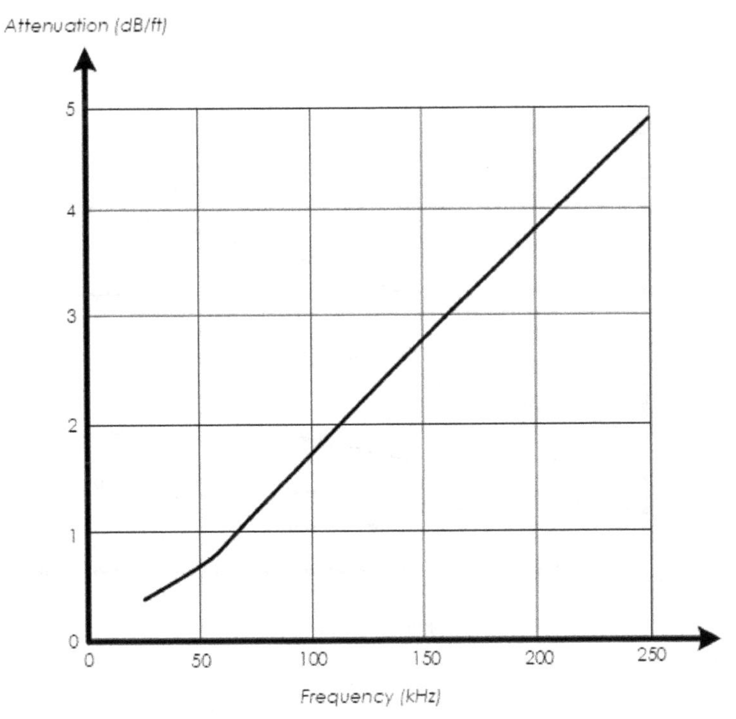

Figure 6.17 – Attenuation of an acoustic wave as a function of its frequency for the worst case of relative humidity.

Attenuation also depends on relative humidity, as shown in Figure 6.18.

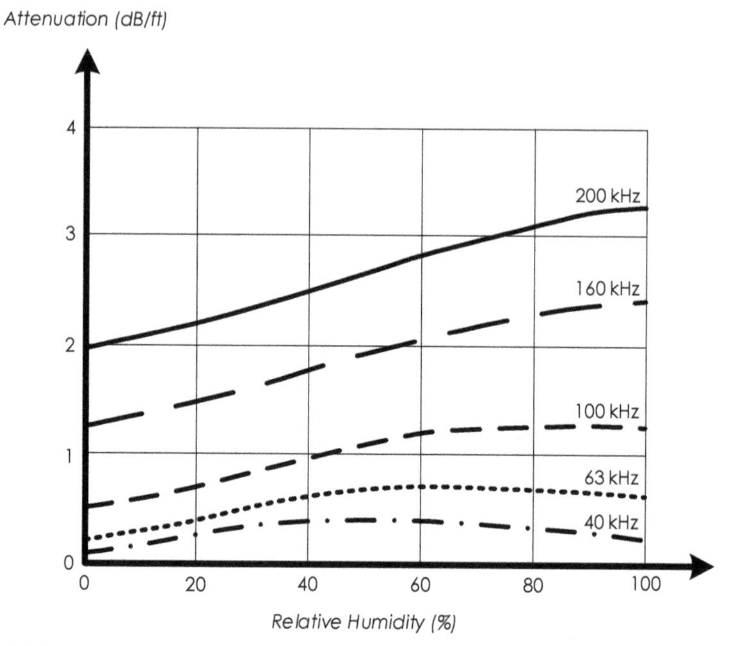

Figure 6.18 – Attenuation of an acoustic wave as a function of relative humidity at different frequencies.

The characteristic acoustic impedance of material reflects the opposition that this material offers to the displacement of particles by sound waves and is defined as:

$$Z_0 = c\rho, \tag{6.10}$$

where ρ is the material density and c is the speed of sound in that material. The characteristic acoustic impedance of air (420 Pa.s/m) is very different from the impedance of water (1.5 MPa.s/m). This makes the acoustic phenomena that take place in the air very different from those that take place in water.

When a wave hits perpendicular to the boundary between two materials with different impedances, part of the energy is reflected, and part is transmitted. The transmitted wave is given by

$$\frac{Transmitted\ Wave}{Incident\ Wave} = \frac{4Z_1Z_2}{(Z_1+Z_2)^2}, \tag{6.11}$$

where Z_1 and Z_2 are the acoustic impedances of both materials. The signal reflected at interfaces given by

$$\frac{Reflected\ Wave}{Incident\ Wave} = \frac{(Z_1 - Z_2)^2}{(Z_1 + Z_2)^2}. \tag{6.12}$$

Figure 6.19 shows, as an example, the case for water and steel, which have acoustic impedances of 1.5 MPa·s/m and 45 MPa.s/m, respectively. Notice that only 12% of the transmitted signal penetrates the steel.

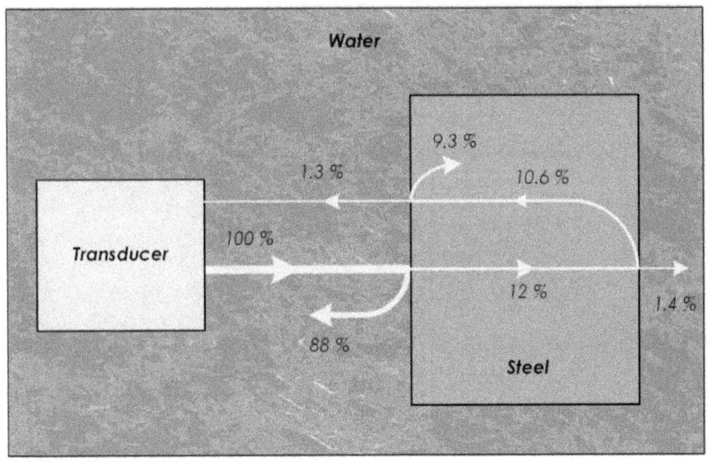

Figure 6.19 – Example of the loss of signal due to the interface between water and steel.

6.5 Ultrasound Transducers

In general, ultrasound transducers can act as transmitters or receivers. These transducers can be built using piezoelectric materials. Before 1950 magnetostrictive materials were used rather than piezoelectric materials. Only when ceramic piezoelectric materials appeared did it become more advantageous to use them in ultrasound transducers.

The emitters convert a voltage into a deformation which in turn displaces the air (or another element), causing a wave. The receivers perform the opposite. The pressure variation of the environment deforms the piezoelectric material, which in turn produces a voltage.

Figure 6.20 shows the typical construction of an ultrasonic transducer. The piezoelectric materials are placed within an encapsulation against one of the faces, which is free to vibrate (wear plate). On the other side of the piezoelectric element, a material is placed whose purpose is to absorb the sound waves generated inside the transducer.

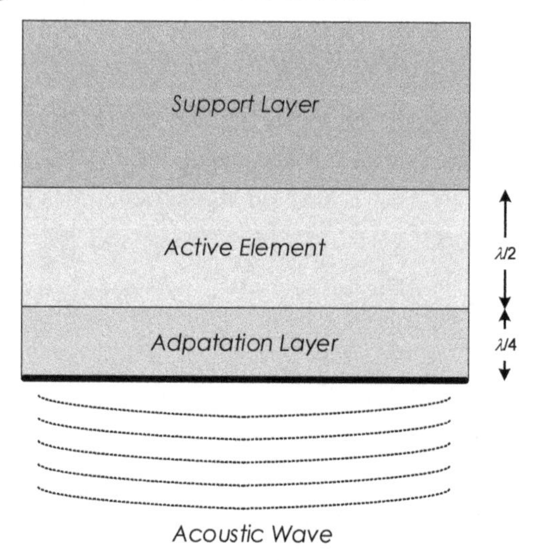

Figure 6.20 – Diagram showing the assembly of an ultrasound transducer with the thickness of each layer as a fraction of the wavelength.

The resonant frequency of an ultrasonic transducer depends on the thickness of the piezoelectric element. The thinner the element, the greater the frequency of resonance. The corresponding wavelength, in general, is twice the thickness (Figure 6.20).

The outer layer is used to perform an impedance matching between the piezoelectric element and the outside environment. An optimum impedance adaptation happens when this layer has a thickness of ¼ of the wavelength. This causes the waves that are reflected at the interface between the adaptation layer and the environment and between this layer and the piezoelectric material to be in phase when they leave the transducer leading to constructive interference.

The material used for the impedance matching layer is chosen to have an acoustic impedance characteristic with an intermediate value between the impedance of the piezoelectric material and the material of the external environment.

The impedance of the material placed on the back of the transducer is of the same piezoelectric material to be able to absorb the waves that propagate into the transducer.

The sound produced by the transducer does not originate from a single point but most of the surface of the piezoelectric material. The typical intensity of the acoustic wave is shown in Figure 6.21.

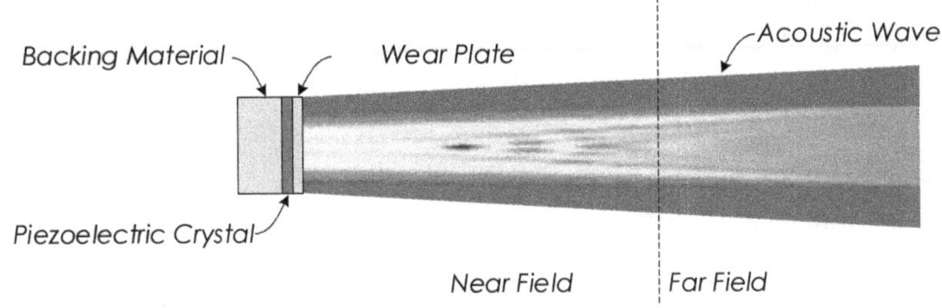

Figure 6.21 – Acoustic wave intensity of an ultrasound transducer. Lighter colors represent higher intensities.

The sound waves come from different points through the surface of the transducer, so the beam intensity is affected by constructive and destructive interference of several waves. These interferences lead to large variations in intensity near the transducer. This region is named the **near-field zone**.

The sound waves combine to form a virtually uniform wavefront at the end of the near field zone. The distance of these points to different points on the surface of the piezoelectric material is virtually the same, which makes the sum of several "basic waves" always constructive.

This next zone is called the **far-field zone**. For detection of flaws in materials, for example, the zone immediately following the end of the near field zone is the most appropriated and where the intensity is greater.

The radiation diagram of an ultrasonic transducer plots the transducer sensitivity as a function of spatial angle. This diagram depends on the operating frequency and acoustic characteristics, size, and shape of the piezoelectric element. This diagram is identical if the transducer operates as a transmitter or receiver.

The ultrasonic transducers can be constructed with different radiation patterns from omnidirectional up to narrow beams. For a transducer with a circular surface radiative, the width of the beam depends on the ratio between the diameter of the surface and the corresponding wavelength of the operating frequency. The larger the diameter of the transducer, the narrower the beam is. For example, for a diameter two times greater than the wavelength, the angle is 30°. If this ratio goes to 10 times, then the angle goes to 6°. In most applications, it is desirable to have a narrow beam, and therefore the diameter of the transducers is large compared with the wavelength.

The beamwidth is defined as the value of the angle for which the acoustic pressure decreases by 3 dB.

In ultrasonic emitters, the sound pressure produced is usually specified in decibels relative to a reference pressure of 20 μPa[1] (0.0002 μbar). This specification is called SPL (Sound Pressure Level), given by the formula

$$SPL_{dB} = 20 \cdot \log\left(\frac{p}{p_0}\right), \tag{6.13}$$

and corresponds to the acoustic pressure wave at a certain distance from the transducer and for a given rms value of the applied voltage.

In the case of ultrasonic receivers, sensitivity is used to express the performance of the transducer. The sensitivity is the ratio of the output voltage in rms and the pressure in the incident acoustic wave of the receptor. The sensitivity is generally expressed in decibel, relative to a certain reference value, typically 1 V/μbar.

Note that the acoustic pressure of the sound wave is proportional to $1/d$ where d is the distance traveled. On the other hand, the sound wave suffers attenuation as it propagates:

$$A = A_0 \cdot e^{-\alpha \cdot z}. \tag{6.14}$$

The attenuation coefficient (α) is generally proportional to the square root of the frequency.

---

[1]This value is the one commonly accepted as the minimum value detectable by the Human ear.

6.6 Measuring Distance Using Ultrasound (Sonography, SONAR, etc...)

6.6.1 *Introduction*

Ultrasonic waves can be used to measure the distance between two points. In Nature, one finds bats that use ultrasound for navigation and hunting. Today there are various man-made systems that use ultrasound to measure the distance and have applications in mapping, determining the geometry of objects, detection of obstacles, and as an aid for car parking, among others. Bosch, for example, has developed a system used for the first time in the Citroen C4 Picasso, which measures the size of the parking place.

The waves are emitted by a transmitter and received by a receiver. The time between emission and reception (Δt) is used to estimate the distance, knowing the propagation velocity (v),

$$d = v \cdot \Delta t. \tag{6.15}$$

There are three typical configurations:

- Emitter and receiver face to face;
- Emitter and receiver side by side;
- The same transducer as the transmitter and the receiver.

These are depicted in Figure 6.22.

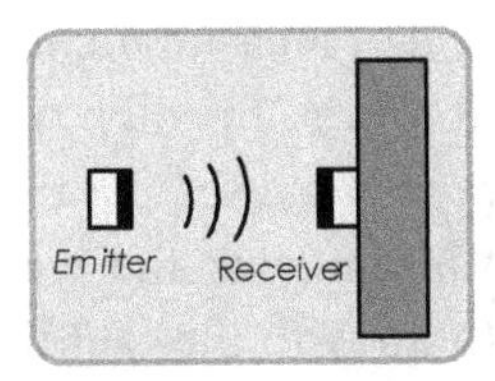

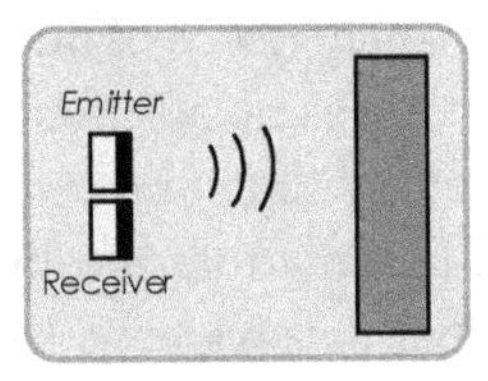

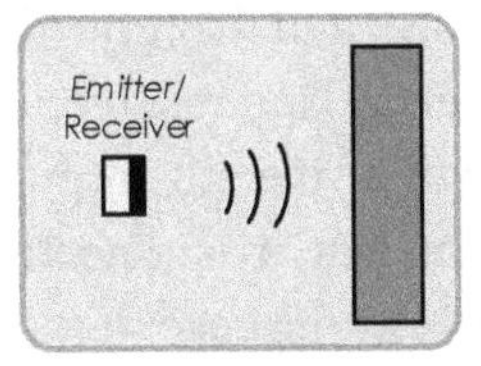

Figure 6.22 – Illustration of the three basic configurations used to measure distance using ultrasounds: face to face (left), side by side (middle), and common emitter and receiver (right).

The use of ultrasound allows measuring distances of a few centimeters to a few meters. They achieve a greater range than capacitive or inductive sensors but a smaller range than sensors using electromagnetic waves. It has, however, the advantage over these last ones of being easier to use.

Regarding the type of signal, one can distinguish two cases:

- Continuous Transmission;
- Pulsed Transmission.

In the case of **Continuous Transmission**, one typically uses unmodulated sinewaves and amplitude modulated sinewaves or frequency modulated sinewaves whose frequency changes in discrete jumps. Note that in this case, it is necessary to use a transmitter and a receiver separated from each other because it needs to simultaneously send and receive the acoustic wave.

In the case of **Pulsed Transmission**, one uses bursts of sinewaves with constant frequency or variable frequency.

6.6.2 *Determining the Distance from the Flight Time*

Determining the time of flight is sometimes difficult due to the nature of the narrow band of the ultrasound piezoelectric transducers typically used, and which causes delay and distortion in the sent and received signals.

Using a Decision Level

Different techniques are used to determine the time of flight, such as comparing the received signal with a threshold, as illustrated in Figure 6.23.

This technique has the disadvantage that it depends on the level of the received signal. This level is directly affected by the distance traveled by the wave and the reflective characteristics of the object of measurement (if the sender and receiver are in the same position). Moreover, the high time growth of the signal produced by the ultrasonic receiver causes the time instant that it is detected different from the time instant that the wave arrives at the receiver. Another disadvantage has to do with the fact that the signal to noise ratio has to be high enough so that it is possible to distinguish whether the signal output of the ultrasonic receiver is due to the acoustic wave from the issuer or if it is only noise present in the environment (in the range of ultrasound) or introduced by the receiver.

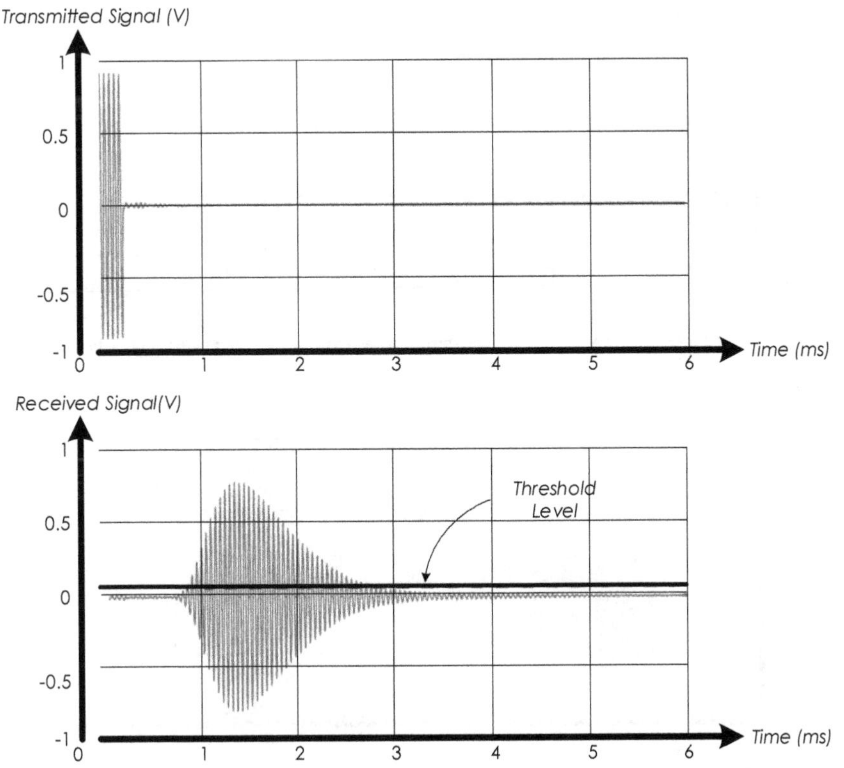

Figure 6.23 – Illustration of the estimation of the time of flight of an ultrasonic wave using a threshold.

This technique, however, has the advantage of being very fast, simple to implement, and having a low cost.

Using Cross-Correlation

Another technique used is the cross-correlation between the emitted signal and the received signal. Mathematically, the correlation is given by

$$\phi[\tau] = \sum_{n=1}^{N} f[n] \cdot g[n+\tau] \; com \; 1 \leq \tau \leq N. \tag{6.16}$$

In Figure 6.24, there is an example of the result of the cross-correlation between the emitted signal (f) and the received signal (g) with different delays.

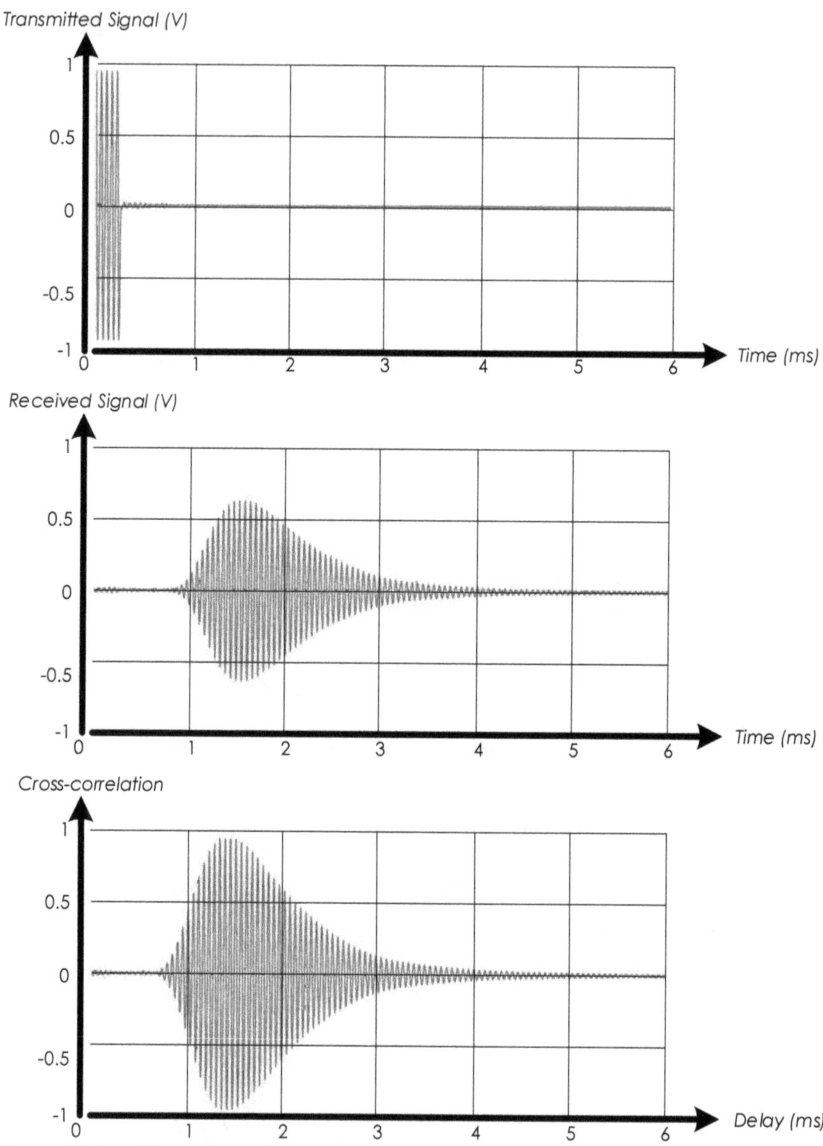

Figure 6.24 – Illustration of the cross-correlation between the emitted signal (f) and received signal (g) for different delays.

This technique is good in terms of determining the time of flight as it uses all the information contained in the signals. The accuracy of the result depends on the sampling frequency used but may be improved by using parabolic interpolation [3]. The accuracy also depends on the quality of the received signal, which can be low if there are objects that interfere with the path of the acoustic wave.

This technique works well with low levels of signal-to-noise ratio, which allows the determination of larger distances. However, it requires relatively high computing power.

Use of a Moving Window

Another technique related to the decision level technique is the moving window technique. A window is applied to the samples of the received signal to examine a small set of samples at a time. For each filtering by the window, the percentage of samples that have a voltage higher than a given threshold is determined. The window is then advanced one sample at a time (hence the name "moving"). The first time this percentage exceeds a certain value, it is considered that the acoustic wave emitted by the transmitter has reached the receiver. The exact moment when one considers that this happens can be determined with different criteria, namely:

- the instant of the first sample of the window;
- the instant of the first sample that exceeds the level of decision;
- the instant of the sample in the center of the window.

This technique is more immune to noise than the decision level technique. Its performance depends, however, on the window size, the value of the levels of decision, and the strategy employed to determine the time of flight.

Use of Received Signal Modeling

The received signal depends on the shape of the transmitted signal, the characteristics of the transducers used, the distance traveled, and the type of reflective surfaces (if applicable). It is possible to define a mathematical model that considers these factors and that describes the received signal mathematically. This model necessarily must have several parameters like:

- the duration of the signal;
- the arrival time;
- the central frequency;

- the amplitude;
- the phase.

The mathematical function is then fitted to the experimental values to determine the value of the parameters and to minimize the mean squared error between the model and experimental data. One must pay attention, however, to the choice of the initial value of the parameters and to the ability of the algorithm no to be trapped in a local minimum of the error.

Despite being computational quite complex, this technique can achieve high resolution in the estimated distance.

Measuring Phase Shift

Another technique is to continuously transmit a sinusoidal signal and measure the phase shift between the received and transmitted signals (Figure 6.25). This phase shift is related to the flight time.

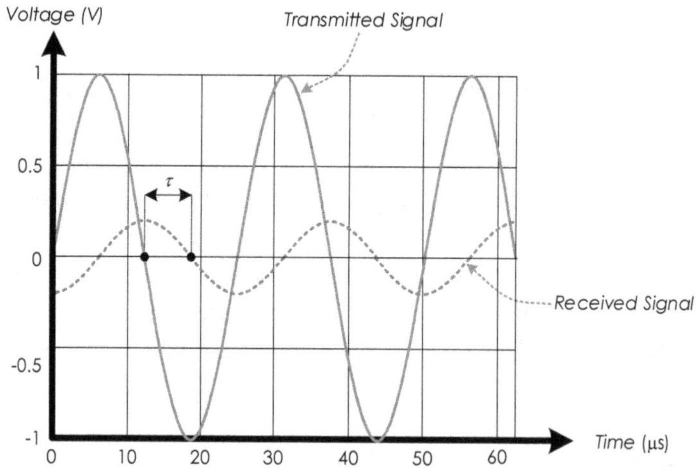

Figure 6.25 – Illustration of the phase difference between transmitted and received sinewaves.

This technique is simple and fast. One can reach a good resolution without a minimum threshold distance that can be measured. It is also very immune to noise. It has, however, the disadvantage that there is an ambiguity in the determination of distances greater than one wavelength because each time the acoustic wave travels a range corresponding to a wavelength, the initial phase of the sinusoid increases 360°, which mathematically corresponds going back to 0 (Figure 6.26). It is impossible to distinguish, for example, a phase shift of 370° from one of 10° (because 370° − 360° = 10°).

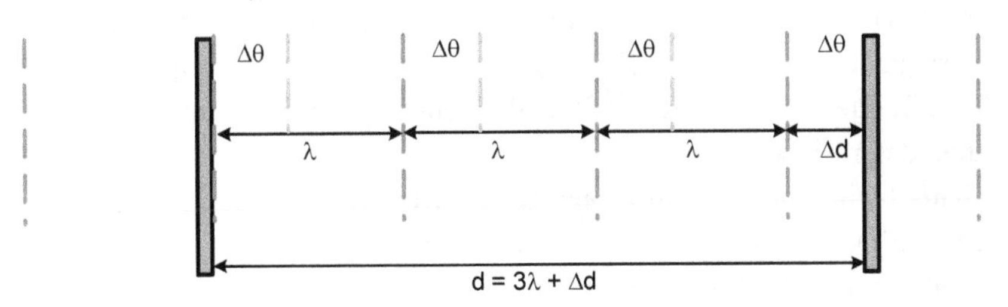

Figure 6.26 – Illustration of the relationship between phase and distance.

Using Amplitude Modulation

Another technique is to modulate the amplitude of a sinusoid with another (lower frequency) sinusoid. The received signal will also have that modulation, and it is possible to determine the time of flight from the phase shift between the envelope of the transmitted and received signals (rather than the signal itself).

This technique requires a transducer with relatively high bandwidth and requires rather complicated electronics.

Using Frequency Modulation

Instead of modulating a sinewave in amplitude is possible to modulate the waveform in frequency so that the frequency of the transmitted signal increases linearly with time (Figure 6.27).

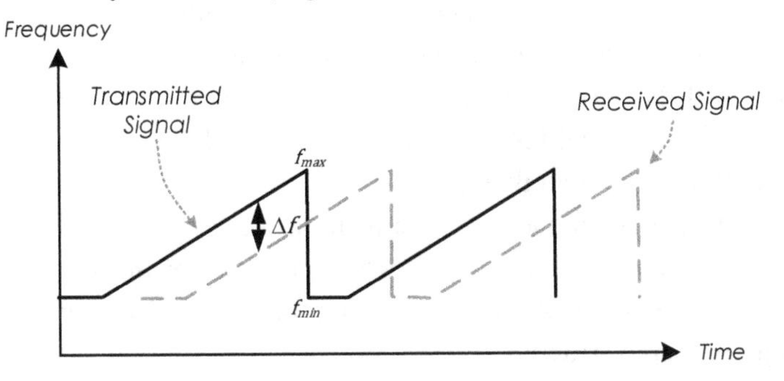

Figure 6.27 – Illustration of the change in frequency of the transmitted and received signal in the case of linear frequency modulation.

The received signal will be a sinusoid in which the value of the frequency is delayed in time. The flight time is determined from the frequency difference of the two signals. This is done by mixing the two signals. The result will be two sinusoids with the sum and difference of the frequencies. Low-pass filtering allows the isolation of the sinusoid, whose frequency is the result of the difference. The frequency of the signal is then determined.

This technique is quite robust to external disturbances and has very high resolution. It has the disadvantages of the complexity necessary for the electronics involved and a relatively high bandwidth required for the transducers.

6.6.3 *Echography*

Echography, or ultrasonography, is used to obtain images from inside the human (and animal) body, including internal organs, muscles, tendons, joints, veins, and arteries. It uses pulsed ultra-sonic acoustic waves, with a typical frequency between 2 and 18 MHz, which are created with a piezoelectric transducer and enter the body where, due to the different acoustic impedances of tissues, it suffers a series of reflections which are measured by the same transducer. A water-based gel is used between the transducer to facilitate the adaptation of impedances and thus optimize the amplitude of the wave that enters the body.

In order to create an image of the structures within the body, multiple transducers are used side by side, working simultaneously. Flight times are used for determining the depth of the tissue that caused the reflection. The higher the acoustic impedance difference between tissues, the higher is the echo. If the acoustic wave reaches gases or solids, most part of the wave is reflected, making it virtually impossible to get a picture of structures that are behind them.

The use of high frequencies due to its lower wavelengths allows for greater resolution in the image. However, the higher the frequency, the greater the wave attenuation in tissue, making it more difficult to obtain an image of deeper structures.

The use of an array of transducers allows one to obtain a two-dimensional image (the depth is the second dimension) (Figure 6.28).

For a three-dimensional image, a matrix of piezoelectric transducers is used.

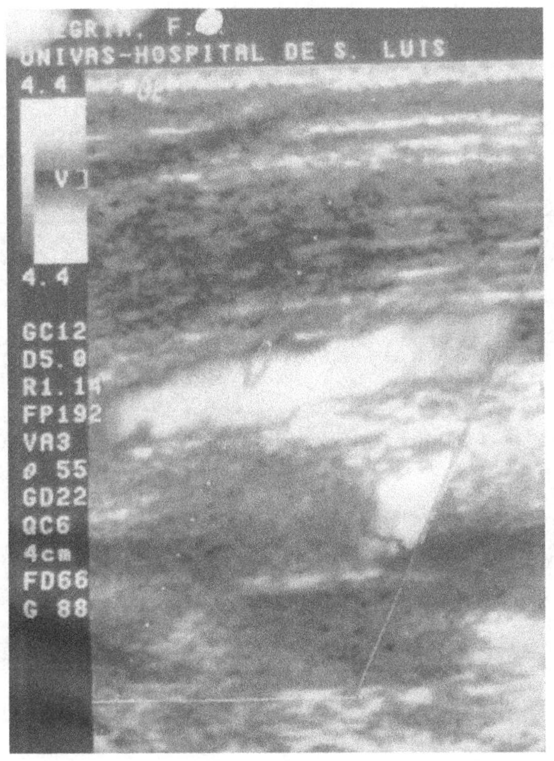

Figure 6.28 – Example of a bi-dimensional image obtained using ultrasounds.

The Doppler ultrasound can be used to measure the velocity of blood in veins and arteries. The Doppler effect is the change in frequency of a wave when reflected from a moving object. Objects approaching increase the frequency (the sounds become more acute), and objects that go away decrease in the frequency (the sounds are more severe). In modern Doppler ultrasound, the frequency variation of the acoustic wave is not measured. Instead, signal pulses are used, and the phase difference between consecutive pulses is measured. The frequency variation is then estimated from the rate of phase variation. The use of pulses instead of a continuous signal allows the simultaneous determination of the distances from the points that originated the echoes. Figure 6.29 shows an example of an image obtained by ultrasound Doppler. The color is used to represent the speed of blood flow.

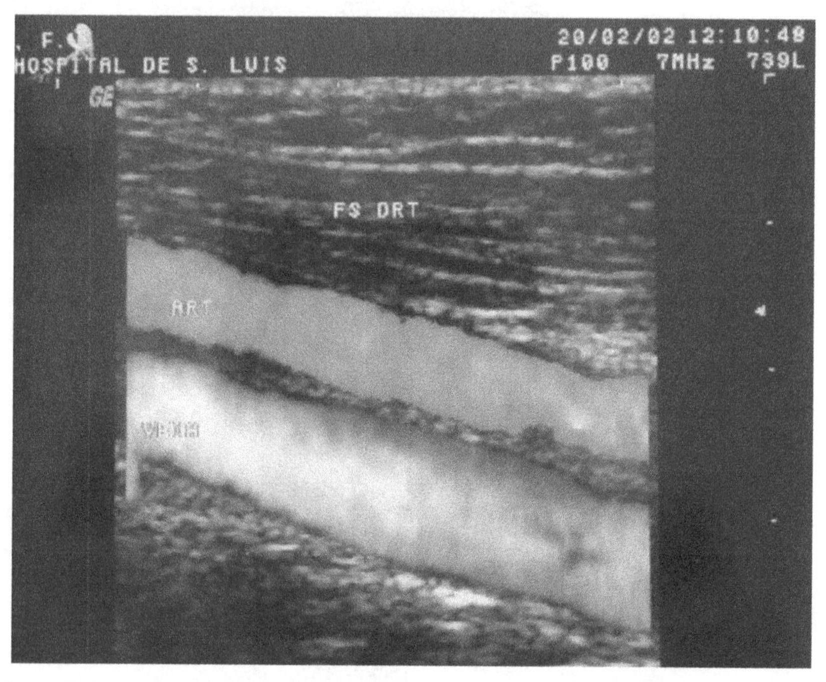

Figure 6.29 – Example of a Doppler Echography. Color represents the velocity of blood flow.

Echography is very good at determining the interface between the different organs, especially in the case of the interface between solids and spaces filled with liquid. It also allows one to obtain images in real-time in which the operator can interactively choose the best place to look. Other advantages are the low cost and size when compared with other diagnostic methods such as computed tomography or magnetic resonance imaging.

The disadvantages of ultrasound as a means of medical diagnosis are the difficulty in visualizing behind bone structures, as in the case of the brain, or behind areas where there is air, like the lungs and intestine; this requires a skilled operator to be possible to obtain good images.

6.7 Fluid Actuators

6.7.1 *Introduction*

Fluid actuators are actuators for producing movement from fluid under pressure. In the case of pneumatic actuators, this fluid is usually air, nitrogen, or another inert gas. In the case of hydraulic actuators, this fluid is usually mineral oil, water, or a combination of both (Figure 6.30).

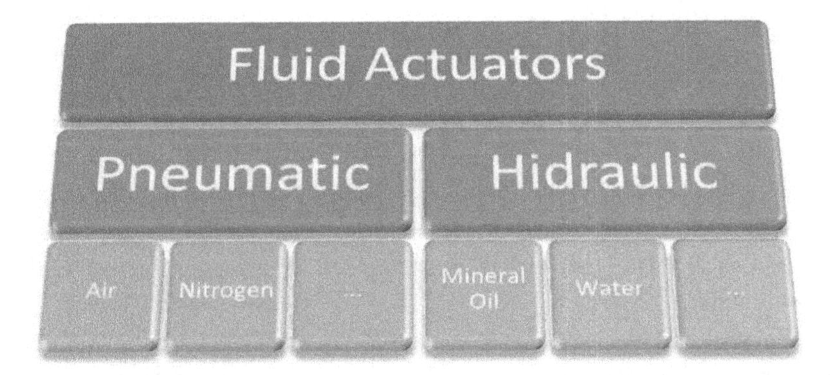

Figure 6.30 – Illustration of the classification of fluid actuators into pneumatic and hydraulic, including the most used fluids.

Fluid actuators are better than electric ones for applications in which the most important thing is the force (hydraulic) or speed (pneumatic). Some areas which are used for fluid actuators are cars (assisted steering and brakes), industry (lifting weights, presses), construction (excavators, cranes), and robotics (robotic limbs).

In presses, for example, it is often necessary to exert a constant force without any movement. In these cases, the hydraulic actuators are very advantageous because they allow one to exercise that force (or binary) without energy consumption. The electric actuators (motors) require, in these conditions, a high current to maintain strength even when stopped.

Other advantages of fluid actuators in relation to electric are their smaller size and lower noise. The pumps necessary to create the pressure in the pneumatic circuit may be located away from the actuator.

The hydraulic actuators, due to the incompressibility of the fluid used, are preferentially employed when it is necessary to tight position control. Moreover, the pneumatic actuators are used in environments where the presence of oil is unwanted, as is the case of the food industry.

6.7.2 *Most Used Quantities*

The pressure is the effect that arises when a force is applied to a surface. The pressure (p) is the amount of force (F) exerted perpendicularly to a surface per unit area (A),

$$p = \frac{F}{A}. \tag{6.17}$$

The unit in the International System (SI) for pressure is the Pascal (Pa) corresponding to 1 newton per square foot (N/m^2). Other units are also normally used in the field of fluid actuators, which are pound per square inch (psi — pound per square inch) and *bar*[2]. A pascal bar corresponds to 10^{-5} and $14{,}5 \times 10^{-5}$ psi. The instrument used to measure pressure is the pressure gauge.

The pressure is not always measured in absolute terms relative to vacuum but often relative to atmospheric pressure. An example of this is the pressure in automobile tires, where the typical pressure is around 32 psi. This value is relative to atmospheric pressure, which has a value of 14.7 psi. The absolute air pressure in the tire is therefore 46.7 psi. In the case of fluid actuators, one normally uses the relative pressure and not the absolute pressure.

The reason why fluids can be used to transmit power can be explained using Pascal's Law: "The pressure exerted anywhere in a confined fluid is transmitted in all directions through the fluid." This pressure, when exerted on a surface, gives rise to a force perpendicular to this surface.

Figure 6.31 illustrates the operation of Pascal's Law in a physical system. The force F_1 exerted on the smaller cylinder with surface A_1, in the left give origin to a pressure $p = F_1/A_1$. This is the same pressure that is exerted on all surfaces of the container containing the liquid, including the surface of the larger cylinder to the right. Since this cylinder has a

---

[2]The word "*bar*" comes from the greek *βάρος* (baros) which means weight. This unit was introduced by Sir Napier Shaw in 1909 having been internationally adopted in 1929.

larger area, $p = F_1/A_1$, the force exerted bottom-up, will have a value of $F_2 = p \times A_2$. Since area A_2 is greater than area A_1, the force exerted on the cylinder on the right is greater than the force exerted on the left cylinder.

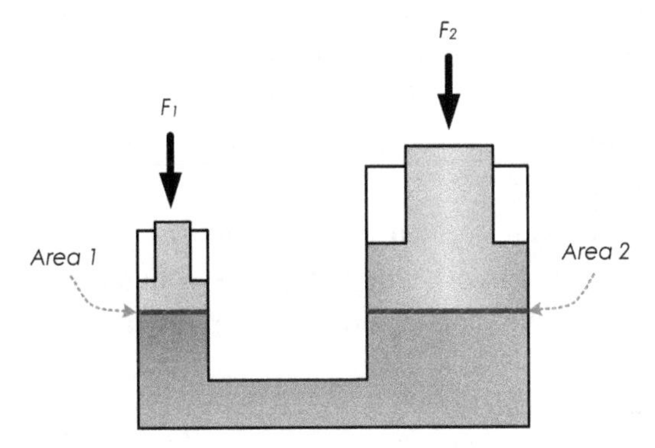

Figure 6.31 – Illustration of the Pascal's Law.

The relationship between forces is thus

$$F_2 = F_1 \frac{A_2}{A_1}. \tag{6.18}$$

If the cylinder on the right has a load coupled to it with a weight equal to F_2, the system will be at rest.

Note that the use of a force less than the weight of a body to sustain it does not violate any physical law since what matters is the preservation of the work done. The work (W) is given by the product of the force and the distance. If the force F_1 exerted on the left cylinder leads to a displacement d_1, then the right cylinder will suffer a smaller displacement, d_2, such that

$$W = F_1 \cdot d_1 = F_2 \cdot d_2. \tag{6.19}$$

An alternative interpretation in terms of work is using the volume of fluid displaced. Inserting (6.18) into (6.19) yields

$$A_1 \cdot d_1 = A_2 \cdot d_2. \tag{6.20}$$

This means that the volume of fluid displaced by the cylinder from the left, $V_1 = A_1 d_1$, is equal to the volume of fluid displaced by the cylinder on the right, $V_2 = A_2 d_2$.

Power (P) is the value of the work done during a certain time interval (Δt),

$$P = \frac{W}{\Delta t}. \tag{6.21}$$

It is measured using the international system in watt (W). A more common unit in fluid actuators horsepower[3] (cv). One horsepower is, by definition, the weight (in pounds) that a horse can lift a foot (a foot-long unit used by the British) for one second. This weight is 550 pounds. Thus 1 hp = 550 lb.ft / s = 745.7 W.

The power of a pump is given by the product of flow rate (in liters per second) and pressure (pascal). The pumps also have the specification of their capacity, and the volume expelled at each rotation.

6.7.3 *Hydraulic Circuit*

Figure 6.32 shows the components of a basic hydraulic circuit. An electric motor is used to drive a pump which draws the fluid from a reservoir and supplies it to the rest of the circuit.

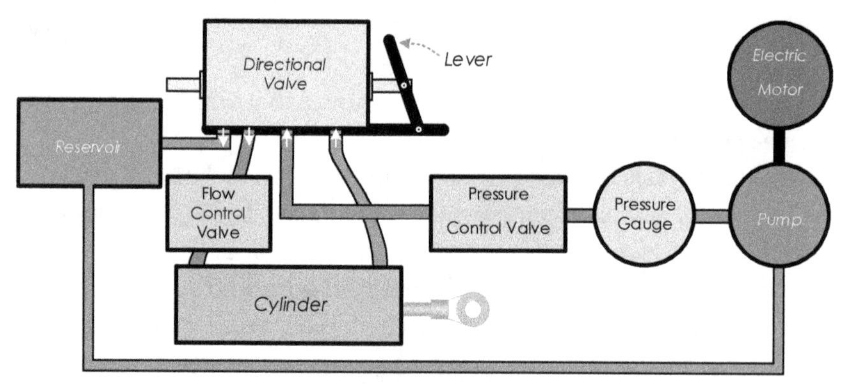

Figure 6.32 – Basic hydraulic circuit.

Note that the pump does not create pressure in the fluid. The pump creates a flow. The pressure is created by the opposition that the rest of the circuit provides to the fluid path. A low or zero value of the output pressure

[3]The horsepower was initially used to compare the performance of the steam engines with that of horses.

of the pump does not necessarily indicate a malfunction in the pump. It may sometimes mean, for example, that there is a leak in the rest of the circuit. Diagnosing a fault in the pump would require a flow meter connected to the pump's output.

The fluid delivered by the pump passes through a pressure control valve, one-directional control valve, one flow control valve, finally reaching the actuator, generally designated as cylinder due to its shape.

There are different types of cylinders that can be classified into single-acting cylinders and double-acting cylinders. Figure 6.33 shows the formation of different variants of a single-acting cylinder. In the two examples above, the fluid under pressure causes the piston to move out of the cylinder, while in the two examples below, this pressure causes the piston to move inward.

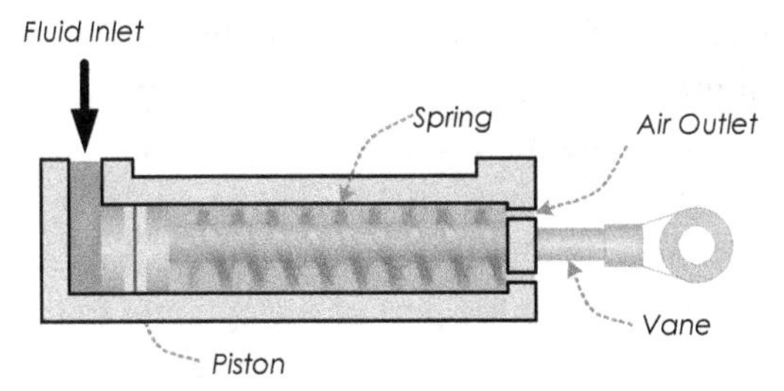

Figure 6.33 – Simple effect cylinder.

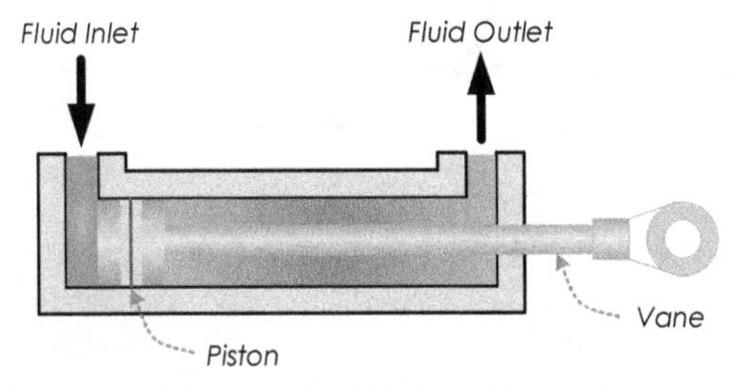

Figure 6.34 – Double cylinder with one piston.

In these examples, a spring is used to return the piston to the starting position when the pressure fluid diminishes. Figure 6.34 shows a double-acting cylinder with one piston.

The valves allow determining the direction of movement, speed of action, and the maximum strength of the system. The pressure control valve allows you to adjust the pressure that works in the rest of the circuit. This valve normally does not let the fluid pass due to a sphere pressed by a spring against fluid ingress (Figure 6.35).

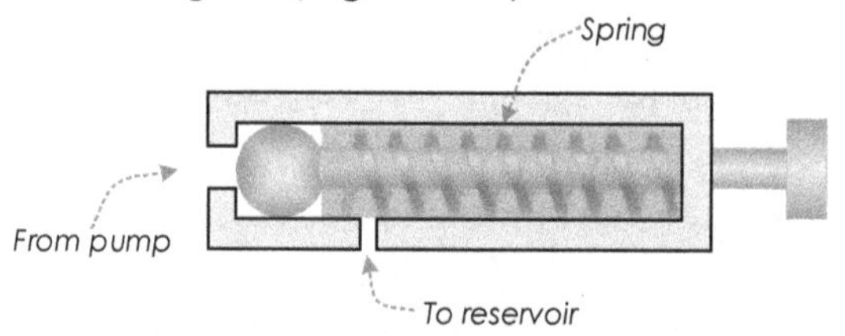

Figure 6.35 – Pressure control valve used in a hydraulic circuit.

When the pressure of the fluid is greater than a certain threshold, the force it exerts on the ball is greater than the force exerted by the spring, causing the ball to move. This lets the fluid reach the valve exit and proceeds to the reservoir, which contains the fluid that is not being used to maintain circuit pressure. This threshold can be adjusted through a screw that compresses the spring depending on its position.

A directional valve, in turn, allows you to control which of the two ports of the cylinder is the fluid injected (Figure 6.36). At the same time, it directs the fluid from the other port of the cylinder to the reservoir.

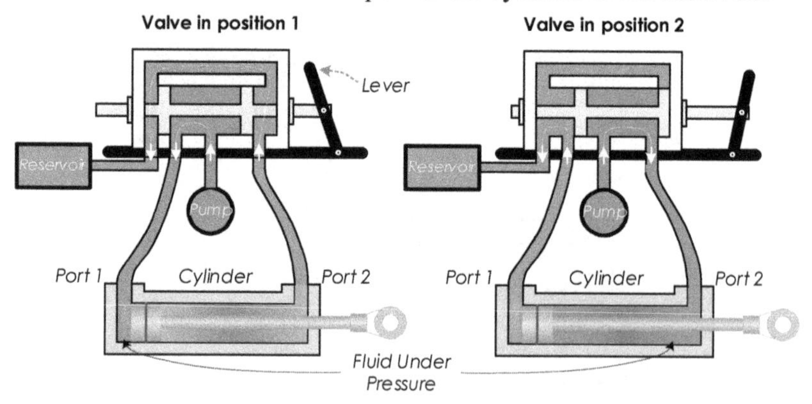

Figure 6.36 – Direction control valve in two different positions.

The flow control valve, in turn, adjusts the flow of fluid reaching the cylinder through a screw that controls the volume through which the volume can pass.

6.7.4 *Pneumatic Circuit*

A pneumatic circuit has the same type of components as a hydraulic circuit with some minor differences. Figure 6.37 shows a basic pneumatic circuit in which the fluid is air. In this case, contrary to what happens with the hydraulics circuit, it is not necessarily a closed circuit where the fluid circulates. The air enters the circuit coming from the surrounding atmosphere, and it is blown into it when the circuit is in excess.

Power for the circuit generally comes from an electric motor that provides mechanical power to a pump which in this case is usually called the **compressor**. The compressor consumes air from the atmosphere through a filter to remove particles that could damage the components of the pneumatic circuit. This air is sent to a reservoir that accumulates it by increasing its pressure, and where the rest of the circuit can be constituted by one or more actuators. A pressure sensor connected to the reservoir is used to connect the electric motor when the pressure is less than desired and to disconnect when the pressure is correct.

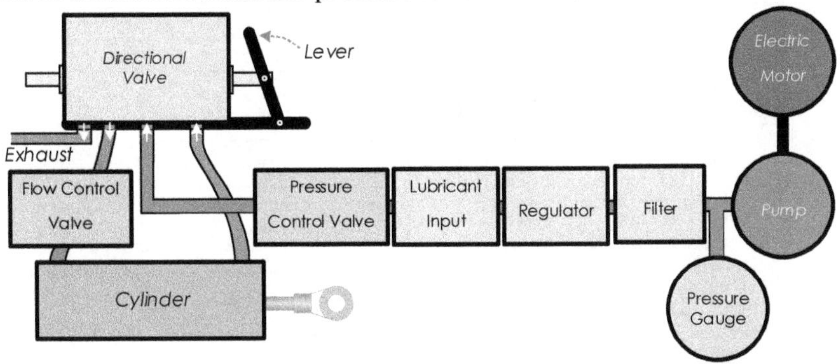

Figure 6.37 – Basic pneumatic circuit.

The air overpressure in the reservoir then passes through a regulator that allows controlling the air pressure in the rest of the circuit. Then, the air passes through a lubricator adding little drops of oil serving to lubricate the other circuit components. The air then reaches the directional control valve and hence the actuator. Then the air present in the actuator returns to the directional control valve and is released to the atmosphere.

Position control of a pneumatic actuator in a system is difficult due to the compressibility of air. The speed, however, is controlled by a flow control valve placed at the output of the actuator to maintain the air pressure on both sides of the actuator piston and thus allow better control of speed.

6.7.5 *Compressors*

There are various types of compressors that can be classified into two categories: positive displacement and dynamic (Figure 6.38). The positive displacement compressors operate by trapping the air in each volume and then decreasing this volume to increase the air pressure. Dynamic compressors, for their part, do not compress the air but instead, they push the air.

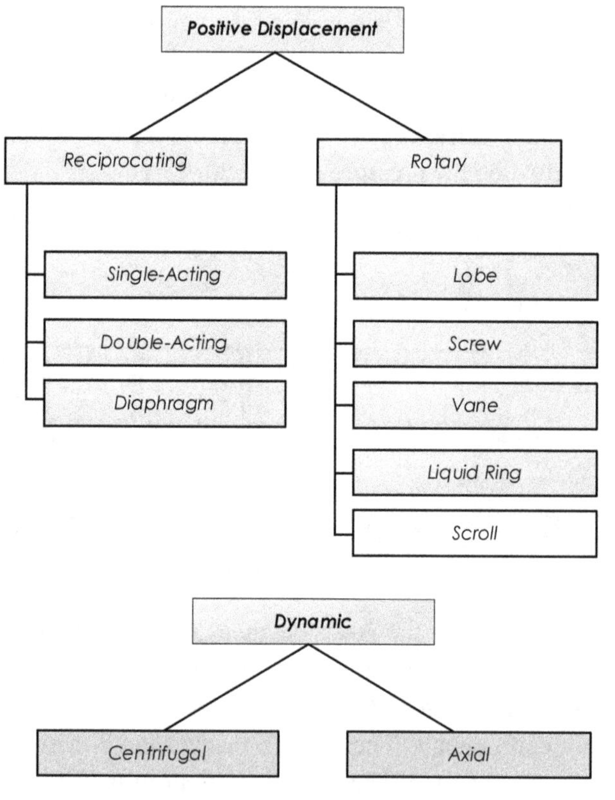

Figure 6.38 – Compressor types.

Figure 6.39 shows a compressor piston in which the air inside a cylinder is compressed by the action of the piston that is pushed by the shaft of the compressor.

Figure 6.40 shows a positive displacement compressor of the rotary type: the screw compressor. The rotation of the screws makes the air move to the opposite end of the compressor, where the volume is less.

The two rotors are shown in more detail in Figure 6.41.

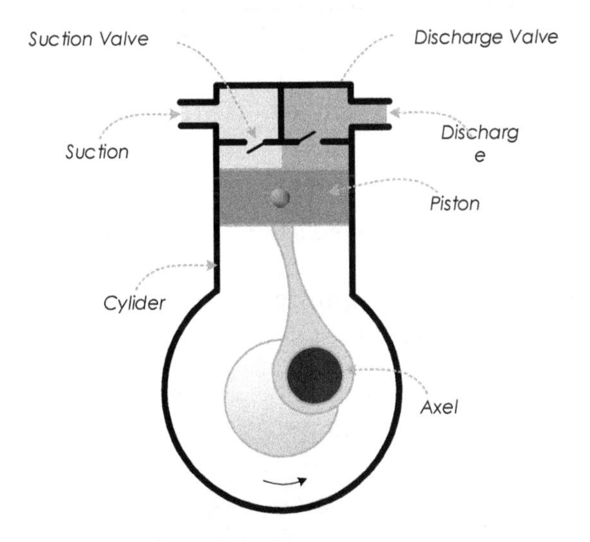

Figure 6.39 – Piston compressor.

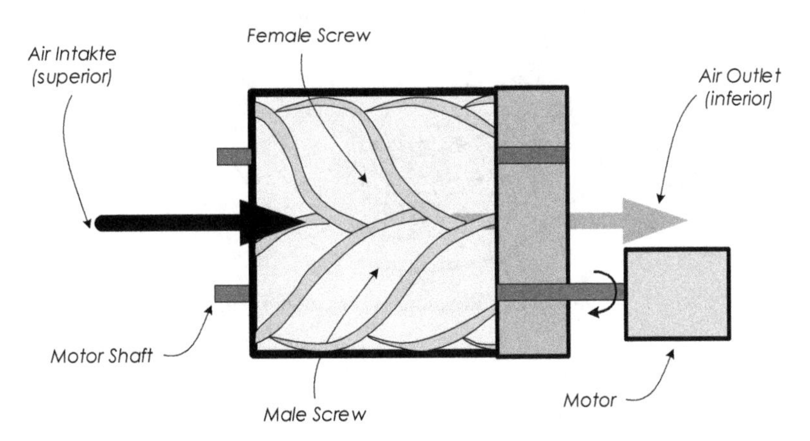

Figure 6.40 – Rotational screw compressor.

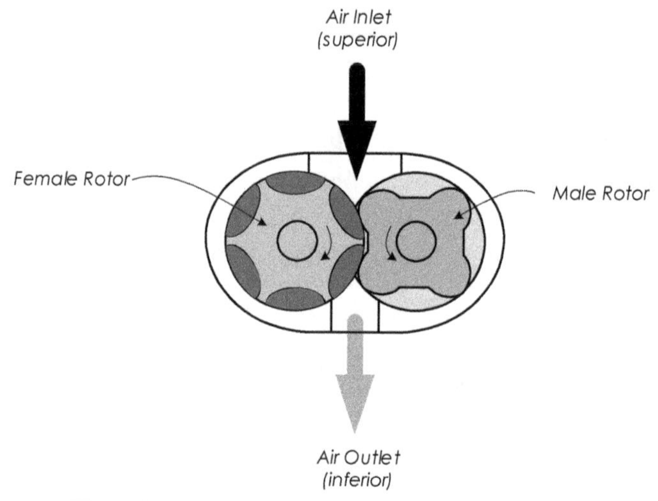

Figure 6.41 – Top view of a rotational screw compressor.

Figure 6.42 also shows a rotating compressor, but instead of using a screw, it used two vanes (like two straws).

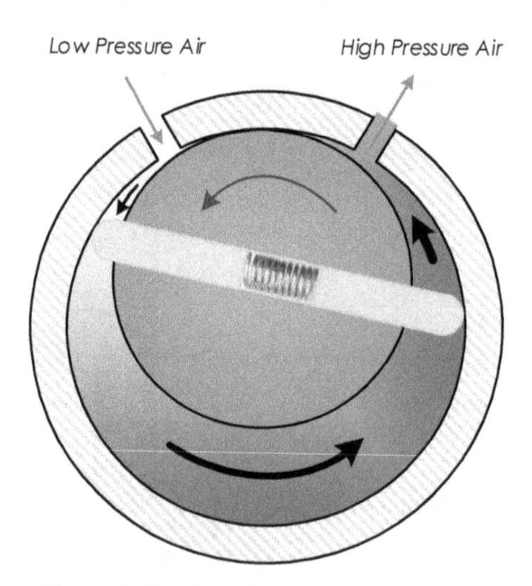

Figure 6.42 – Rotational vane compressor.

Finally, Figure 6.43 shows a centrifugal dynamic compressor.

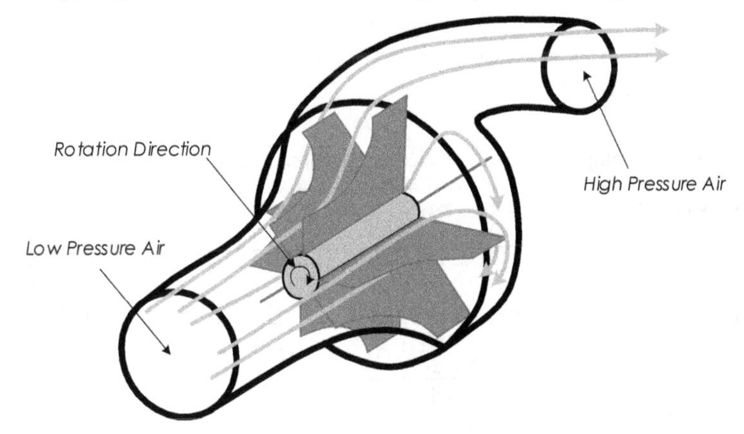

Figure 6.43 – Centrifugal compressor.

The double-acting piston compressor (compresses air when the piston moves in both directions) is the most efficient type of compressor, although the initial cost of installation and maintenance is high. The rotary screw compressor has a low cost but is inefficient. The centrifugal compressor is the only one capable of reaching a power higher than 600 hp, but it has a high initial cost.

6.7.6 *Hydraulic / Pneumatic Comparison*

Pneumatic actuators work at a higher speed than hydraulic actuators, ensuring a clean environment. They are also more reliable, insensitive to temperature, low cost, and easy to install and maintain. There is also no risk of fire due to the use of flammable fluid.

It does, however, have some disadvantages over hydraulic circuits in that they are more expensive than hydraulic actuators, produce more noise, react more slowly to control, have limited strength, and do not allow very precise positioning due to the compressibility of air.

The hydraulic actuators, in turn, allow very high force and binary, operate at a moderate rate, have a good weight/ power ratio, and allow one to keep the position, exerting a high force without energy consumption.

These actuators have, however, disadvantages in terms of cost and are susceptible to leakage, which leads to problems of cleaning the oil. This prevents them from being used in the food, pharmaceutical, and textile industries.

Table 6.2 shows the comparison between electric, pneumatic, and hydraulic actuators.

Table 6.2 – Comparison between electric, pneumatic, and hydraulic actuators.

PARAMETER	ELECTRIC	PNEUMATIC	HYDRAULIC
Speed	High	Very High	Medium/High
Control	Easy	Difficult	Reasonable
Torque at low speed	Low/Medium	Low	High
Precision	Good with gearboxes	Bad	Good
Environmental Issues	Eventual Electrical Arcs	Clean	Danger of leaks
Costs	Low	Low	High

6.8 Questions

1. What percentage of an acoustic wave passes from water to steel? Consider that the characteristic acoustic impedances of water and steel are 1.5 MPa·s/m and 45 MPa·s/m, respectively.

2. A sinusoidal signal with a 14 V rms value and 40 kHz frequency is applied to an ultrasonic emitter. An ultrasound receiver is placed 2m away. What is the rms value of the voltage at the receiver terminals? Emitter characteristics: SPL = 119 dB referred to 20 μPa, for a distance of 30 cm, an rms voltage of 10 V, and a frequency of 40 kHz. Receiver characteristics: 65 dB sensitivity referred to 1 V/μbar for a frequency of 40 kHz.

3. It is intended to use a double-acting hydraulic cylinder to move a load with a mass of 1500 kg on a plane with a 30° inclination. The cylinder has a 95% yield. Consider that the frictional force is independent of the speed and has a value of 10% of the force exerted by the load in the direction of the plane. What diameter should the cylinder piston have if it is inserted into a hydraulic circuit with a pressure of 50 bar?

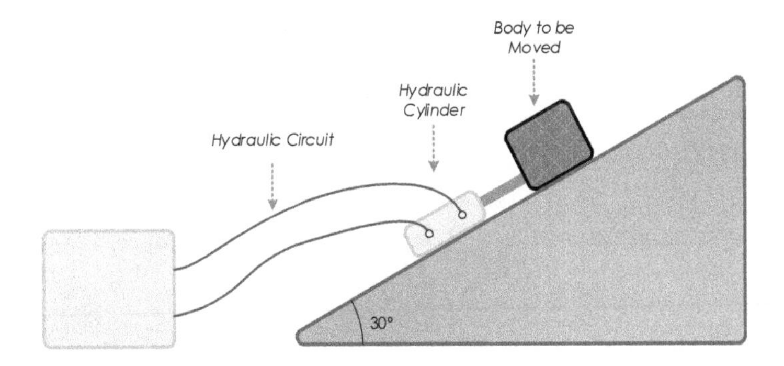

4. Describe the piezoelectric effect.
5. Describe the following modes of operation of a piezoelectric material related to the way it is cut: Transverse, Longitudinal, and Tangential.
6. How is the typical frequency response of a sensor based on the piezoelectric effect? In what kind of applications are the different zones of this frequency response used?
7. Give at least three examples of applications where ultrasonic acoustic waves are normally used.
8. Do all materials show a piezoelectric effect?
9. Why can't ultrasound be used to measure the distance between satellites in space?
10. What are the possible configurations in terms of the location of the emitter and receiver of a distance measurement system using ultrasound?
11. Which of the possible configurations in terms of the location of the emitter and receiver of a distance measurement system using ultrasound allows to measure greater distances?

    a. Emitter and reflector face to face.
    b. Emitter and reflector side by side.
    c. Use of the same transducer as an emitter and reflector.
    d. Everyone can measure the same distance.
    e. None of the above.

Chapter 7

Devices Based on Thermal Phenomena

7.1 Introduction

The temperature of a body is a parameter that describes the kinetic energy of the particles thereof. The higher the agitation of the particles, the higher their kinetic energy and therefore the higher the temperature. In the case of an ideal gas, for example, the temperature is directly proportional to the average kinetic energy of the molecules,

$$T = \frac{2}{3} \cdot \frac{n}{R} \cdot \overline{E}_k, \tag{7.1}$$

where n is the Avogadro's number ($6.02214179 \times 10^{23}$ mol^{-1}), R is the constant of ideal gases (8.314472 JK^{-1}mol^{-1}) and $\overline{E}_k$ is the average value of kinetic energy of the molecules. This works when you have a very high set of molecules, having each molecule different kinetic energy, not only for having different masses but also for having different speeds.

There are three thermoelectric effects that are used to make sensors and actuators involving thermal effects: the Thomson effect, the Peltier effect, and the Seebeck effect. These effects are due to the movement of electrons in the material. This movement not only gives origin to an electric current but also to a thermal current, i.e., a transport of heat from one point to another material.

7.2 Thomson Effect

Consider an aluminum bar that is heated at one end and cooled at the other end, as shown in Figure 7.1.

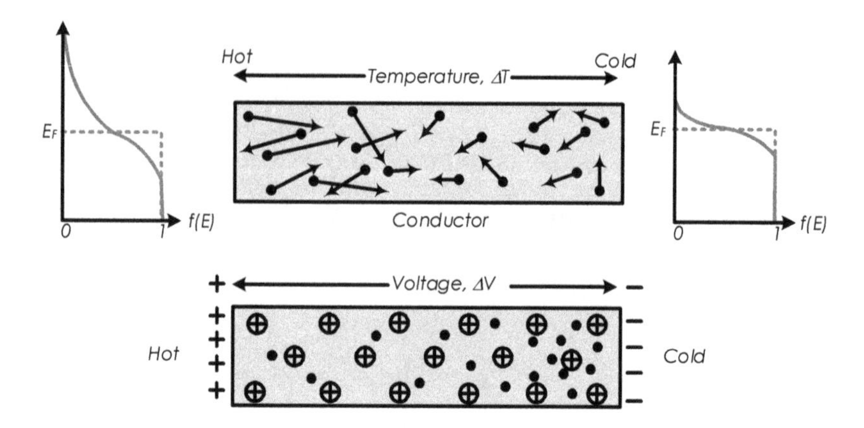

Figure 7.1 – Kinetic energy of electrons in a material whose extremities are at different temperatures.

There are electrons from the hot side traveling to the cold side and vice-versa. The electrons on the hot side have higher energy and, therefore, higher speed than the electrons on the cold side. Therefore, the number of electrons which will travel to the cold side will be greater than the number of electrons which will travel in the opposite direction giving rise to an excess of electrons on the cold side and a deficit of electrons (compared to ion metal structure) in the hot side. This difference in concentrations of electrons relative to positive ions means that there is an overall negative charge on the cold side and an overall positive charge on the hot side. There is thus an electric field inside the material and a consequent electromotive force due to the difference in temperature. This electric field will exert a force on the electrons that oppose their motion due to diffusion; a balance is established between the action of the difference of temperature and the electric field created. This is called the Thomson Effect.

A coefficient is used (thermocouple coefficient) to express the magnitude of this effect. It relates the potential variation with the temperature change in a material,

$$\alpha = \frac{dV}{dT}, \tag{7.2}$$

where dV is, by convention, the potential difference between the cold side and hot side. In the case of aluminum, used to illustrate the Thomson effect, the thermoelectric coefficient is negative. There are, however, materials such as iron, where the coefficient is positive (Table 7.1).

Table 7.1 – Thermoelectric Coefficients of some materials. Values are taken from [2].

METAL	THERMOELECTRIC COEFFICIENT AT 0°C	THERMOELECTRIC COEFFICIENT AT 27°C
Na		−5
K		−1.8
Al	−1.6	
Mg	−1.3	
Pb	−1.15	−1.3
Pd	−9	−9.99
Pt	−4.45	−5.28
Mo	4.71	5.57
Li	14	
Cu	1.7	1.84
Ag	1.38	1.51
Au	1.79	1.94

A positive value means that there are electrons that migrate from the cold side to the hot side. This happens because, in addition to having to consider the migration of electrons of conduction, which behave as free electrons, one must consider the interaction of the electrons with positive ions of the structure as well as vibrations of the structure itself.

Be aware that the value of the thermoelectric coefficient is not independent of temperature, as seen in Figure 7.2 for different materials.

7.3 Peltier Effect

The Peltier effect occurs in the interface between two different materials. Considering Pauli's exclusion principle, there are at most 2 electrons in each orbital of an atom (Figure 7.3). In each of these orbitals, electrons are associated with an energy that is negative by convention. Its value has to do with the amount of energy you need to supply to move the electron from its orbital position to somewhere outside the atom (actually, to an infinite distance from the atom). The work that is required to be made to remove electrons from atoms depends on the material itself and the temperature. In the example in Figure 7.3, which is represented the energy levels of the two metals (A, B), it is observed that it is necessary to spend more energy to remove an electron from the atoms of the metal B than A. This means that, on average, the free electrons in metal A have higher energy than metal B (because less energy is used to withdraw atoms).

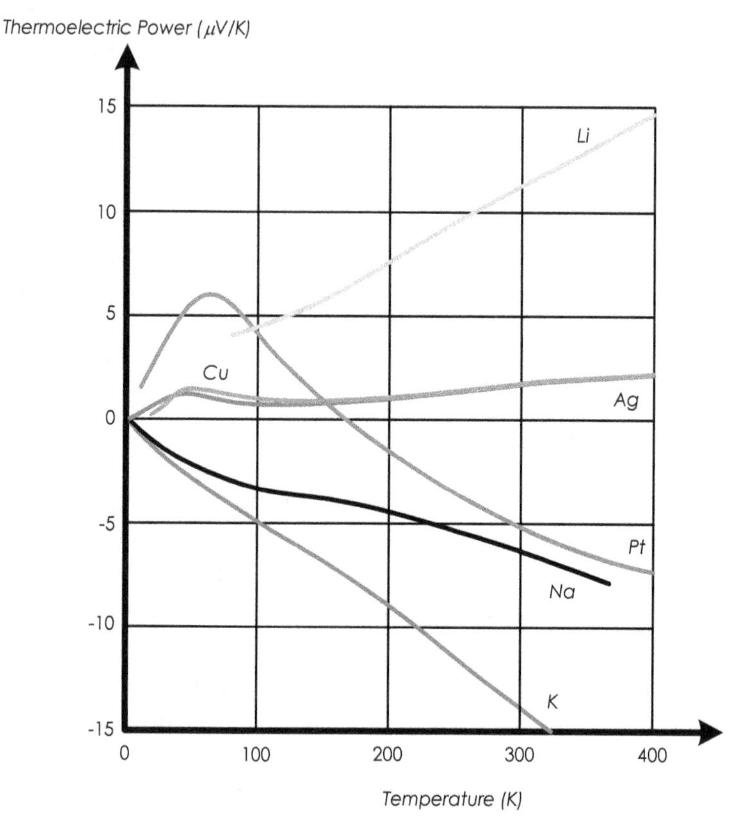

Figure 7.2 – Dependence of the thermoelectric coefficient with temperature in some materials.

This results in the diffusion of electrons from metal A to metal B, which means that there is an excess of positive charge on the metal A side and an excess of negative charge on the metal B side. Therefore, a potential difference is created in the interface of the two metals just because they have different electronic constitutions — contact potential. This voltage is proportional to the difference between the value of the work W_A and W_B that is needed to free electrons from atoms of different metals. This can be seen as a battery that supplies an electromotive force to the exterior (Figure 7.4).

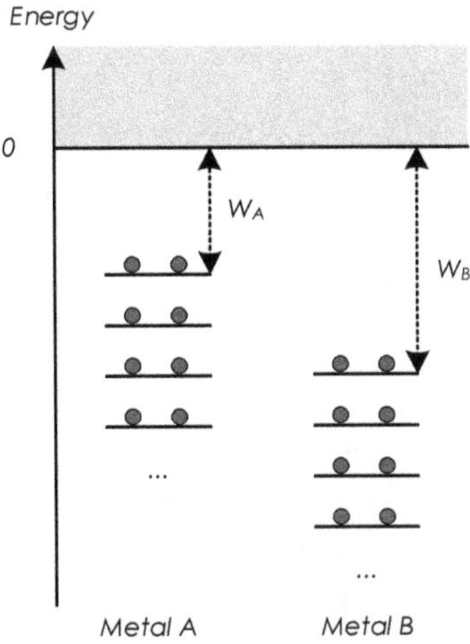

Figure 7.3 – Illustration of the different energy of the electrons in the different orbitals of an atom and the work necessary to remove them from it.

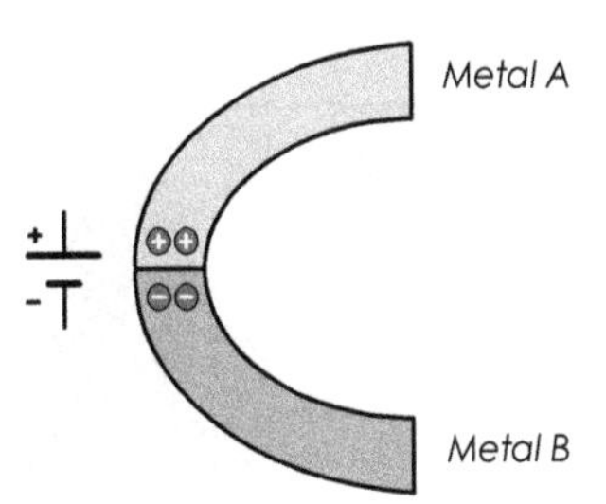

Figure 7.4 – Electromotive force due to the contact of two different metals.

If two different metals are joined together, as illustrated in Figure 7.5, the electromotive force is equal in each junction. This causes the voltage connected to this circuit to be zero.

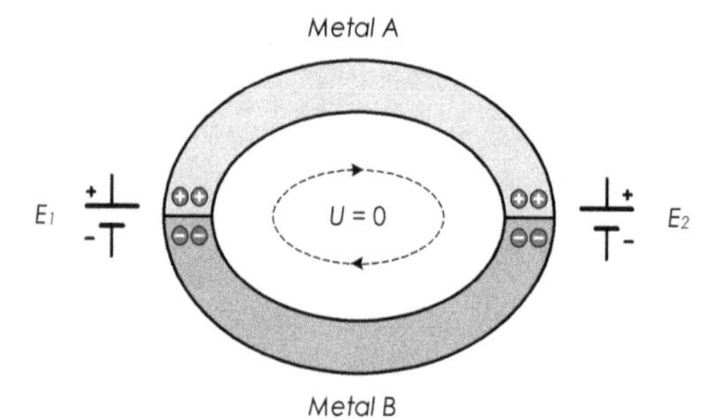

Figure 7.5 – Electromotive force due to the contact of two different metals connected in a close circuit.

The work necessary to remove electrons from atoms depends on the temperature in a way that depends on the type of material. The difference of the value of the work in two different metals, therefore, depends on the temperature, which causes the electromotive force that arises in the interface of metals to depend on temperature. If the temperature of the interface between the two metals is different, as illustrated in Figure 7.6, the voltage in the circuit ceases to be 0.

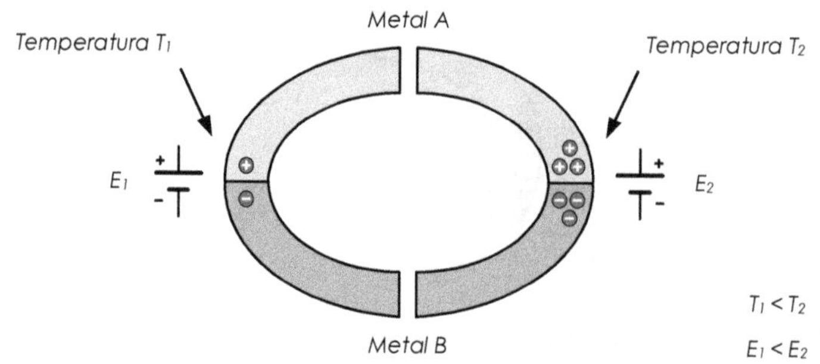

Figure 7.6 – Electromotive force due to two different metals connected in two places and at different temperatures.

7.4 Seebeck Effect

The Seebeck effect is a combination of the Peltier and Thompson effect. When we have two different metals forming two interfaces, as illustrated in Figure 7.7, there are four simultaneous electromotive forces — two due to the Thomson effect (E_3 and E_4) due to the fact that there is a temperature gradient in the materials and two due to the Peltier effect (E_1 and E_2) because the two junctions are of two different metals.

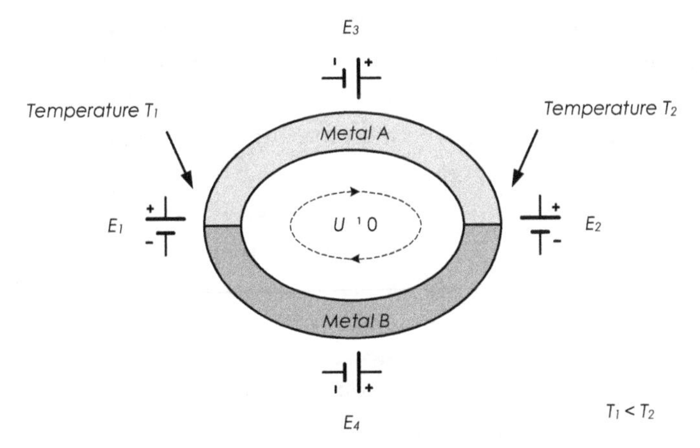

Figure 7.7 – Electromotive forces due to the Thomson and Peltier effects.

The electric voltage that is generated in the circuit made by the two metals is given by

$$U = E_2 + E_4 - E_1 - E_3. \tag{7.3}$$

This voltage may have a higher or lower value depending on the difference in temperature and the type of materials used.

7.5 Thermocouple

The Seebeck effect is used to construct temperature sensors. Using two different metals in contact with each other and maintaining one of the junctions at a reference temperature, it is possible to determine the temperature of the other junction from the measured voltage (Figure 7.8).

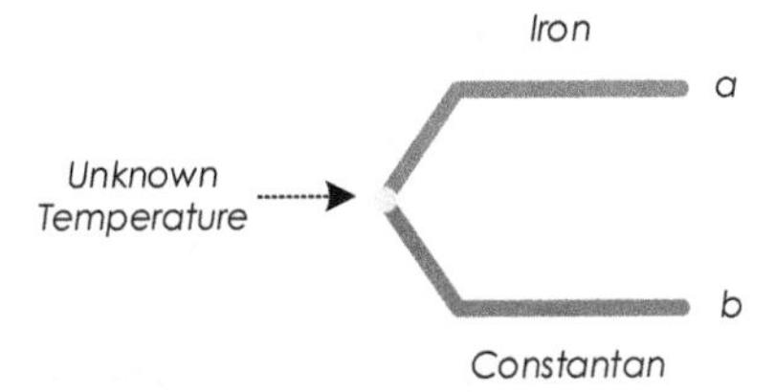

Figure 7.8 – Example of a Thermocouple.

The temperature between the terminals a and b is proportional to the difference in temperature T_x and the temperature T_{ab} (the temperature at which the terminals a and b are).

There are two problems when making a temperature sensor. The first has to do with the measurement of the voltage between terminals a and b. If a voltmeter is used with cables made of copper, for example, they form two new junctions. You have, therefore, three different junctions at unknown temperatures (Figure 7.9).

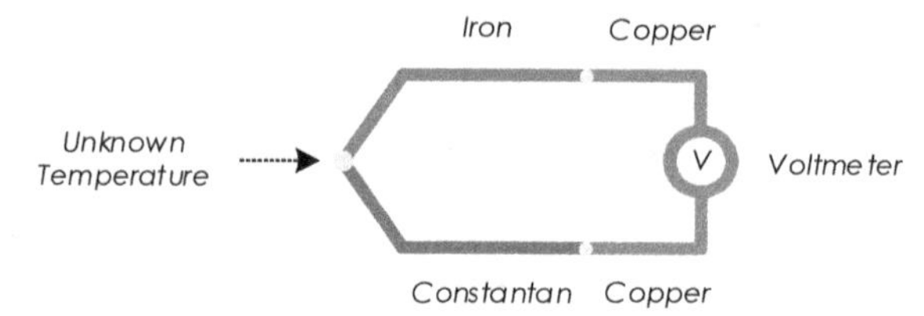

Figure 7.9 – Example of the three junctions that are formed when a voltmeter is connected to a thermocouple.

One solution is to add a second identical thermocouple, connected on the opposite way and maintained at a known temperature (Figure 7.10). The thermocouples are attached so that the two junctions created by the connection to the voltmeter (Copper/Iron) are identical. The voltage in these two junctions, due to the Peltier effect, cancels one another if they are kept at the same temperature, which is accomplished with an isothermal block (Figure 7.11).

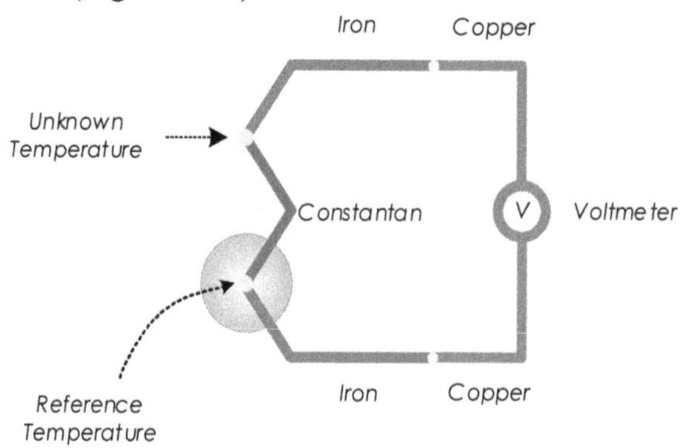

Figure 7.10 – Use of two identical thermocouples to make a temperature sensor.

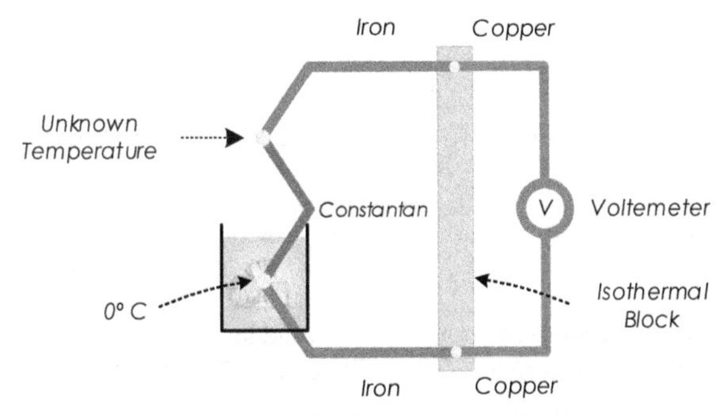

Figure 7.11 – Use of an isothermal block (yellow) to maintain the two junctions at the same temperature.

However, there is still the difficulty of requiring a reference temperature. This is traditionally performed using an ice bath to keep the junction at 0°C, which is not very practical to measure the temperature and today is only used in calibration laboratories. The solution typically used is to measure the temperature of the reference junction with another temperature sensor (a thermistor, for example), as illustrated in Figure 7.12.

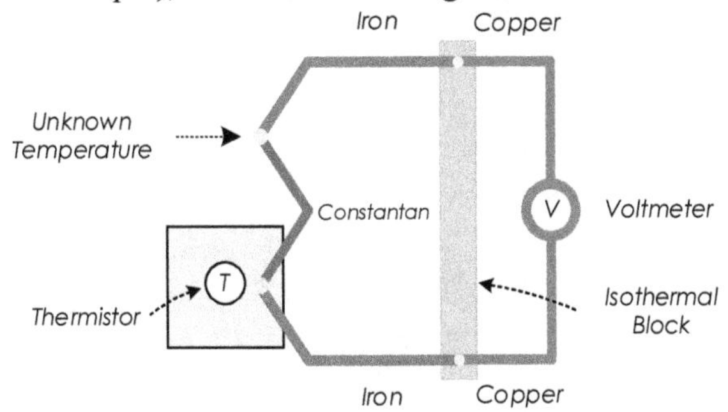

Figure 7.12 – Use of a second temperature sensor to measure the temperature of the reference junction.

Since we measure the reference temperature of the sensor with another sensor, it is not necessary to use the reference junction. Simply connect the voltmeter to the thermocouple using an isothermal block and measure its temperature (Figure 7.13).

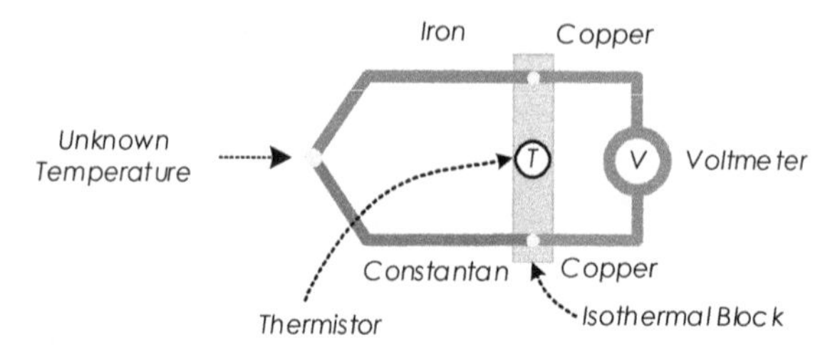

Figure 7.13 – Use of a second temperature sensor and an isothermal block to measure the temperature of the junctions formed by the voltmeter cables and the thermocouple.

Using the indication of the voltmeter (V) and the value measured for the reference temperature (T_{ref}) it is possible to obtain the unknown temperature (T_x). Figure 7.14 illustrates the procedure. The value of the reference temperature is used to look in the calibration curve for the corresponding value of the thermocouple voltage (V_{ref}). The measured voltage (V) is added to it (leading to V_x) and the calibration curve is used to obtain the unknown temperature.

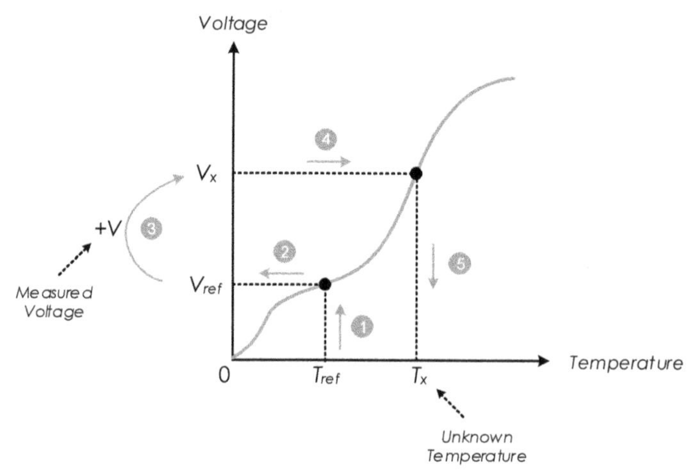

Figure 7.14 – Illustration of how the unknown temperature is obtained from the measured reference temperature and voltmeter voltage together with the thermocouple calibration curve.

The main advantage of the use of thermocouples for temperature measurement is the capacity to measure high temperatures and to withstand harsh environments (subject to corrosion, for example). There are different types of thermocouples, depending on the metals which are used in their construction (Table 7.2).

Table 7.2 – Some thermocouple types.

MATERIALS	**SENSITIVITY AT 25°C (μV/°C)**	**TEMPERATURE RANGE (°C)**	**NAME**	**COMMENT**
Copper and Constantan	40.9	−270 to 600	T	
Iron and Constantan	51.7	−210 to 760	J	
Chromel and Alumel	40.6	−270 to 1300	K	Most used, cheap
Chromel and Constantan	60.9	−200 to 1000	E	Most sensitive
90% Pt/10% Rh and Platinum	6	0 to 1550	S	
87% Pt/13% Rh and Platinum	6	0 to 1600	R	
82%Ni/18%Mo and 99.2%Ni/0.8%Co	50	Up to 1400	M	Used in vacuum furnaces
Nicrosil and Nisil	39	−270 to 1300	N	Most stable
70%Pt/30%Rh and 94%Pt/6%Rh	6	Up to 1800	B	
95%W/5%Re and 74%W/26%Re	10	−196 to 2329	C	High temperature
97%W/3%Re and 75%W/25%Re	10	−196 to 2329	D	High temperature
W and 74%W/26%Re	10	−196 to 2329	G	High temperature

Naturally, the use of different materials leads to different sensitivities, as seen in Figure 7.15.

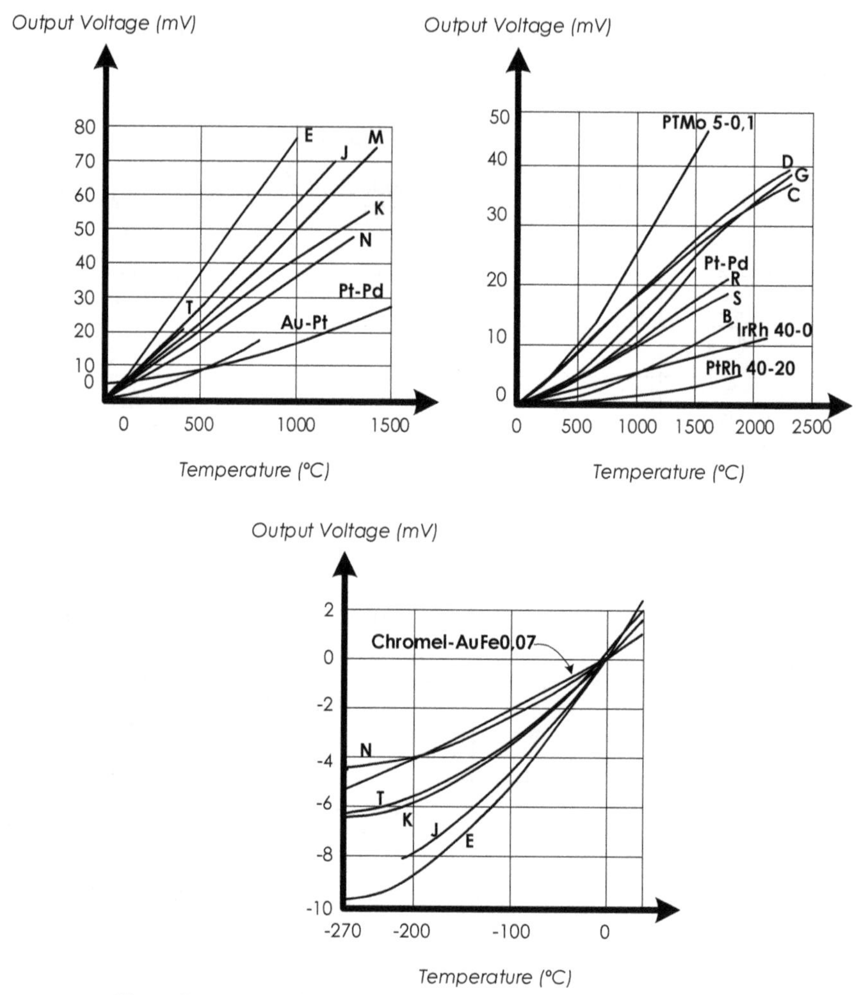

Figure 7.15 – Transfer functions of different thermocouple types.

Figure 7.16 shows a thermocouple connected to a multimeter. This can be done in multimeters that have this function.

Figure 7.16 – Photograph of a thermocouple connected to a multimeter.

Depending on the application in question, the use of a thermocouple may be a good choice. Below are listed some advantages and disadvantages of thermocouples as temperature sensors [8]:

Advantages

- Small size allows fast response time
- Low cost
- Extensive range of measurement
- They are robust, enabling the use in environments with severe vibrations

Disadvantages

- They are likely to corrode
- Special extension cables are needed
- They are less stable than RTDs in measuring medium and high temperatures
- Need to measure the temperature of the reference junction

7.6 Peltier Cell

A Peltier cell is a device based on the Peltier effect used for heating and cooling. It is an actuator that acts in the surrounding environment by increasing or decreasing the temperature. The degree of heating or cooling is controlled by the amount of current applied to the cell.

The Peltier effect exists in the interface between two different materials. However, to enhance its use in refrigeration usually cells formed by a semiconductor material sandwiched between two metallic materials are used. Figure 7.17 represents the energy of the valence and conduction bands of an n-type semiconductor placed between two metals of the same type. The two metals are connected to a battery which raises the energy bands of the metal connected to the negative terminal.

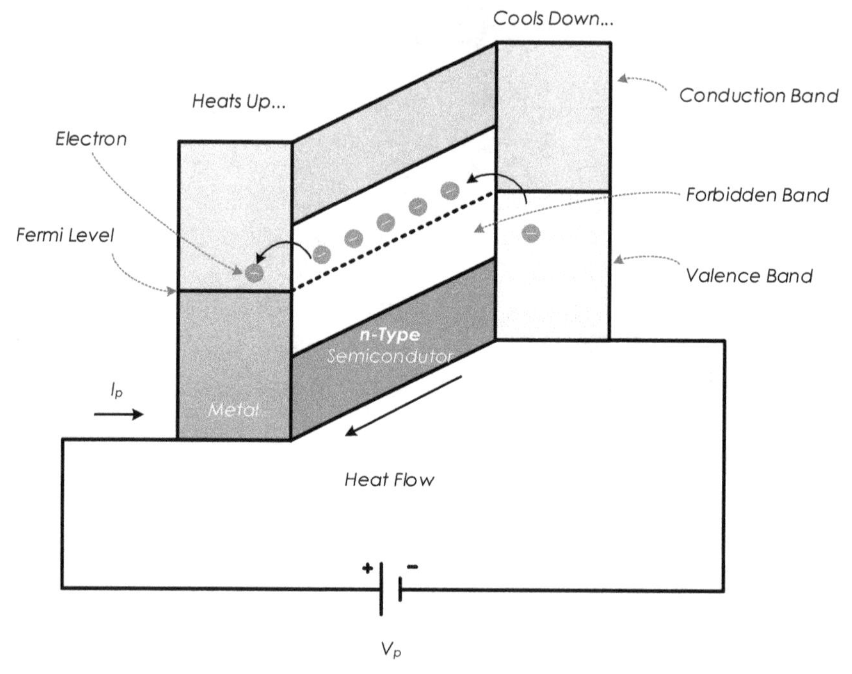

Figure 7.17 – Illustration of the Peltier effect in an n-type semiconductor placed between two metals.

In metals, there is no forbidden zone between the valence band and the conduction band. The electrons from the valence band can easily gain energy and move along the material. In semiconductors, there is a forbidden zone between the valence band and the conduction band. To have a current, it is necessary that the electrons have enough energy to reach the conduction band energy of the semiconductor.

As illustrated in Figure 7.17, the more energetic electrons of the metal from the right have enough energy to move to the conduction band of the semiconductor energy. These electrons pass through the semiconductor material (by diffusion) until they reach the left metal. As a result, the temperature of the material in the right lowers because it loses its most energetic electrons. However, the temperature of the material in the left increases because it receives the most energetic electrons that were in the rightmost metal.

Figure 7.18 shows a similar case but with a p-type semiconductor. The operation is similar to holes instead of electrons.

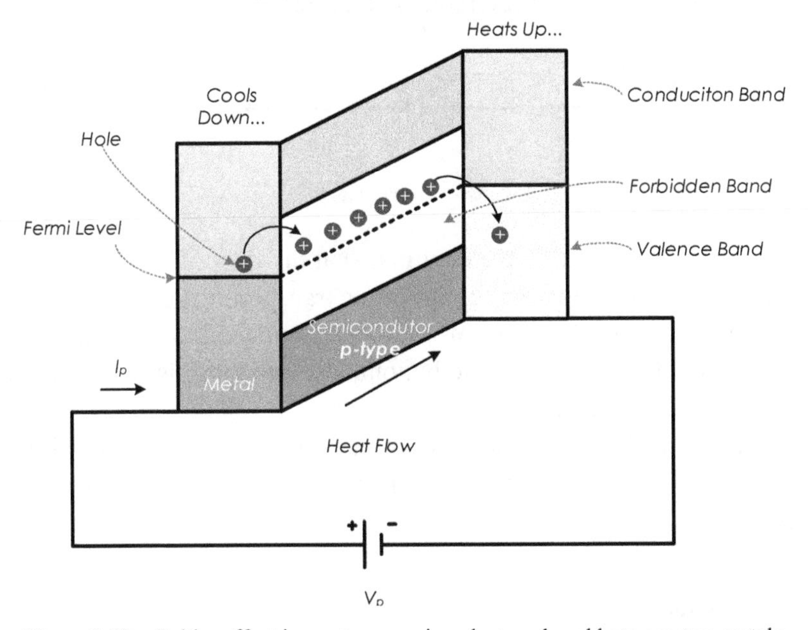

Figure 7.18 – Peltier effect in a p-type semiconductor placed between two metals.

A Peltier cell is made of two semiconductors, one n-type and another p-type, placed between metals as illustrated in Figure 7.19.

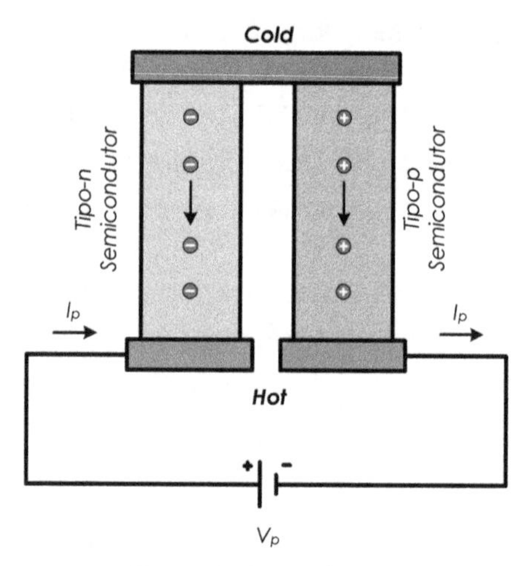

Figure 7.19 – Peltier cell.

A current is imposed in the circuit so that the flow of electrons in the n-type semiconductor is from top to bottom transporting heat from the top to the bottom[1] metal plate. In a p-type semiconductor, there is a displacement of holes from top to bottom, also transporting heat from the upper to the lower plate.

The transfer of thermal energy from the upper to the lower metal, due to the Peltier effect, is proportional to the applied current and can be expressed by

$$Q_P = \Pi \cdot I_P, \tag{7.4}$$

where Π is the Peltier coefficient which is related to the thermoelectric coefficient (α) by

$$\Pi = \alpha \cdot T, \tag{7.5}$$

where T is the temperature in degrees Kelvin.

There is, however, another source of thermal energy transfer, namely due to heat conduction from the hot plate to the cold plate given by the Fourier law

$$Q_c = k \cdot \Delta T, \tag{7.6}$$

---

[1]Electrons have a negative charge and therefore moving in the opposite direction of the electric current direction.

where k is the thermal conductance. This thermal energy is transported from the hot side to the cold side, that is, contrary to what happens with the Peltier effect.

Finally, there is also heating of the material due to the Joule effect. This heating depends on the square of the current value:

$$Q_J = R_P \cdot I_P^2, \tag{7.7}$$

where R_P is the electrical resistance of the material. One can consider that half of that thermal energy is transferred to the hot side and a half to the cold side.

So, for the heat removed from the cold side, one has

$$Q_F = -\Pi \cdot I_P + \frac{1}{2} R_P \cdot I_P^2 + k \cdot \Delta T, \tag{7.8}$$

meaning the Peltier effect removes heat from the cold side while the Joule effect and thermal conduction add head in that side (Figure 7.20). Recall that, by convention, a negative value means loss of thermal energy, and a positive value means a gain of thermal energy.

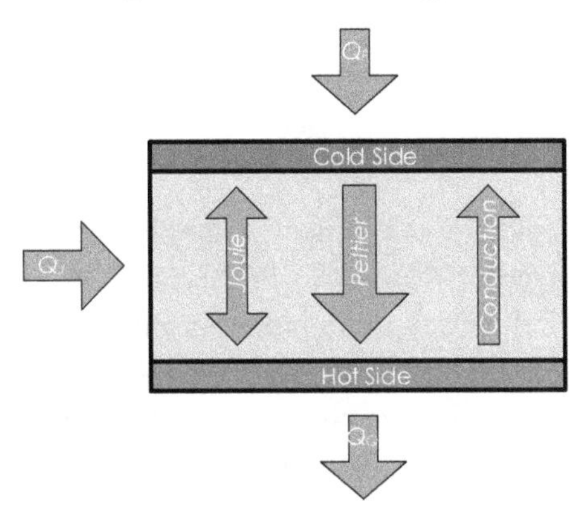

Figure 7.20 – Heat transport in a Peltier cell.

On the hot side, on the other hand, the Peltier and Joule effect leads to increased heat, and the thermal conductivity leads to a decrease in energy:

$$Q_Q = \Pi \cdot I_P + \frac{1}{2} R_P \cdot I_P^2 - k \cdot \Delta T. \tag{7.9}$$

While the thermal energy carried by the Peltier effect is proportional to the electric current, the energy produced by the Joule effect is proportional to the square of the electric current. This limits the amount of cooling that can be achieved by increasing the amount of current used. There is an optimum value of current which leads to the lowest temperature on the cold side of the cell (Figure 7.21).

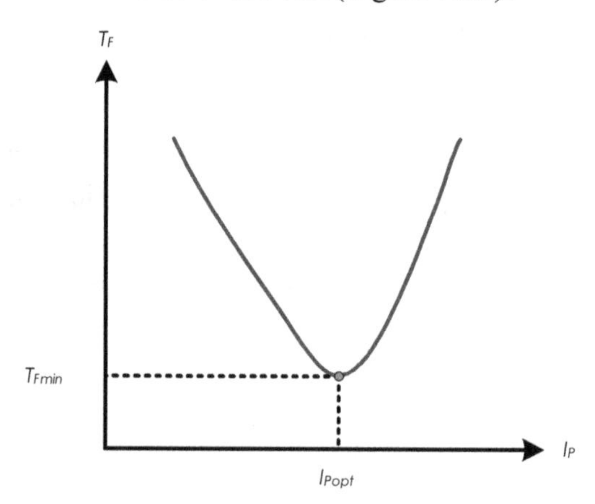

Figure 7.21 – Temperature in the cold side (T_F) of a Peltier cell as a function of the applied current (I_P).

A factor of merit is defined for Peltier cells given by

$$Z = \frac{\alpha^2 \sigma}{k}, \tag{7.10}$$

where α is the thermoelectric coefficient, σ is the electrical conductivity, and k is thermal conductivity. The units of Z are K^{-1}. A good material for the construction of a Peltier cell has a high thermoelectric coefficient and high electrical conductivity but a low thermal conductivity. The materials that have these characteristics are the semiconductors (Figure 7.22), in particular, bismuth telluride (Bi_2Te_3), having a merit factor of about 0.003.

Several cells are associated in a single module to obtain a greater cooling power, as illustrated in Figure 7.23. They are connected electrically in series and thermally in parallel.

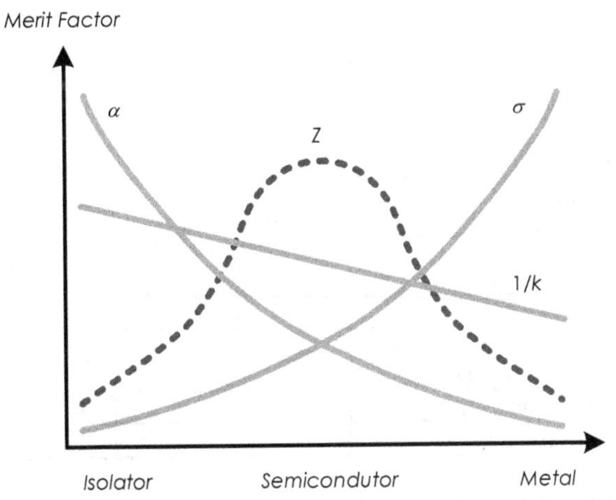

Figure 7.22 – Dependence of the Merit Factor on the type of material used in Peltier Cells.

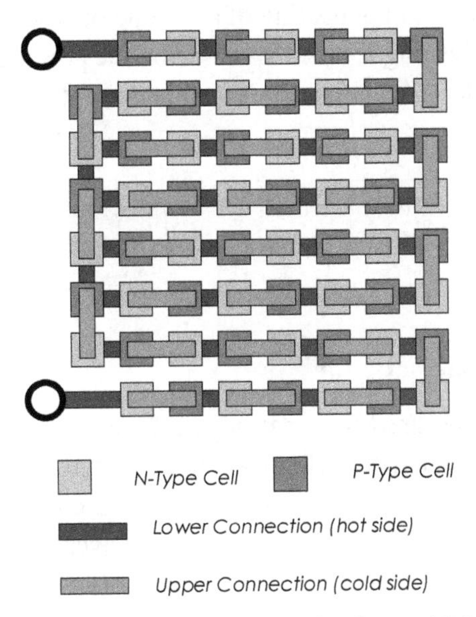

Figure 7.23 – Peltier module made of several Peltier cells.

The operation of a Peltier module consists simply of applying a DC voltage or current whose value determines the temperature difference that is intended between the hot side and the cold side. You must be careful not to exceed the maximum specifications given by the manufacturer.

Table 7.3 shows an example of specifications of a thermoelectric module of Melcor, model CP1.4-17-045. It should be noted the temperature difference depends on the value of the temperature of the hot side.

Table 7.3 – Datasheet of a thermoelectric module by Melcor, model CP1.4-17-045.

FEATURE	NORMAL TEMPERATURE	HIGH TEMPERATURE
Hot Side Temperature (°C)	25	50
Q_{max} (W)	9.2	10.9
Max. Temperature Difference (°C)	65	74
Maximum Current (A)	8.5	8.5
Maximum Voltage(V)	2.06	2.19
Electric Resistance (Ω)	0.21	0.23

7.7 Joule Effect

The Joule effect consists of the heating of a conductor when crossed by an electric current. This heating can be expressed by the power dissipated (P) as a function of the electrical resistance (R) and current (I):

$$P = R \cdot I^2. \tag{7.11}$$

The dissipation of power leads to an increase in temperature. This increase can be determined from the thermal resistance (R_θ):

$$\Delta T = P \cdot R_\theta. \tag{7.12}$$

7.8 Guckel Thermal Actuator

The Joule effect can be used to construct an actuator that causes linear movement. The thermal actuator of Guckel [12] is an example of this type of actuator (Figure 7.24). It is made of a U-shaped conductive material with one arm wider than the other. When this driver is traversed by a current, it will heat up due to the Joule effect. Since the heat is proportional to the electrical resistance and the thinner arm has greater resistance to the passage of current, it will heat up more. This will cause an expansion of the material in that arm, and the actuator will bend.

This type of actuator is particularly used in MEMS devices (micro-electromechanical systems). Many are connected together to increase the force made. These actuators produce more power than the electrostatic ones but are slower.

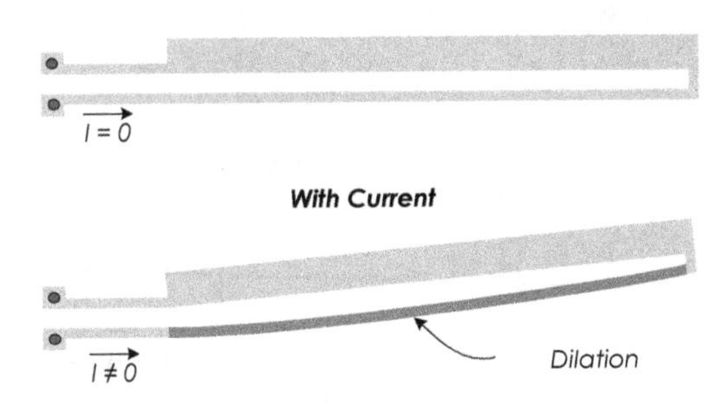

Figure 7.24 – Guckel thermal actuator.

7.9 Hot Wire Anemometer

The Joule effect and convection heat are used in the hot wire anemometers sensors that are used to measure the velocity of a liquid or gas. A conductive wire made typically of tungsten or platinum is inserted into the fluid whose speed is to be measured, and an electric current is imposed so that the wire heats it by Joule effect.

The increase of the temperature in the wire is offset by a decrease in temperature caused by the movement of the flow that tends to cool the wire (if its temperature is lower than the wire). Thus, there is a balance between the electrical power input, given by Ri^2, and power due to heat transfer from the wire to the fluid given by $hA_W(T_W - T_f)$ where h is the heat transfer coefficient, A_W is the area of the wire in contact with the fluid, T_W is the temperature of the wire and T_f is the fluid temperature:

$$R \cdot i^2 = h \cdot A_W \cdot (T_W - T_f). \tag{7.13}$$

This equation is the combination of Eqs. (176) and (177) where $1/R_\theta = h \cdot A_W$. Moreover, the thermal resistance is a function of fluid velocity

$$\frac{1}{R_\theta} = a + b \cdot v^c, \tag{7.14}$$

where a, b and c are constants obtained by calibration and depend on the shape of the wire and the type of fluid. We must also consider that the electrical resistance of the wire depends on the temperature, as seen in Eq. (4.4):

$$R = R_0[1 + \alpha(T_W - T_{W0})]. \tag{7.15}$$

Variable R_0 represents the value of the resistance at the temperature T_{W0}. Combining Eqs. (7.13), (7.14) and (7.15) we get

$$v = \left[\frac{\frac{I^2 R_0[1+\alpha(T_W - T_{W0})]}{A_W(T_W - T_f)}}{b}\right]^{\frac{1}{c}}. \tag{7.16}$$

By measuring the temperature of the fluid and the wire independently, it is possible to estimate the velocity of the fluid.

There are two modes of operation: constant current and constant temperature. In the first case, the current injected into the wire is maintained at a constant value. This form has the disadvantage that if the flow suddenly stops, the wire temperature can rise too much, which can lead to its destruction. In the second case, an electronic circuit that automatically adjusts the current to maintain a constant temperature wire is used. This second mode is the most used.

The hot wire anemometer has the advantage of having an optimal spatial resolution and a high-frequency response (up to 400 kHz) but has the disadvantage of being fragile, too expensive, and needs to be periodically recalibrated due to the accumulation of debris on the wire.

7.10 Questions

1. Describe the Thomson, Peltier, and Seebeck phenomena.
2. Explain how a thermocouple works.
3. What are the advantages and disadvantages of using thermocouples as temperature sensors?
4. Why is it not possible in a Peltier cell to lower the temperature to 0 K?
5. Consider a voltage divider consisting of a thermistor and a fixed resistance of 4 Ω. What is the maximum voltage that can be used in the voltage divider so that there is no self-heating of the thermistor of more than 0.1°C to a temperature of 200°C? Thermistor characteristics: Electrical resistance at 200°C: 1 Ω., Thermal resistance: 0.2°C/W.

6. The appearance of a potential difference between the ends of a metal kept at different temperatures is due to what effect?

   a. Seebeck effect.
   b. Peltier effect.
   c. Thomson effect.
   d. Matteucci effect.
   e. None of the above.

7. What is the advantage of a Guckel thermal actuator when compared to an electrostatic actuator?

8. How does a hot wire anemometer work?

Chapter 8

Devices Based on Electromagnetic Radiation

8.1 Quantities and Units

Electromagnetic radiation can be conceptualized as a set of photons or as waves propagating in space at a velocity (c) of 3×10^8 m/s. Actually, any elementary particle or even any set of particles at the macroscopic level can be seen as a wave. This duality arises from the way they are used to explain the observable phenomena in nature. There are phenomena that are easier to explain if we use the concepts of particles, such as the photoelectric effect, while there are others, such as interference, that it is easier to explain using the concept of waves.

The photons, like other particles, are characterized by their energy (E). In terms of waves, the photons can be characterized by their frequency (f)[1]. The relationship between the photon energy and the frequency of the wave associated is

$$E = h \cdot f, \tag{8.1}$$

where h is Planck's constant and has the value $6{,}67 \times 10^{-34}$ Js.

The wavelength is related to frequency using

$$\lambda = \frac{c}{f}. \tag{8.2}$$

Electromagnetic radiation can be classified according to the frequency in different bands, as shown in Figure 8.1.

---

[1]In physics it is common to use the Greek letter nu (ν) to represente frequency instead of the letter f typically used by engineers.

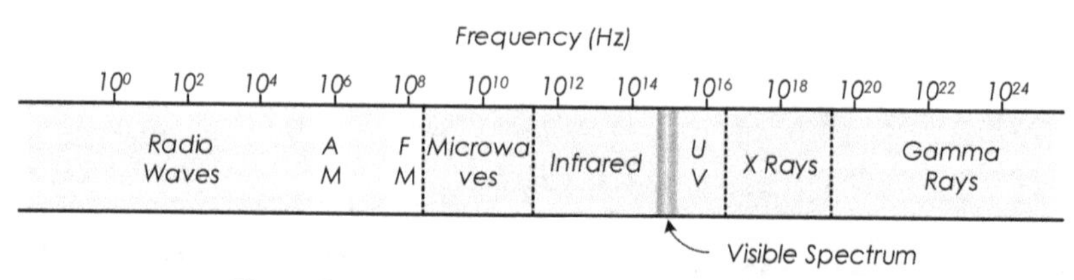

Figure 8.1 – Electromagnetic radiation spectrum.

The electromagnetic radiation of all frequencies is used in different applications, from communications and medicine to astrophysics and cooking.

There are different magnitudes used to express the amount of electromagnetic radiation. These quantities are divided into two major classes used in different domains: photometry and radiometry. The first case is used solely for radiation in the visible spectrum, while the second case is used across the whole electromagnetic spectrum.

Table 8.1 lists the quantities and units used in photometry.

Table 8.1 – Quantities and units used in photometry.

QUANTITY	SYM.	UNIT	UNIT SYMBOL	DESCRIPTION
Luminous energy	Q_v	lumen *x* second	lm·s	Radiation energy
Luminous flux	F	lumen (cd·sr)	lm	Luminous power that leaves a light source
Luminous intensity	I_v	candela (lm/sr)	cd	Luminous power that leaves a light source per unit of solid angle
Luminance	L_v	candela per square meter	cd/m^2	Luminous intensity that leaves an elementary area
Illuminance	E_v	lux (lm/m^2)	lx	Quantity of light that impinges into an elementary area
Luminous emittance	M_v	lux (lm/m^2)	lx	Quantity of light emitted by a surface
Luminous efficacy		lumen per watt	lm/W	Relationship between luminous flux and radiant flux

The luminous flux (F) is the luminous power that comes from a light source. This quantity is typically used in lamps whose aim is lighting and therefore are made to radiate visible light in practically all directions.

The luminous energy (Q_v) relates to the energy radiated during a certain time interval and is obtained from the luminous flux by multiplying it by the desired time interval (if the flow is constant). This magnitude is not used as a source of light/radiation parameter because it depends not only on the light source (but also on time).

The luminous intensity (I_v) is the light power coming out of a light source per unit solid angle. A solid angle is an angle defined in two dimensions (Figure 8.2).

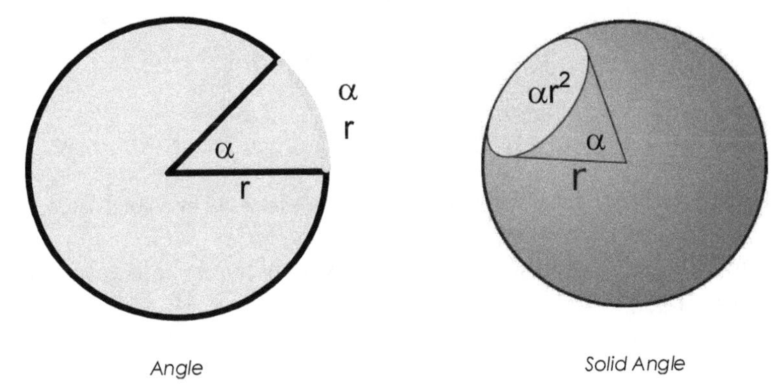

Figure 8.2 – Illustration of the unit of solid angle (steradian).

An angle α is the ratio of the length of the arc defined by this angle and the radius of the circle. Similarly, the solid angle is the ratio of the surface area of a sphere defined by a cone centered on the sphere and the radius of the sphere.

The light intensity is used for point light sources such as LEDs. For example, in the case of the blue LED TLHB4400, Vishay Semiconductors is the typical light intensity of 15 mcd (Table 8.2).

Table 8.2 – Extract from the datasheet of the blue led by Vishay Semiconductors, model TLHB4400.

PARAMETER	TEST CONDITION	SYM.	MIN.	TYP.	MAX.	UNIT
Forward voltage	$I_F = 20$ mA	V_F		3,9	4,5	V
Reverse voltage	$I_R = 10$ μA	V_R	5			V
Luminous intensity	$I_F = 20$ mA	I_K	6.3	15		mcd
Dom. wavelength	$I_F = 10$ mA	λ_d		466		nm
Angle of ½ sensitivity	$I_F = 10$ mA	φ		±30		°
Peak wavelength	$I_F = 10$ mA	λ_p		428		nm

Figure 8.3 illustrates the concept of light intensity from a point source. Considering a light tube defined by the light rays from a point light source, the power running through area A is the same that crosses surface B despite it having a larger area. This happens because the light scatters. The luminous intensity depends only on the light source and not on the distance.

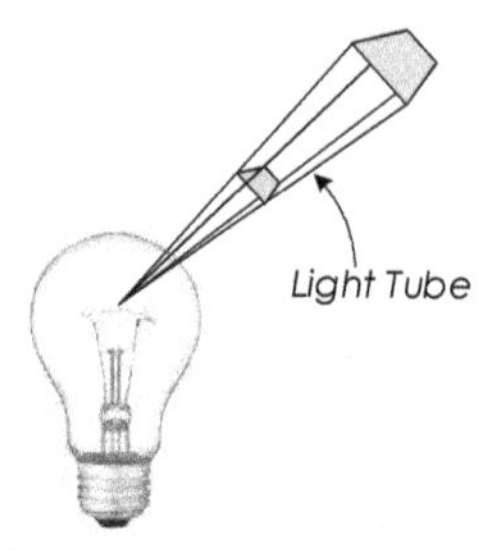

Figure 8.3 – Illustration of the concept of luminous intensity by a point source.

Figure 8.4 illustrates the relationship between electromagnetic quantities related to radiation.

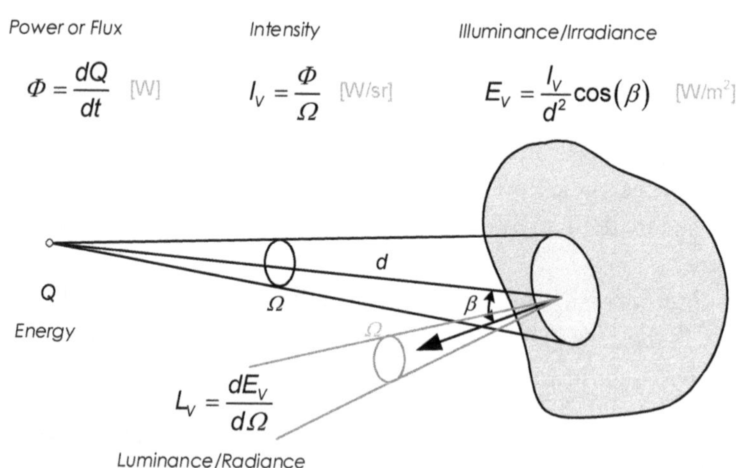

Figure 8.4 – Illustration of the different quantities related to a point source of electromagnetic radiation.

Table 8.3 shows typical Illuminance values on Earth's surface for different conditions.

Table 8.3 – Typical Illuminance values on Earth's surface for different conditions.

CONDITION	ILLUMINANCE
Direct Sunlight	100 000 lux
Cloudy Day	10 000 lux
Twilight	10 lux
Full Moon	1 lux
Moonless Night	0.01 lux

Figure 8.5 shows typical illuminance values of different surfaces.

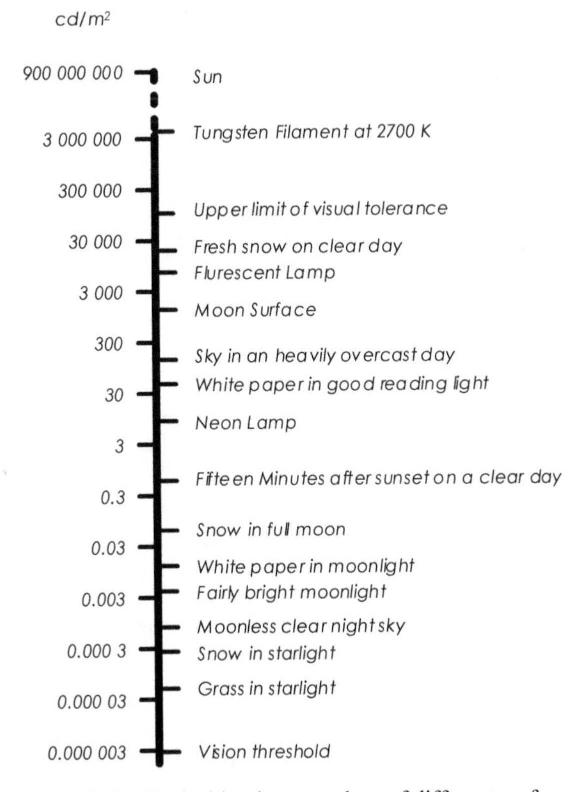

Figure 8.5 – Typical luminance values of different surfaces.

Table 8.4 lists the quantities and units used in radiometry. The values of the quantities in the field of radiometry can be converted to their respective values in the photometric area considering the sensitivity of the "perfect eye" (Figure 8.6).

Table 8.4 – Quantities and units used in radiometry.

QUANTITY	SYM.	UNIT NAME	UNIT	DESCRIPTION
Radiated Energy	Q	joule	J	Energy
Flux or radiated power	Φ	watt	W	Energy per unit time (power)
Radiant Intensity	I	watt per steradian	W/sr	Power per unit solid angle
Radiance	L	watt per steradian per square meter	W/sr/m²	Power per unit solid angle per area
Irradiance	E	watt per square meter	W/m²	Power incident on a surface
Emittance	M	watt per square meter	W/m²	Power emitted by a surface

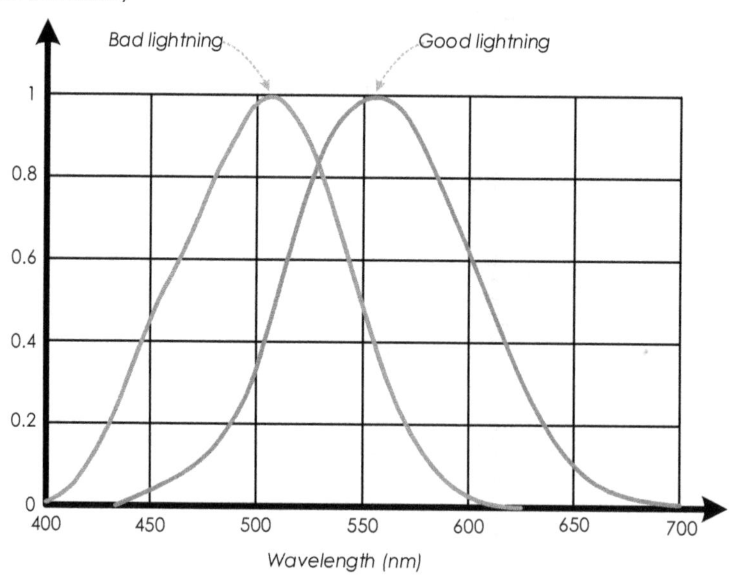

Figure 8.6 – Luminous efficiency of the Human eye in conditions of good lighting (blue) and bad lighting (red) according to the Commission Internationale de l'Éclairage (CIE).

The luminous flux (F), for example, can be obtained from the radiant power spectral density, $J(\lambda)$ and from the luminous efficiency, $\overline{y}(\lambda)$, given by the curve in Figure 8.6, using

$$F = 683 \cdot \int_0^\infty J(\lambda) \cdot \overline{y}(\lambda) \cdot d\lambda. \quad (8.3)$$

The numeric value 683 was chosen so that the lumen value used could be kept the same as that used before 2019 that was defined based on the luminous flux emitted by 1 square centimeter of platinum at its solidification temperature.

The radiated energy indicates the total amount of energy that a radiation source produces.

The radiant power specifies the power of the radiation emitted by a light source, and radiant intensity specifies the radiant power per steradian. These parameters are found typically in devices that emit radiation outside the visible spectrum, such as infrared LEDs. An example is found in Table 8.4, where you can observe that the typical radiant power is 10 mW and the typical radiant intensity is 5 mW/sr.

Table 8.5 – Extract of the datasheet of infrared led from Vishay Semiconductors, model CQY37N.

PARAMETER	TEST CONDITION	SYM.	MIN.	TYP.	MAX.	UNIT
Radiant intensity	$I_F = 50$ mA, $t_p \leq 20$ ms	I_e	2.2	5	11	mW/sr
Radiant power	$I_F = 50$ mA, $t_p \leq 20$ ms	φ_e		10		mW
Temperature coefficient of φ_e	$I_F = 50$ mA	$TK\varphi_e$		−0.8		%/K
Angle of half sensitivity		φ		±12		°
Peak wavelength	$I_F = 50$ mA	λ_p		950		nm
Spectral bandwidth	$I_F = 50$ mA	$\Delta\lambda$		50		nm
Rise time	$I_F = 1.5$ A, $t_p/T = 0{,}01$, $t_p \leq 10\mu s$	t_r		400		ns
Fall time	$I_F = 1.5$ A, $t_p/T = 0{,}01$, $t_p \leq 10\mu s$	t_f		450		ns
Virtual source diameter				1.2		mm

There are situations where you want to work with a given wavelength. In these cases, you use spectral densities such as spectral density of the radiated power and express it by an analytical function or through a chart of the spectral density versus wavelength.

The radiated power, Φ, is obtained from the spectral density of the radiated power, $J(\lambda)$, integrating all the wavelengths:

$$\Phi = \int_0^{\infty} J(\lambda) \cdot d\lambda. \tag{8.4}$$

8.2 Electroluminescence

Luminescence is defined as the emission of optical radiation (ultraviolet, visible, or infrared) as a result of electronic excitation. This effect is known as electroluminescence when the excitation is caused by an electric field or current.

The electroluminescence can be triggered by various processes, including intrinsically by avalanche or tunnel effect or by injection processes. The latter process is of particular interest and happens when you inject minority charge carriers into a p-n junction where the phenomenon of radiative transition occurs.

Radiative transition is defined as the transition between two states of a molecular entity in which the difference of energy between these states is related to the emission or absorption of photons.

The emission or absorption of photons can occur in three different ways:

- When an electron moves from a filled valence state to an empty state in the conduction band by absorbing a photon band.
- When a photon triggers the emission of another photon by an electron going from a filled state in the conduction band to an empty state in the valence band.
- When an electron in the conduction band returns spontaneously to an empty state in the valence band by emitting a photon.

The transition of an electron from the conduction band to the valence band accompanied by the release of a photon is called radiative recombination (Figure 8.7).

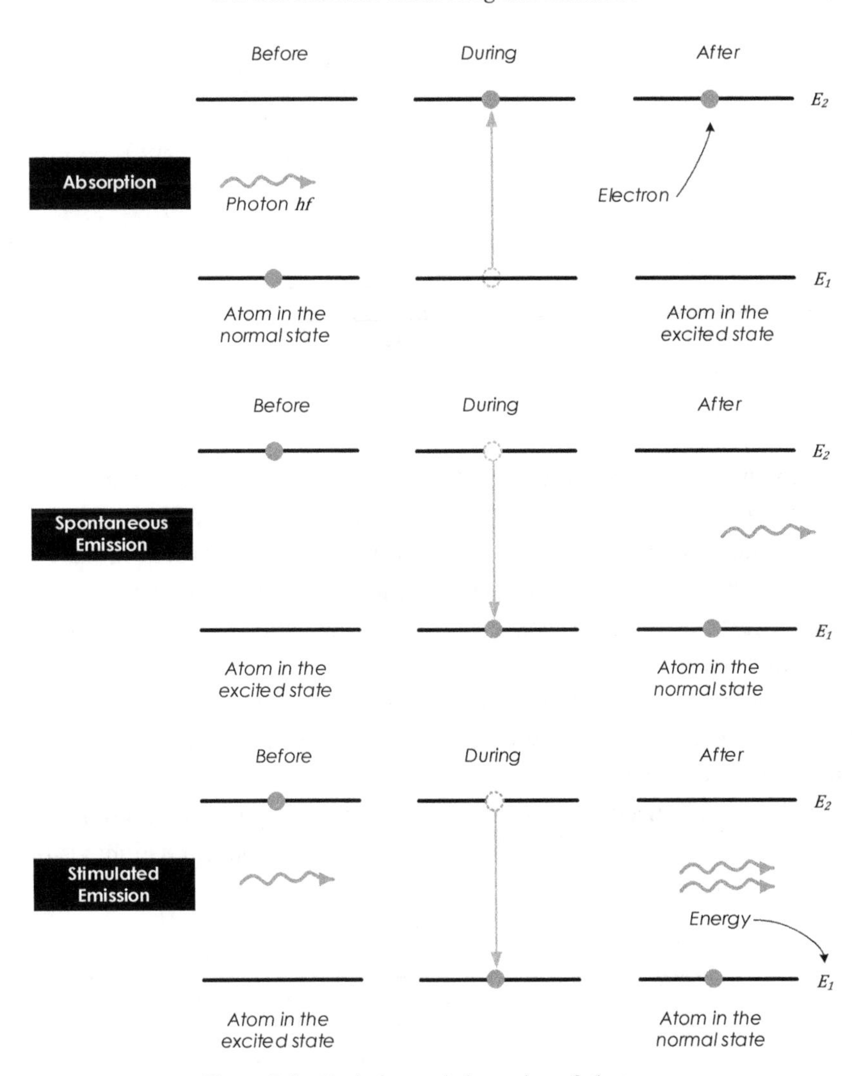

Figure 8.7 – Emission and absorption of photons.

8.3 Photovoltaic Effect

The photovoltaic effect is the absorption of energy by the electrons in the valence band of a material that thus acquires enough energy to move to the conduction band (Figure 8.8).

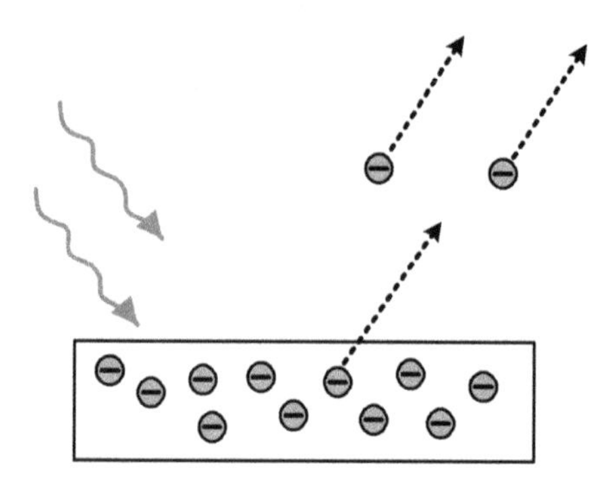

Figure 8.8 – Illustration of the Photovoltaic effect.

Photons can only promote the release of electrons from the valence band if they have energy ($h\nu$) higher than the bandgap energy in the case of semiconductors. This means that for a given material, only electromagnetic radiation above a certain frequency can cause the release of electrons. The excess energy that photons possess is transformed into kinetic energy of the released electron.

An increase in light intensity corresponds to an increase in the number of photons and not in an increase of their energy and thus gives rise to further freed electrons.

This effect is used in solar panels (photovoltaic panels), photodiodes, photoresistors, image sensors, photoelectric spectroscopy, and night vision devices.

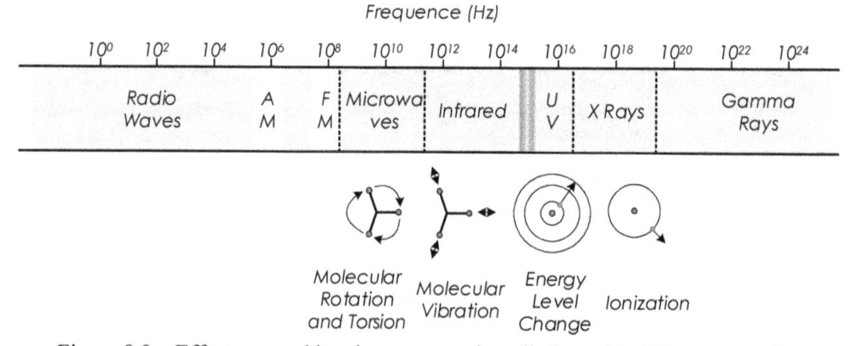

Figure 8.9 – Effects caused by electromagnetic radiation with different energies.

Figure 8.9 exemplifies the effects electromagnetic radiation has on matter according to its energy. It is noted that only visible light can cause the promotion of the electrons in the energy levels of the atoms (and the matter). Infrared radiation, for example, gives origin to an increase in the agitation of the molecules and is therefore related to the temperature of the bodies.

8.4 LED

The LED (light-emitting diode) is a P-N semiconductor junction that emits visible light when energized by electroluminescence. This light is not monochromatic such as a laser but consists of a narrow spectral band as shown in Figure 8.10 taken from the specification sheet of the LED TLCR5100 from Vishay Semiconductors and represents the relative light intensity (the maximum value) as a function of wavelength.

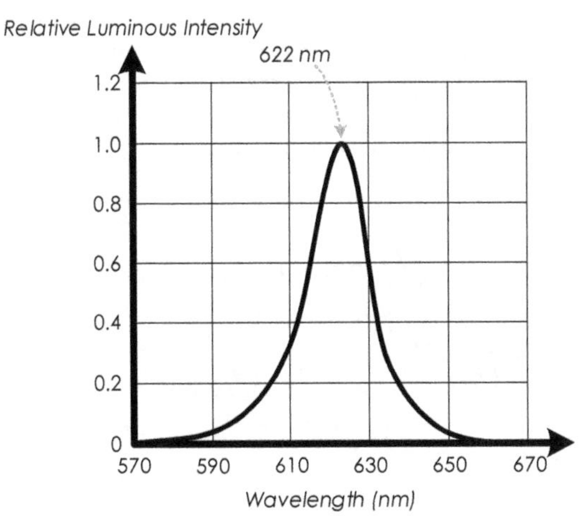

Figure 8.10 – Relative luminous intensity as a function of wavelength for a red led from Vishay Semiconductors, model TLCR5100.

Referring to the specifications provided by the manufacturer to this led (Table 8.6), it appears that the peak light intensity (peak wavelength) occurs at the wavelength of 622 nm.

Note that also, in this case, the manufacturer specifies the value of the dominant wavelength that is typically between 611 and 622 nm. The dominant wavelength is related to the sensitivity of the human eye. This sensitivity is not the same for all wavelengths. The dominant wavelength is the value of the wavelength of a monochromatic light that produces the same sensation of the color light source in question.

The spectral width is defined at the level of 50% of the maximum luminous intensity and is 18 nm for the case of the LED shown in Table 8.6.

Table 8.6 – Datasheet of a red led from Vishay Semiconductors, model TLCR5100.

PARAMETER	TEST CONDITION	SYM.	MIN.	TYP.	MAX.	UNIT
Forward voltage	$I_F = 50$ mA	V_F		2.1	2.7	V
Reverse voltage	$I_R = 10$ µA	V_R	5			V
Luminous intensity	$I_F = 50$ mA	I_V	4300	11000		mcd
Dominant wavelength	$I_F = 50$ mA	λ_d	611	616	622	nm
Spectral bandwidth at 50% $I_{rel\ max}$	$I_F = 50$ mA	$\Delta\lambda$		18		nm
Angle of half sensitivity	$I_F = 50$ mA	φ		±9		°
Peak wavelength	$I_F = 50$ mA	λ_p		622		nm

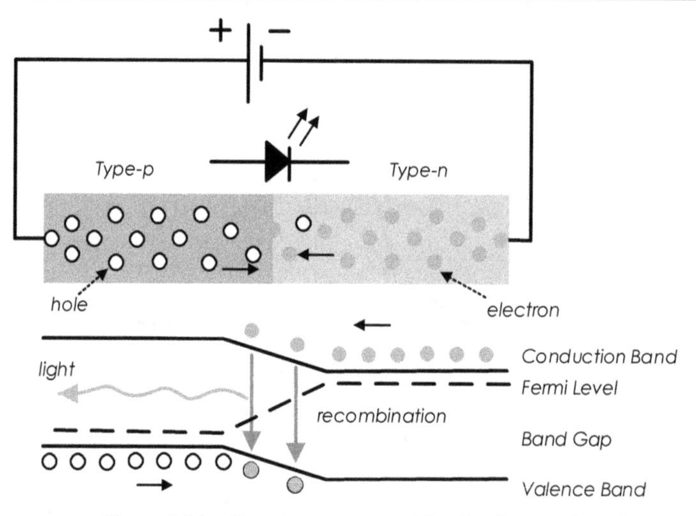

Figure 8.11 – Spontaneous recombination in a p-n junction.

The light emitted by a LED is produced by the energetic interactions of the electron. In either P-N forward-biased junction within the structure, near the junction, recombination of holes and electrons occurs. This recombination requires the energy possessed by this electron, which until then was free to be released, which occurs in the form of heat or light photons. The wavelength of the emitted light depends on the size of the bandgap of the semiconductor. When you directly polarize the junction, you promote the displacement of electrons and gaps to the junction zone, where they recombine, leading to spontaneous emission (Figure 8.11).

Different colors are achieved by using different materials having different forbidden bandwidths (Table 8.7).

Table 8.7 – Materials used to make LEDs with different colors.

COLOR	WAVELENGTH (nm)	SEMICONDUTOR MATERIAL
Infrared	$\lambda > 760$	Gallium Arsenide (GaAs) Aluminium Gallium Arsenide (AlGaAs)
Red	$610 < \lambda < 760$	Aluminium Gallium Arsenide (AlGaAs) Galium Arsenide Phosphide (GaAsP) Aluminium Galium Indium Phosphide (AlGaInP) Galium Phosphide (GaP)
Orange	$590 < \lambda < 610$	Galium Arsenide Phosphide (GaAsP) Aluminium Galium Indium Phosphide (AlGaInP) Galium Phosphide (GaP)
Yellow	$570 < \lambda < 590$	Galium Arsenide Phosphide (GaAsP) Aluminium Galium Indium Phosphide (AlGaInP) Galium Phosphide (GaP)
Green	$500 < \lambda < 570$	Indium Gallium Nitrade (InGaN) Fosfato de Gálio (GaP) Aluminium Galium Indium Phosphide (AlGaInP) Aluminium Galium Phosphide (AlGaP)
Blue	$450 < \lambda < 500$	Zinc Selenide (ZnSe) Indium Galium Nitrade (InGaN)
Violet	$400 < \lambda < 450$	Indium Galium Nitrade (InGaN)
Purple	Several types	Dual Blue/red
Ultraviolet	$610 < \lambda < 760$	Diamond (C) Aluminium Nitrade (AlN) Aluminium Galium Nitrade(AlGaN)
White	Broad Spectrum	Blue/UV Diode with Yellow Phosphor

The light intensity is proportional to the current flowing through the junction. There is, however, a limit to the amount of current that can be used due to the Joule heating effect. In Table 8.8, we can observe that for the LED TLCR5100 from Vishay is specified the maximum value of continuous current of 50 mA. It also specifies a maximum current value of 1 A that can be maintained for no more than 10 μs.

Table 8.8 – Datasheet of the LED TLCR5100 from Vishay Semiconductors showing maximum operating values.

PARAMETER	TEST CONDITION	SYM.	VALUE	UNIT
Reverse voltage		V_R	5	V
DC forward current	$T_{amb} \leq 85°C$	I_F	50	mA
Surge forward current	$t_p \leq 10$ μs	I_{FSM}	1	A
Power dissipation	$T_{amb} \leq 85°C$	P_V	135	mW
Junction temperature		T_j	125	°C
Operating temperature range		T_{amb}	−40 to +100	°C
Storage temperature range		T_{stg}	−40 to +100	°C
Soldering temperature	$t \leq 5$ s	T_{sd}	260	°C
Thermal resistance junction/ambient		R_{thJA}	300	K/W

The voltage drop across the LED depends on its type. In the case of the Vishay Semiconductors LED TLCR5100, this voltage drop (V_F) is typically 2.1 V (Table 8.8). This value is important for an adequate power supply of the electronic circuit of a LED (Figure 8.12). The most common situation is the use of a continuous voltage source (E). To obtain maximum light intensity in the steady state, it is necessary to limit the current (I_F) that runs through the LED. This is usually done by placing a resistor in series with the value of

$$R = \frac{E - V_F}{I_F}. \quad (8.5)$$

Figure 8.12 shows a typically led polarization circuit.

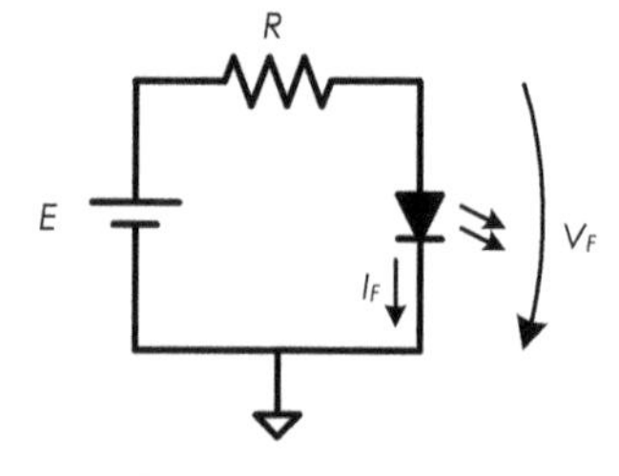

Figure 8.12 – Polarization of a LED with a DC voltage source.

8.5 Lighting Using LEDs

LEDs can be used for lighting in homes or cars, for example. LED-type lamps use several LEDs together to produce considerable light intensity. As the LEDs emit light in some direction, it is necessary to use a diffuser to disperse that light in several directions (Figure 8.13).

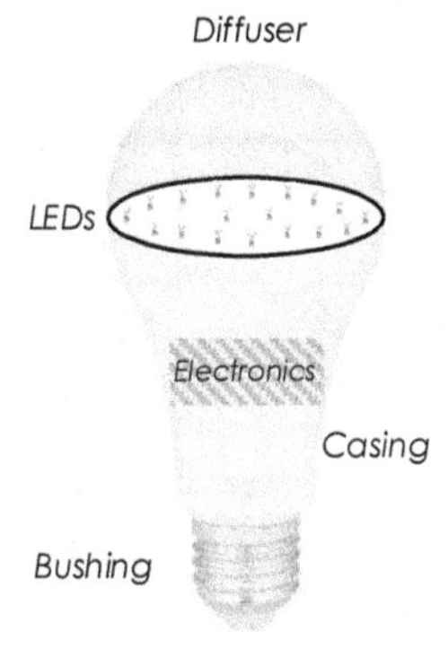

Figure 8.13 – Illustration of a LED light bulb.

To power these LEDs from the electrical power grid in homes that have an AC voltage of 110 V or 220 V, it is necessary to include a rectifier circuit inside the lamp that transforms this voltage into a continuous voltage that is used to power the LEDs connected in series. In addition, the encapsulation must guarantee the dissipation of the heat generated by the LEDs and electronics.

8.6 Liquid Crystal Panel

A liquid crystal panel is a two-dimensional actuator that emits light whose intensity can be controlled at each point independently. It is thus possible to create an image of light just like the one we see, for example, on televisions.

We will first discuss how the intensity of light is controlled at a given point and then how to achieve a color display. Finally, we look at how the color of individual pixels is controlled independently to create an image.

The white light source is a cold cathode fluorescent lamp that is always on. In front of it, a material is placed that lets light through depending on an electrical control signal. It is thus possible to have a controlled light intensity. This material is a liquid crystal.

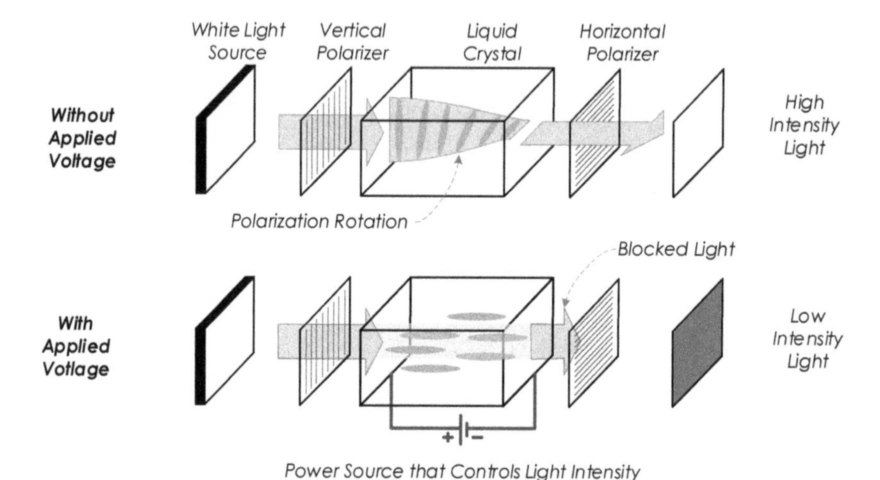

Figure 8.14 – Luminous flow in each point of the liquid crystal panel. The different elements are drawn apart for illustration purposes.

A liquid crystal is a state of matter between the solid-state and the liquid state. In a solid, the molecules maintain their position and orientation, and in a liquid, that position and orientation are variable. In a liquid crystal, the molecules maintain their orientation, but their position can vary. In this application, the liquid crystal molecules have an elongated structure with a direction that varies continuously from one side to the opposite side. The light that falls on that first side will see its polarization rotated along the path to the opposite side by a total of 90° (Figure 8.14).

The liquid crystal panel, therefore, has a white light source that emits light through a vertical polarizer. This light passes through the liquid crystal that rotates its polarization, making it horizontal. In this case, all light passes through the horizontal polarizer after the liquid crystal and exits the panel, as shown in the upper part of Figure 8.14. However, if a voltage is applied to the liquid crystal, this causes the elongated molecules to reorient so that the polarization of the light does not change when it passes through the crystal, as shown at the bottom of Figure 8.14. This maintains the vertical polarization. The result is that the light coming out of the crystal is not able to pass through the horizontal polarizer that is at its exit, and no light emerges from the panel (in the ideal case).

In order to have a color screen, three chromatic filters are used that transform white light into blue, red, and green light depending on the filter they cross. The combination of these three components in different amounts can produce any color (Figure 8.15). For this technique to work, it is necessary that the size of each filter is quite small, that is, smaller than the spatial resolution of the human eye so that our brain registers a single resulting color.

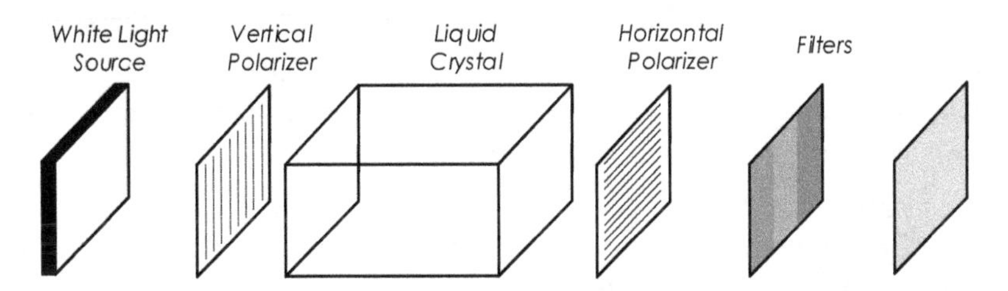

Figure 8.15 – Illustration of the use of chromatic filters to create a color panel.

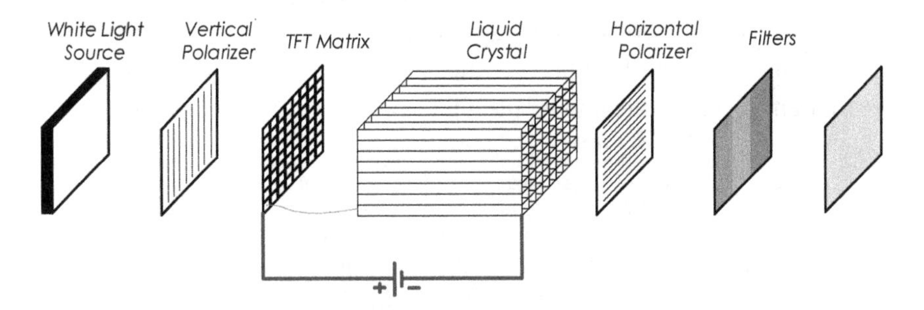

Figure 8.16 – Illustration of a liquid crystal panel with several pixels.

In order to build a two-dimensional screen, it is necessary to use a liquid crystal matrix in which each element (pixel) is controlled by an independent voltage source. This is accomplished by the TFT ("Thin Film Transistor") matrix, as shown in Figure 8.16.

This type of panel is quite inexpensive, although it has a limited viewing angle. Another disadvantage is that the darker color that can be produced is not entirely black.

This type of panels underwent, at the time, an improvement that consists of a change in the white light source. Instead of using a fluorescent lamp, a set of LEDs was placed behind or on the sides of the panel, which made them thinner and with less energy consumption. Another advantage that was obtained was the possibility of obtaining a darker "black color" since the LEDs can have their intensity quickly decreased while the fluorescent lamps must always be on with the same intensity. These panels came to be known as "LED panels."

8.7 OLED Panel

OLED panels ("Organic Light Emitting Diode") work in a similar way to LEDs. Instead of using semiconductor materials doped with impurities, they use organic materials, usually polymers. In order to create an image, several elements arranged in a matrix form are used.

As shown in Figure 8.17, electrons flow, by the action of the power supply, from the cathode to the anode through the organic layers. The electrons are injected into the emissive layer and removed from the conductive layer. The electrons in the atoms in the emitting layer have more energy than the electrons in the atoms in the conducting layer. When electrons pass from one organic layer to another, they lose energy that is released in the form of a photon. The released photons pass through the transparent layers of the anode, the TFT matrix, and the substrate.

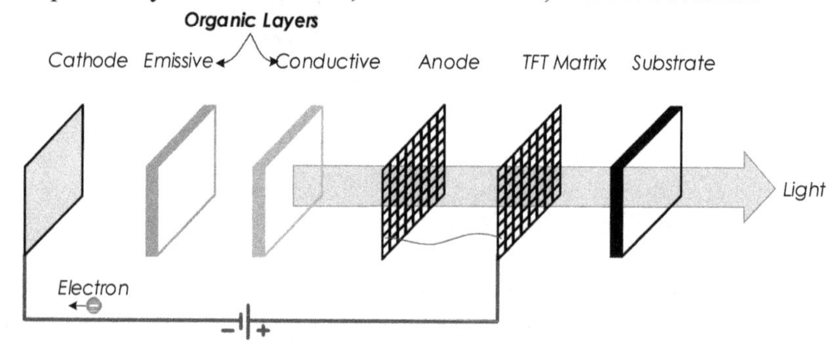

Figure 8.17 – Illustration of the constitution of an OLED panel. The various elements are separated for illustration purposes only.

The type of material used in the organic layers determines the energy levels of the orbitals of its atoms and consequently the energy of the photon created when an electron passes from the emissive layer to the conductive layer. In order to create a color OLED panel, three pairs of organic layers with different energy levels are used to create three different colors that, combined in different proportions, allow the creation of a wide range of colors.

The main advantage of OLED panels over LCD panels is their low energy consumption as they do not require an additional light source. They are also very thin and can be built with high dimensions, and can even be flexible. Another advantage is its greater viewing angle.

The large OLED panels still suffer from problems in the deposition of organic layers that make them more expensive. In applications that use smaller panels, such as, for example, mobile phones or car displays, they are already more economical and consequently are already used in more products.

8.8 Photoresistor

A photoresistor is a device made of a material whose electric resistivity depends on the light intensity of the incident light. This is due to the photoelectric effect, which increases the number of electrons in the conduction band of a material leading to a larger electric current for the same applied voltage, i.e., a lower electrical resistance.

Photoresistors are usually made of cadmium sulfide (CdS) or cadmium selenide (CdSe). They are semiconductors that have a forbidden bandwidth of 2.42 eV, which is in the range of visible light (from 1.65 to 3.1 eV). There are other materials also used, such as silicon (1.12 eV) or germanium (0.66 eV), that are sensitive to radiation with lower frequency, such as near-infrared radiation.

One of the disadvantages of photoresistors is the non-linear relation between electrical resistance and luminance, as shown in Figure 8.18.

The sensitivity is defined by

$$\gamma = \frac{\log\left(\frac{R_a}{R_b}\right)}{\log\left(\frac{E_b}{E_a}\right)} \tag{8.6}$$

where R_a and R_b are resistance values corresponding to the illuminance E_a and E_b which are typically 10 lux and 100 lux, respectively.

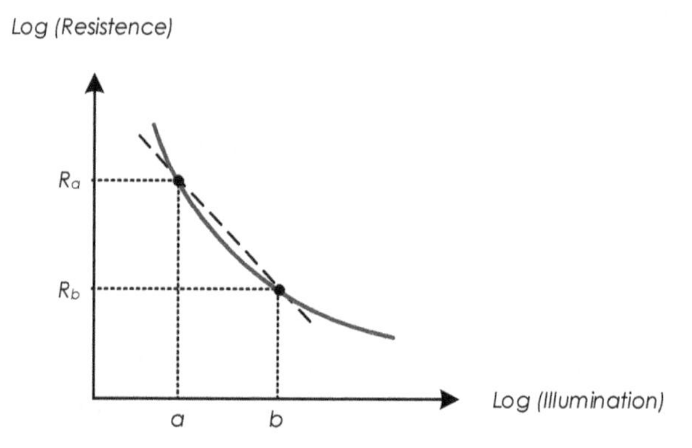

Figure 8.18 – Change of resistance with illumination.

8.9 Photodiode

A photodiode is a semiconductor device used to convert light into an electrical signal. There are several types of photodiodes, and all are based on a junction between doped semiconductors (*extrinsic semiconductors*). They are based on the photoelectric effect, i.e., the radiation incident on them makes some of the electrons in the valence band acquire enough energy to jump into the conduction band forming an electrical current. Electromagnetic radiation is thus converted into an electrical current.

8.9.1 *Operating mode*

If a p-n junction is forward biased (*positive side of the battery connected to the p-side*), there is a current flow from the anode (*p-side*) to the cathode (*n-side*). If electromagnetic radiation with the appropriate frequency impinges on the junction, an additional current will be generated in the same direction that will be much smaller relative to the current in the absence of radiation (*dark current*).

If the junction is reversed biased, the current flowing through it, in the absence of radiation, is practically zero. When radiation falls on the junction, electron-hole pairs are created on both sides of the junction, and the electrons which are in the conduction band flow towards the cathode (*n-side*) and the holes towards the anode. There is thus an electric current from the cathode to the anode.

Table 8.9 – Extract from the datasheet of the photodiode BPV10 from Vishay Semiconductors.

PARAMETER	TEST CONDITION	SYM.	MIN.	TYP.	MAX.	UNIT
Forward voltage	$I_F = 50$ mA	V_F		1.0	1.3	V
Breakdown voltage	$I_R = 100$ μA, E = 0	$V_{(BR)}$	60			V
Reverse dark current	$V_R = 20$ V, E = 0	I_{ro}		1	5	nA
Diode capacitance	$V_R = 0$ V, f = 1 MHz, E = 0	C_D		11		pF
	$V_R = 5$ V, f = 1 MHz, E = 0	C_D		3.8		pF
Open circuit voltage	$E_A = 1$ klx	V_o		480		mV
	$E_e = 1$ mW/cm^2, $\lambda = 950$ nm	V_o		450		mV
Short circuit current	$E_A = 1$ klx	I_K		80		μA
	$E_e = 1$ mW/cm^2, $\lambda = 950$ nm	I_K		65		μA
Reverse light current	$E_A = 1$ klx, $V_R = 5$ V	I_{ra}	38	85		μA
	$E_e = 1$ mW/cm^2, $\lambda = 950$ nm, $V_R = 5$ V	I_{ra}		70		μA
Absolute spectral sensitivity	$V_R = 5$ V, $\lambda = 950$ nm	$s(\lambda)$		0.55		A/W
Angle of half sensitivity		η		±20		°
Peak wavelength		λ_p		920		nm
Spectral bandwidth		$\lambda_{0,1}$		380 to 1100		nm
Quantum efficiency	$\lambda = 950$ nm	Eta		72		%
Noise equivalent power	$V_R = 20$ V, $\lambda = 950$ nm	NEP		3×10^{-14}		W/√Hz
Detectivity	$V_R = 20$ V, $\lambda = 950$ nm	D		3×10^{12}		cm√Hz/W
Rise time	$V_R = 50$ V, $R_L = 50\ \Omega$, $\lambda = 820$ nm	t_r		2.5		ns
Fall time	$V_R = 50$ V, $R_L = 50\ \Omega$, $\lambda = 820$ nm	t_f		2.5		ns

The value of the current due to radiation is somewhat lower than that obtained in the case of direct polarization (due to a greater depletion region); however, there is no additional high valued current from the junction itself, and it is, therefore, possible to amplify this current to produce a reasonable voltage.

The dark current is the reverse saturation current of the diode and which usually has a very small value, as viewed in the example shown in Table 8.9, in which this current is 1 nA.

The mode of operation described above in which the diode is reverse biased is called **Photoconductive Mode**. The higher the illuminance, the higher the absolute value of the current in the photodiode, as shown in Figure 8.19.

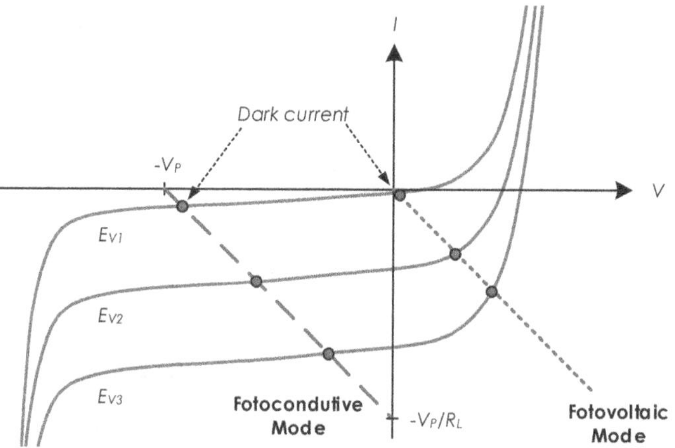

Figure 8.19 – Voltage/current characteristic of a photodiode for different values of illuminance ($E_{V1} < E_{V2} < E_{V3}$). The dashed curve is the load line to a reverse biased circuit for the photodiode (photoconductive mode). The dotted curve is the load line to a circuit without photodiode bias (photovoltaic mode).

The photodiode can be reversely biased using a voltage source (V_P) and series resistance (R_L), as illustrated in Figure 8.20.

This circuit has a line load, as shown in Figure 8.19 (dashed stroke). The value of the voltage source to be used can be as large as the disruptive voltage. In the case of the specifications listed in Table 8.9, it has a value of 60 V (*breakdown voltage*).

Note that the value of the reverse saturation current (I_P) varies fairly linearly with the illuminance (E_V), as shown in Figure 8.21 in the case of the Osram photodiode BP104S with a reverse voltage of 5 V.

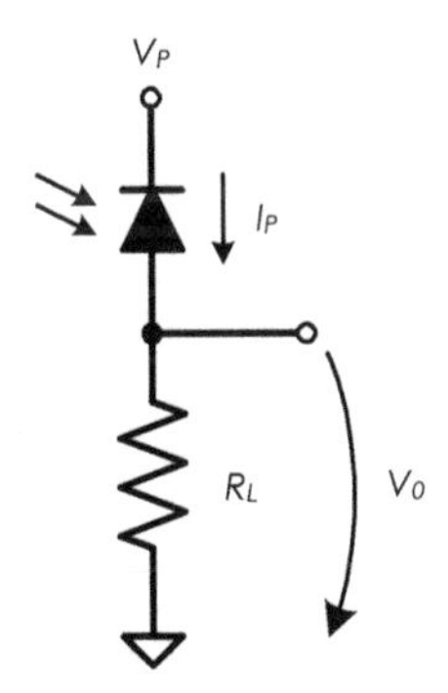

Figure 8.20 – Wiring diagram of a circuit with a reverse-biased photodiode.

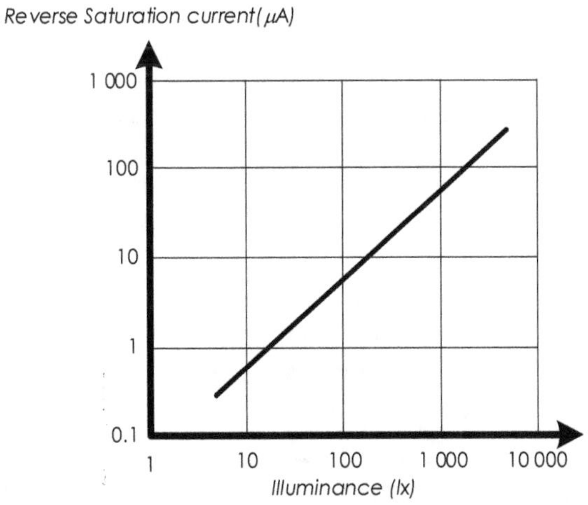

Figure 8.21 – Relation between the current and the illuminance of a photodiode Osram, model BP104S.

The reverse saturation current varies with the inverse bias voltage in the case of the actual photodiode, as shown in Figure 8.22.

The lower the reverse voltage is, the lower the dark current one has. In the limit, one can use the photodiode without a bias. This mode of operation is called the **Photovoltaic Mode**. This mode is used, for example, in the case of solar panels.

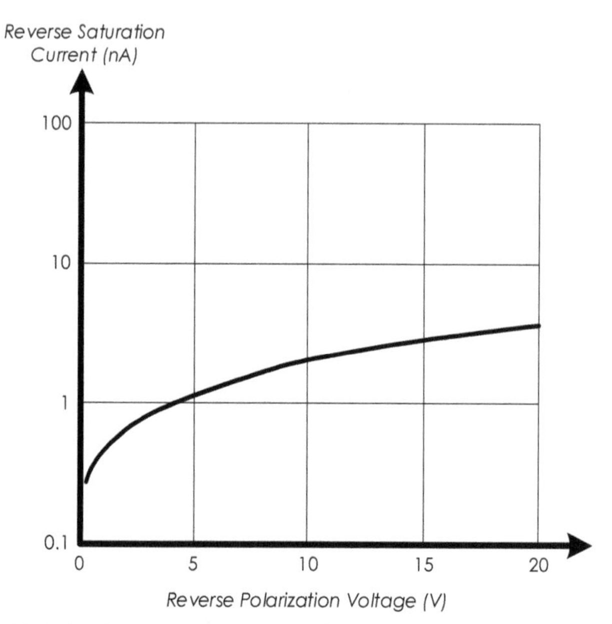

Figure 8.22 – Relation between reverse saturation current and reverse bias voltage of a photodiode Osram, model BP104S.

An example of the assembly can be seen in Figure 8.23.

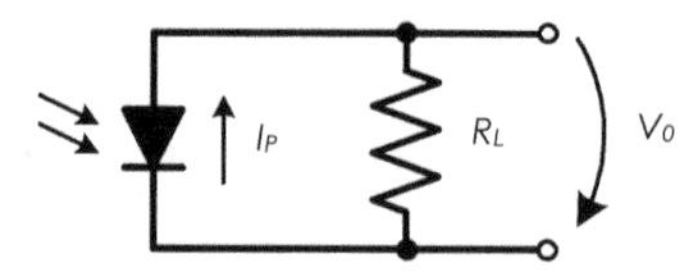

Figure 8.23 – Electrical schematic of a circuit that uses a photodiode in the photovoltaic mode.

This circuit has a load line, as shown in Figure 8.19 (dotted stroke). Notice that the dark current is lower than the photoconductive but also that the relationship between current and illuminance is not as linear. This difference in linearity is illustrated in Figure 8.24.

Another disadvantage of the photovoltaic mode is the highest capacity junction which causes the response time to be longer.

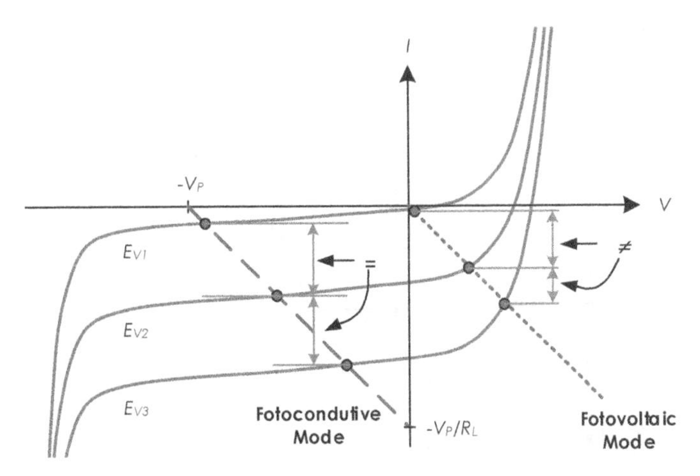

Figure 8.24 – Illustration of the reason for the different linearity between current and illuminance in photoconductive and photovoltaic modes ($E_{V1} < E_{V2} < E_{V3}$).

8.9.2 *Dark Current*

Another aspect to consider is the variation of the dark current with temperature (Figure 8.25). The higher the temperature, the greater the kinetic energy of the electrons, and therefore more electrons can pass from the valence to the conduction band solely due to thermal agitation even without incident electromagnetic radiation.

8.9.3 *Signal Conditioning with a Transimpedance Amplifier*

The assembly represented in Figure 8.20 and Figure 8.23 has the disadvantage of being affected by the load impedance. In order to circumvent this problem, it is common to use operational amplifiers in transimpedance configurations to produce a voltage proportional to the output current of the photodiode.

Figure 8.26 shows a transimpedance amplifier connected to a photodiode operating in the photovoltaic mode (no bias). The differential input voltage of the operational amplifier is practically zero, which causes the bias voltage of the photodiode to be practically zero too.

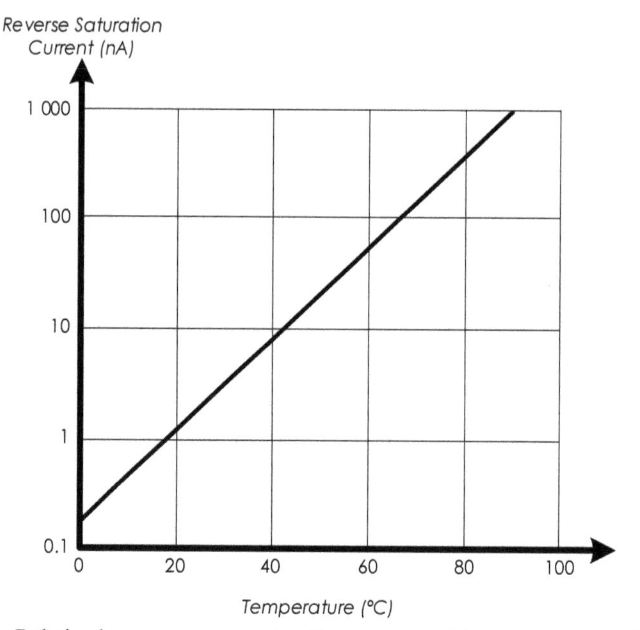

Figure 8.25 – Relation between reverse saturation current and temperature in a photodiode Osram, model BP104S.

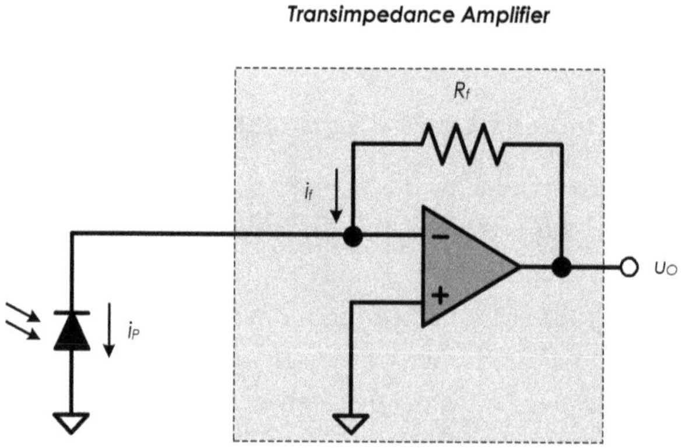

Transimpedance Amplifier

Figure 8.26 – Wiring diagram of a photodiode operating in photovoltaic mode with a transimpedance amplifier.

The output voltage, considering an ideal operational amplifier, is

$$u_0 = R_f \cdot i_f = R_f \cdot I_p. \tag{8.7}$$

Given that the current of the photodiode is very small, of the order of nano-ampere, you must use a very large feedback resistor (R_f), on the order of tens of MΩ or even 1 GΩ. This imposes severe restrictions on the input current of the operational amplifier that must be really low (of the order of pico-ampere).

For the operation in photoconductive mode, the assembly of Figure 8.27 is used, with a constant voltage source to create a reverse bias for the photodiode.

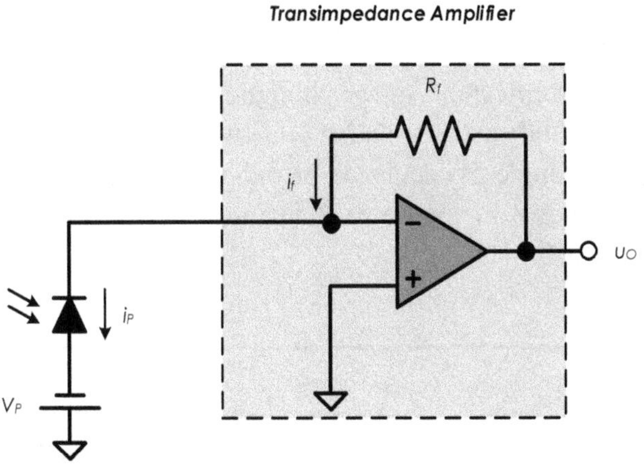

Transimpedance Amplifier

R_f, i_f, u_O, i_P, V_P

Figure 8.27 – Wiring diagram of a photodiode operating in photoconductive mode with a transimpedance amplifier.

The output voltage is given similarly by Eq. (8.7).

8.9.4 *Numerical Example*

To obtain the output voltage one expects in each application, one needs to determine the expected photodiode current (I_p). That is usually achieved by looking up the specification "Reverse Light Current" (it sometimes has a different name). In Table 8.9, for example, the typical value is 70 μA for irradiance (E_e) of 1 mW/cm^2 at a wavelength of 950 nm and a reverse voltage of 5 V.

Next, one needs to compute the actual irradiance one expects. Imagine a LED like model QEC423 from Fairchild, which has a radiant intensity

(I_R) of at least 20 mW/sr for a current of 100 mA. If it is placed 100 cm from the photodiode, the irradiance at the photodiode will be

$$E_e = \frac{I_R}{d^2} = \frac{20\times10^{-3}}{100^2} = 2\ \mu\text{W/cm}^2. \tag{8.8}$$

So, if for 1 mW/cm^2, one has 70 μA of current, for 2 μW/cm^2, one has a current of 0.14 μA.

If one resistor with a value of 10 MΩ is used in the transimpedance amplifier, one will have an output voltage of 1.4 V.

8.9.5 *Directivity*

An important specification of a photodiode is its directivity. The datasheets usually indicate the relative sensitivity to its maximum value as a function of the angle of incidence of radiation using a rectangular or polar graph as shown in Figure 8.28 for the LED from Osram, model BP104S.

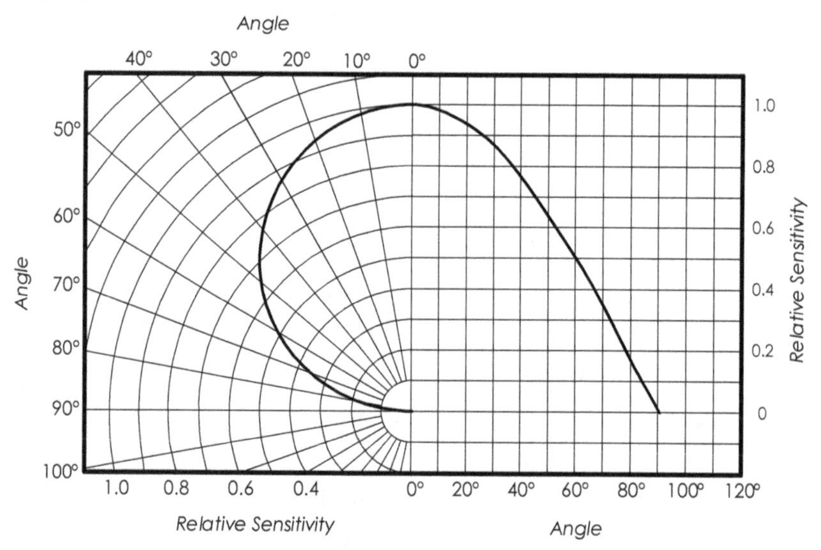

Figure 8.28 – Specification of the directivity of a photodiode (BP104S Osram model) in polar (right) and rectangular (left) coordinates.

It is common to specify the angle for which the sensitivity decreases by half. In the case of specifications listed in Table 8.9, this value is 20° (to each side of the maximum).

8.9.6 *Noise Equivalent Power*

Another important specification for a photodiode is the **Noise Equivalent Power** (NEP). This specification allows one to know the luminous power or minimum detectable radiant power with a given photodiode. It is defined as the power value of incident radiation that produces a current with a value equal to the noise current. This specification is calculated by the manufacturers based on the current noise (n_I), and the sensitivity of the photodiode (S):

$$NEP = \frac{n_I}{S}. \tag{8.9}$$

The equivalent noise power depends on the bandwidth of the measuring system because the noise current is broadband. To remove this dependence on this parameter, the current value used is not the total amount but the spectral density, i.e., the total current value divided by the square root of the bandwidth (ampere per root hertz). Given that the unit of sensitivity is ampere per watt is easily understood that the unit of equivalent noise power is watt per root hertz:

$$[NEP] = \frac{\left[\frac{A}{\sqrt{Hz}}\right]}{\left[\frac{A}{W}\right]} = \left[\frac{W}{\sqrt{Hz}}\right]. \tag{8.10}$$

Given that the sensitivity of the photodiode depends on the wavelength of the incident radiation, the equivalent noise power is specified for a certain wavelength. Like sensitivity, noise equivalent power has a non-linear variation with wavelength. In Table 8.9, one can observe, for example, the case of diode BPV10 from Vishay Semiconductors, where the equivalent noise power is $3\times10^{-14}\ W/\sqrt{Hz}$ at a wavelength of 950 nm and a reverse bias voltage of 20 V.

The dependence of the noise equivalent power on the bias voltage has to do with the origin of the noise current. There are two phenomena that contribute to this noise: thermal and impulsive.

Thermal noise (Johnson noise) is given by

$$I_J = \sqrt{\frac{4 \cdot K \cdot T \cdot B}{R}}, \tag{8.11}$$

where K is Boltzmann's constant, T is the temperature, B the bandwidth, and R is the resistance of the photodiode. This noise is independent of the polarization of the photodiode.

Another type of noise present is impulsive noise (*shot noise*). This type of noise is particularly important in semiconductors since the electric current consists of discrete charges (electrons or holes). For example, a current of 1 nA means that in every nanosecond charge carriers with a total load value 10^{-21} C cross the section of the conductor:

$$I = \frac{dQ}{dt} \rightarrow dQ = I \times dt = 10^{-9} \times 10^{-9} = 10^{-18}\text{ C.} \quad (8.12)$$

This corresponds to about 6.25 electrons per nanosecond since the charge of each electron is 1.60×10^{-19} C. This means that, *on average*, 6.25 electrons traverse a section of the conductor per nanosecond. During one nanosecond, there can be 5 electrons; during another nanosecond, there can be 7, etc. This variation in the number of charge carriers that crosses a section of the conductor is random (following a Poisson distribution) and, therefore, can be regarded as a noise — impulsive noise. Its value is given by

$$I_S = \sqrt{2 \cdot e \cdot I_d}, \quad (8.13)$$

where e is the electron charge and I_d the dark current of the photodiode.

The total noise is the square root sum of the rms value of each of the components of the noise.

$$I_{Total} = \sqrt{I_J^2 + I_S^2}. \quad (8.14)$$

Imagine, for example, that a photodiode with a dark current of 2 nA, a resistance of 500 MΩ, and a sensitivity of 0.5 A/W and is working with a bandwidth of 1 Hz. In this case, it has

$$I_J = 0{,}0056\text{ pA}, I_S = 0{,}025\text{ pA}, I_{Total} = 0{,}026\text{ pA } and$$

$$NEP = 5{,}1 \times 10^{-14}\text{ W.} \quad (8.15)$$

The impulsive noise is the dominant component of the noise current when the diode is reverse biased. When the photodiode is operated in photovoltaic mode, the dark current is almost zero and, therefore, no impulsive noise. This causes the noise equivalent power to be lower, and thus it is possible to detect smaller signals.

8.9.7 *Advantages and Disadvantages of the Operating Modes*

Each mode of operation has its advantages and disadvantages, which are summarized in Table 8.10. The photovoltaic mode has the advantage of not having a dark current and having low noise (only of thermal origin while the photoconductive mode also has impulsive noise), while the photoconductive mode has excellent linearity and short response time.

Table 8.10 – Comparison between the characteristics of photodiodes operating in photovoltaic and photoconductive modes.

PARAMETER	PHOTOVOLTAIC MODE	PHOTOCONDUCTIVE MODE
Polarization	Null	Reverse
Dark Current	Null	Non null
Linearity	Average	Optimal
Noise	Low	Average
Response Time	Average	High

8.10 Pyrometers

A pyrometer is an instrument used to measure the temperature of a body without the need to make contact with that body.

This is possible because anybody with a temperature above absolute zero emits electromagnetic radiation. This radiation has different wavelengths depending on temperature, as seen in Figure 8.29.

Mathematically, the spectrum of radiation from a black body is described by Planck's law,

$$I(\lambda, T) = \frac{2hc^2}{\lambda^5} \times \frac{1}{e^{\frac{hc}{\lambda kT}} - 1}. \quad (8.16)$$

The radiant flux is given by the Stefan-Boltzmann law,

$$j^* = \epsilon \cdot \sigma \cdot T^4 \quad (8.17)$$

where ϵ is the emissivity (1 in case of a black body) and σ is the Stefan-Boltzmann constant which has a value of 5.670400×10^{-8} $Js^{-1}m^{-2}K^{-4}$.

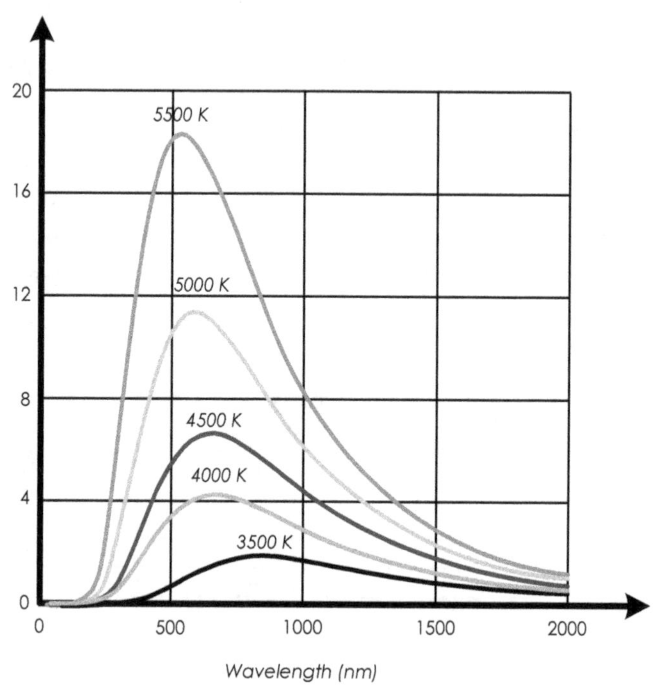

Figure 8.29 – Spectrum of blackbody radiation.

A pyrometer has an optical system and a detector which focuses the radiation from an object on a detector (e.g., a photodiode).

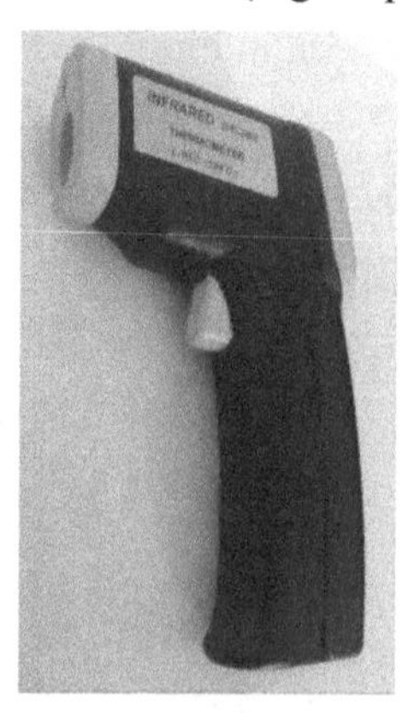

Figure 8.30 – Photograph of a pyrometer.

A pyrometer, although expensive, allows one to measure temperature in moving surfaces or on which we cannot touch due to, for example, a very high temperature. Figure 8.30 shows a picture of a pyrometer being used to measure the temperature of a ventilation duct.

8.11 Source of X-Rays

The X-ray sources are used, for example, in computer tomography apparatus to generate a 3D image of the human body.

The X-ray source is typically a Coolidge tube comprising a tungsten filament (cathode) traversed by a current, which leads to the release of electrons. These electrons are accelerated by a high voltage against the anode made of tungsten, molybdenum, or copper (Figure 8.31).

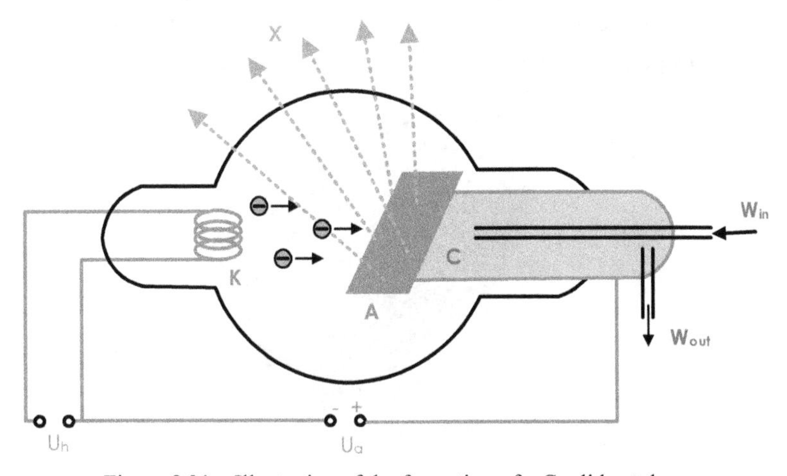

Figure 8.31 – Illustration of the formation of a Coolidge tube.

The clash between the electrons causes the acceleration of other electrons, ions, or nuclei of atoms. About 1% of this energy is released as electromagnetic radiation, that is, photons within the energy band of the X-rays.

X-ray sources are used in medicine to visualize the interior of the human body through radiography or computed axial tomography and to treat certain types of tumors. They are also used as security elements at airports to detect the presence of prohibited objects such as weapons.

8.12 Measurement of the Blood Oxygen Level with a Pulse Oximeter

A pulse oximeter is an instrument used to measure the oxygenation of human blood. In normal conditions, the oxygen saturation in the blood, i.e., the percentage of hemoglobin cells carrying oxygen, should be between 95% and 100%. Monitor the oxygen saturation is important in patients with unstable health (in intensive care units, for example) or to diagnose specific problems that may arise in pilots operating unpressurized aircraft at high altitude or sleep problems as sleep apnea, just to name a few examples.

It is possible to make an instrument capable of measuring the amount of oxygen in the blood rapidly and non-invasively using the physical phenomenon of absorption of the radiation by matter. Figure 8.32 shows a photograph of a pulse oximeter being used in a finger.

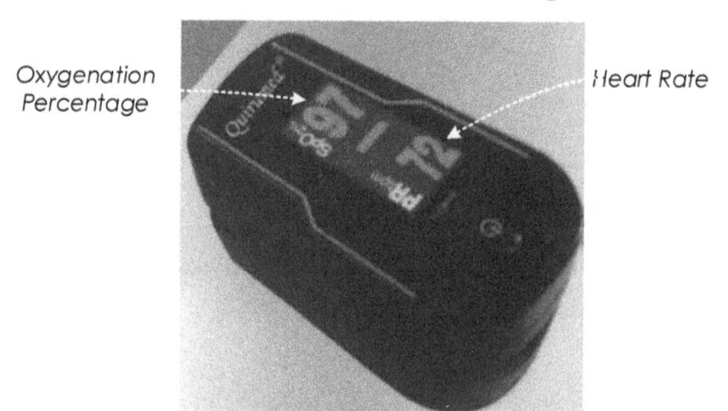

Figure 8.32 – Photograph of a pulse oximeter.

In particular, the light absorption by hemoglobin cells when carrying or not oxygen is different for different wavelengths of light, as shown in Figure 8.33.

Using two LEDs and two photodiodes that work in different wavelengths of light (660 nm and 910 nm), the light goes through a body part that is more or less translucent, such as the earlobe or the finger, is possible to compare the absorption that light suffers due to the traverse of the tissues including the arteries and hemoglobin cells that move in them. As blood flows according to a heartbeat rhythm, it is possible to identify in the signals measured with photodiodes the contribution of the blood by eliminating the contribution of other substances that are traversed by the same light as tissue, bone, venous blood. Table 8.11 shows the different types of waveforms as well as the physiological causes and possible diseases.

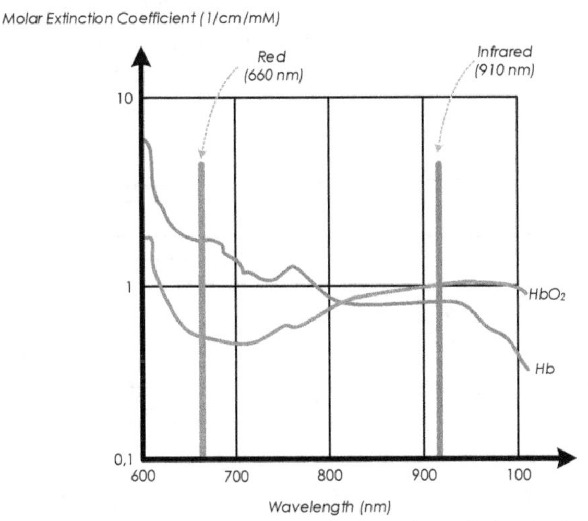

Figure 8.33 – Coefficient of absorption of electromagnetic radiation by hemoglobin (Hb) and oxyhemoglobin (HbO_2) as a function of wavelength.

Table 8.11 – Pulse types, physiological causes, and possible diseases.

PULSE TYPE	WAVEFORM	PHYSIOLOGICAL CAUSE	POSSIBLE DESEASE
Normal	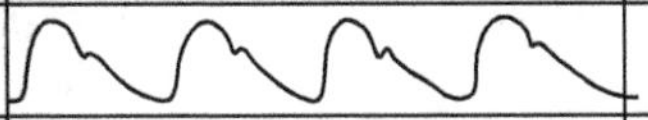	-	-
Small and Weak		Decreased Stroke Volume Increased Peripheral Resistance	Heart Failure Hypovolemia Severe Aortic Stenosis
Large and Bounding	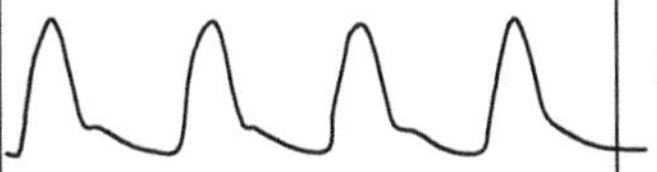	Increased Stroke Volume Decreased Peripheral Resistance Decreased Compliance	Fever, Anemia, Hyperthyroidism, Aortic Regurgitation, Bradycardia, Heart Block
Bisferiens		Increased Arterial Pulse with Double Systolic Peak	Aortic Regurgitation, Aortic Stenosis, and Regurgitation, Hypertropic Cardiomyopathy
Alternans	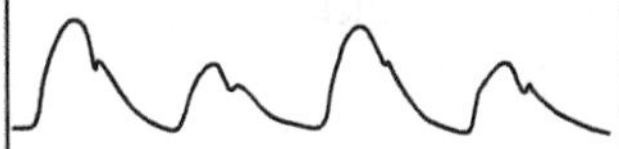	Pulse Amplitude Varies from Peak to Peak, Rhythm Basically Regular	Left Ventricular Failure

8.13 Optical Computer Mouse

The optical computer mouse appeared in 1999. It uses a small video camera to take about 1500 images per second and can work on virtually any surface. The mouse has a small red LED whose light is reflected at the surface where the mouse is set and focuses on a CMOS sensor. Figure 8.34 shows the construction of the mouse.

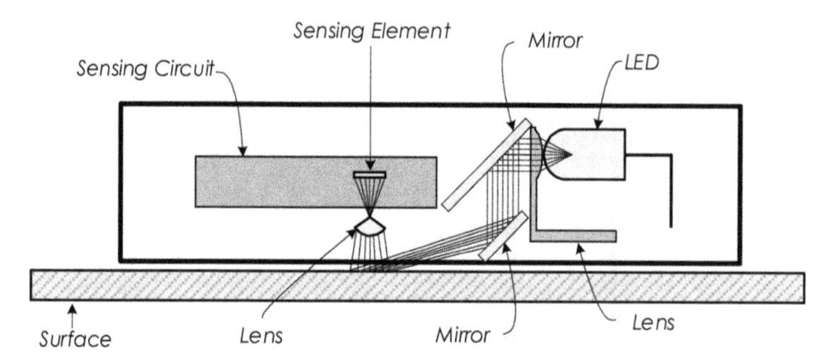

Figure 8.34 – Illustration of the assembly of an optical mouse.

The CMOS sensor sends each image to a digital signal processor, which can detect patterns in the images and determine how these patterns are moving from image to image. Based on the sequence of changes of the patterns in a sequence of images, the digital signal processor determines how much the mouse moved, sending the corresponding coordinates to the computer.

Optical mice have several advantages as compared to sphere mice:

- There are no moving parts which make it more robust and less susceptible to failure due to wear.
- Less affected by dirt because it cannot get inside the mouse and interfere with normal functioning.
- It has a higher resolution which allows a smoother movement of the cursor.
- Do not require a special surface.

8.14 Wii Game Console

In 2006, Nintendo launched a game console called Wii to compete with the Xbox 360 console from Microsoft and Sony's PlayStation 3. One of the revolutionary aspects of this console was to use a remote controller called "Wii Remote" or abbreviated "Wii mote."

The innovative aspects of this remote control have to do with the ability to detect motion. Traditional controllers have buttons that are pressed by the user. In this case, it is possible for the controller to determine where it is being pointed at, allowing, among other things, to control items on the TV screen and simulate the movement of a sword or the aiming of a firearm.

This capability is achieved based on the use of optical and acceleration sensors. An optical sensor (from PixArt) is used together with a suitable filter to capture a two-dimensional image of radiation in the infrared band. The viewing angle of this image sensor is 41°. In conjunction with the remote controller, it is necessary two infrared light sources can be provided by the Nintendo and which consist of a horizontal bar with several infrared LEDs, as seen in Figure 8.35.

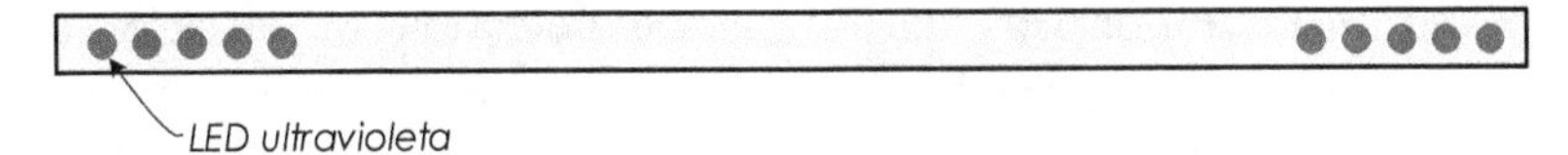

Figure 8.35 – Photograph of the lights bar from Nintendo Wii console.

The position of the infrared lights in the captured image is used to determine in which direction the remote controller is pointing. Note that the bar is stationary, and its position in relation to the remote controller is calibrated at the beginning of the operation.

The distance between the remote control and the light bar is determined from the distance between the lights on the bar, which is constant, and the distance between the two points of light obtained in the captured image. The farther away from the remote control is from the light bar, the smaller the distance between these points of light. This system works up to 5 m. The accuracy in controlling the position of an object in the scene displayed on the television is not as good as in the case of using an optical computer mouse.

Apart from the image sensor, the controller has a three-axis accelerometer from Analog Devices, model ADXL330. This allows it to measure linear acceleration in three axes which are used, for example, to simulate sudden movements like punches and movements of a tennis racket.

The remote controller transmits information about the direction of the infrared lights in the captured image and acceleration measurements in each of the three axes to the console using a Bluetooth wireless connection.

It is also available for use with the remote-control Wii mote a module, called *Wii Motion Plus,* containing a gyroscope that measures the speed of rotation to enable detection of complex movements. The sensor used is the IDG-600 from InvenSense.

8.15 X-Ray Computed Tomography

Computed tomography (CT scan) is a medical imaging technique used to create a three-dimensional image of an object from a collection of images of two-dimensional X-rays taken around a single axis of rotation.

This technique allows us to observe the internal structures of the human body and is therefore very useful in the detection of diseases and disabilities, including internal bleeding, tumors, and fractures. The main advantage over other techniques for the same purpose is that it can distinguish tissues with a variation of density as small as 1%.

The CT scan uses electromagnetic radiation in the band of the X-rays. At these frequencies (3×10^{16} Hz to 3×10^{19} Hz), part of the radiation is absorbed by the human body. This absorption depends not only on the atomic number but also on the density of the material through which the radiation penetrates (Table 8.12). This is used to create an image of the human body where different densities are distinguished and thus different materials, that is, different organs.

Table 8.12 – Values of atomic number and density of different materials relevant for X-ray radiography of the human body.

MATERIAL	ATOMIC NUMBER	DENSITY (kg/m^3)
Fat	6.3	320
Muscle	7.4	910
Lungs	7.4	1000
Air	7.6	1.3
Bone	13.8	1850
Barium	53	3500

In order to detect structures such as veins and arteries, contrast materials are often used, such as barium or iodine, which are injected into the bloodstream at the time of examination.

The 3D image of the human body is made from a collection of parallel 2D images. These 2D images are acquired by placing a source of X-rays on one side of the body and a set of sensors on the opposite side. A collimator is used so that X-rays form a beam that crosses a narrow slice of the human body (Figure 8.36).

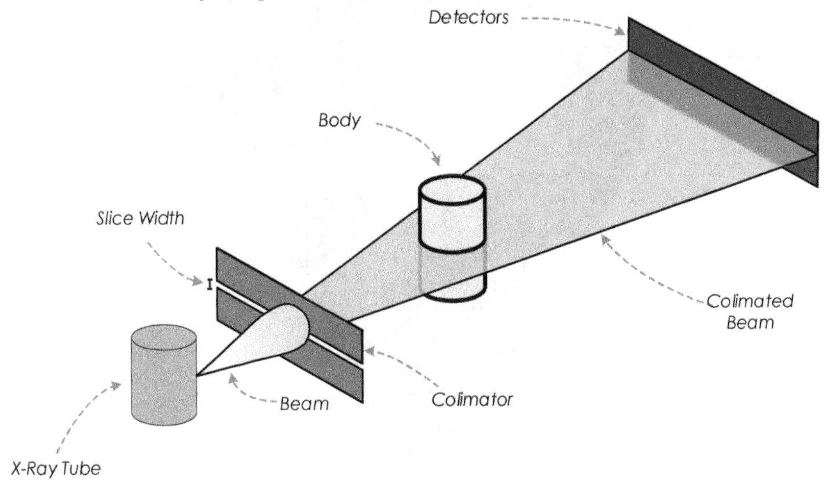

Figure 8.36 – Illustration of the X-ray beam obtained from a collimator.

Both the source and the detector are rotated around the human body to complete one full rotation.

The radiation that reaches the detector (I) depends on the absorption coefficient of the material traversed. If this coefficient is represented by $\mu(x, y)$, where x and y are the coordinates in the plane where the tomographic image is being obtained, then the intensity of the detected radiation is given by

$$I = I_0 \cdot e^{\int_\gamma \mu(x,y) \cdot d\gamma}, \tag{8.18}$$

where I_0 is the intensity of the emitted radiation, and s is the path traversed by the radiation, which in this case is a line segment that joins the source and the X-ray detector. Each detector in the detector array receives a different amount of radiation, as illustrated in Figure 8.37 since the X-ray that reaches it penetrates through a different part of the body.

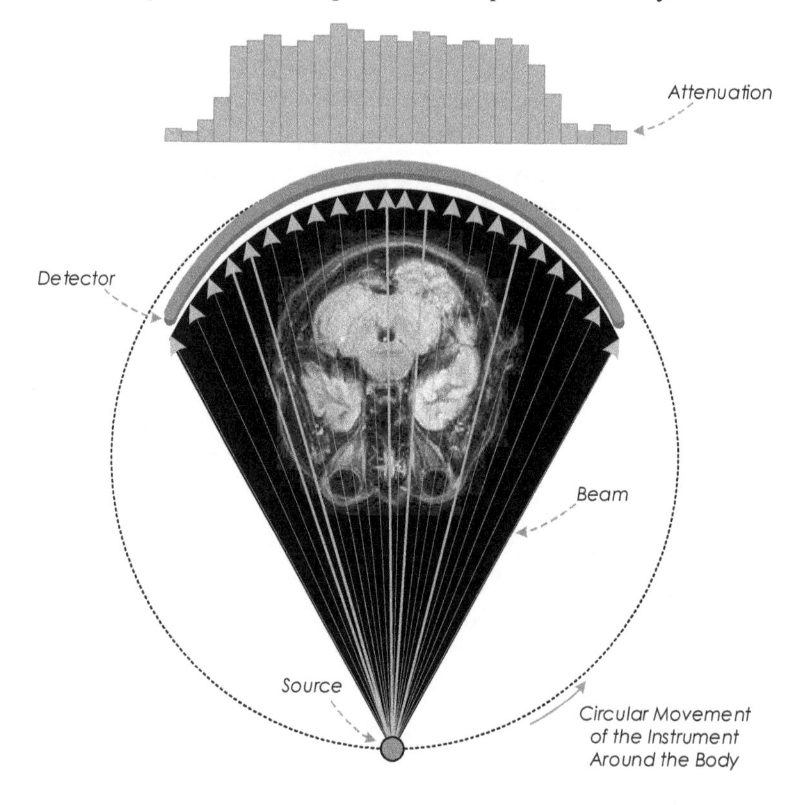

Figure 8.37 – Image attenuation of the X-ray that is detected in each detector.

Figure 8.38 shows an example of the radiation intensity measured in three different views in the case of a hypothetical structure constituted by a small sphere of a material within a cube of another material.

With the data obtained for the different detectors and the source in different positions, it is possible to reconstruct an image containing the value of density of the material traversed by the X-rays.

The image reconstruction is made by taking the intensity values measured in the detectors and "spreading them" through the image along the path traversed by the X-rays in each view (Figure 8.39).

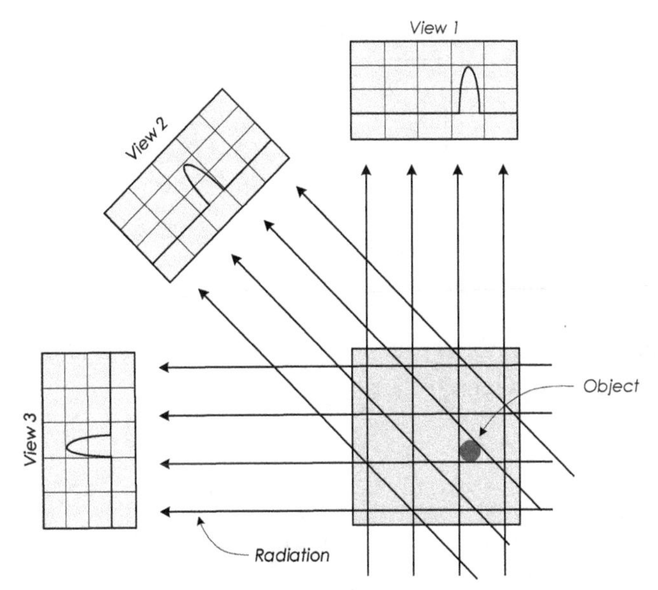

Figure 8.38 – Example of the radiation intensity measured in three different views in the case of a hypothetical structure constituted by a small sphere of a material within a cube of other material.

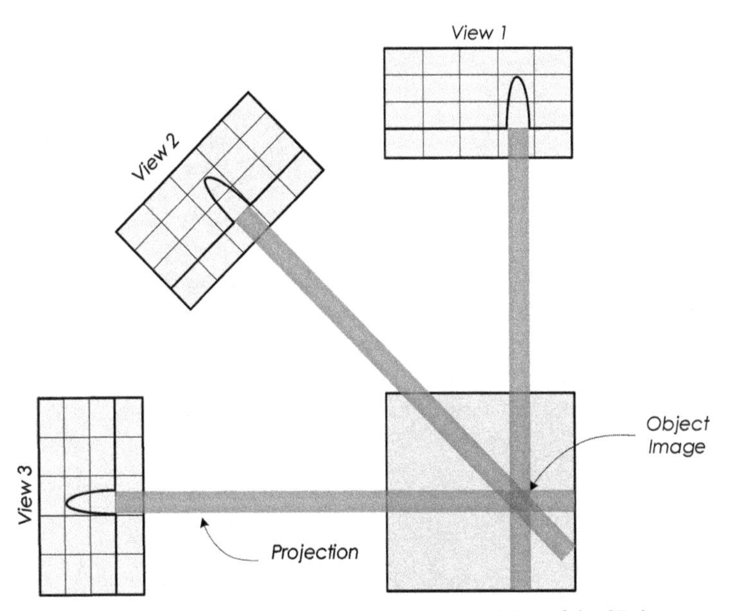

Figure 8.39 – Example of the reconstruction algorithm of the 2D image.

Using three views, we get something like Figure 8.40.

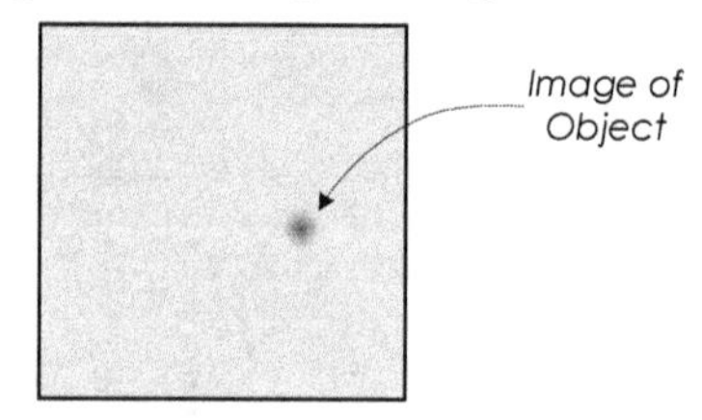

Figure 8.40 – Example of the reconstruction algorithm using three views.

In order to improve the resolution of the final image, data measured in each view is filtered. Figure 8.41 illustrates the result.

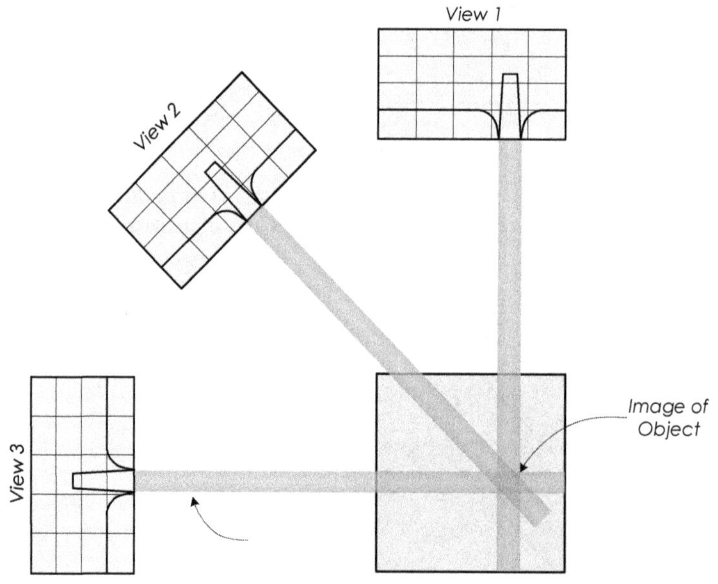

Figure 8.41 – Example of the reconstruction algorithm of the 2D image where filtering is used to improve resolution.

The use of many views allows for a more accurate object reconstruction (Figure 8.42).

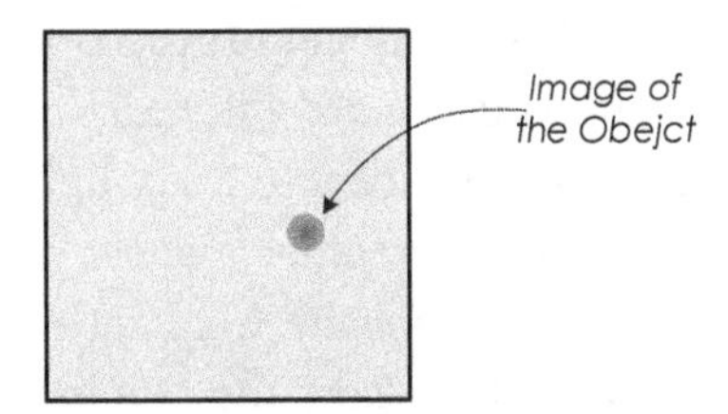

Figure 8.42 – Exemplo do resultado da reconstrução usando muitas vistas e filtragem.

Figure 8.43 shows an actual example of an image obtained from a slice of the human body using computed tomography.

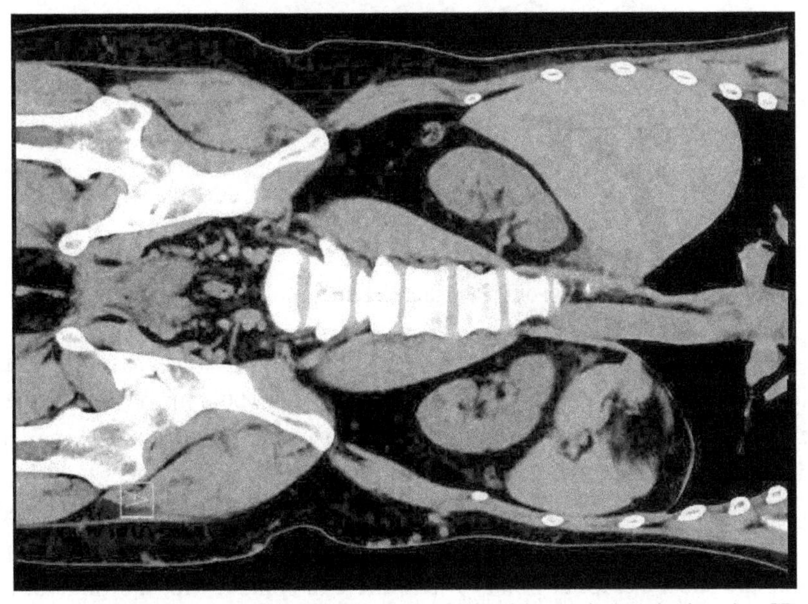

Figure 8.43 – Example of a set of 2D images obtained from a human brain using CT.

8.16 Multi-touch screen

The multi-touch screens allow not only to display an image but also to detect the position of contact with one or more fingers to increase the possibilities in terms of the user interface.

One of the most successful techniques in performing multi-touch screens is the *Frustrated Total Internal Reflection* (FTIR). When the light meets the interface with a material having a lower refractive index, such as, for example, in the case of the glass to air interface, and the incidence on that interface is made with an angle greater than a certain critical angle, this light is totally reflected (*total internal reflection*). This principle is used, for example, in optical fibers to carry light over great distances with minimal losses. If, however, another material is present on the interface, such as the finger of a person, the total reflection ceases to happen (is frustrated), causing light to escape the conducting material (the glass in this case), as shown in Figure 8.44.

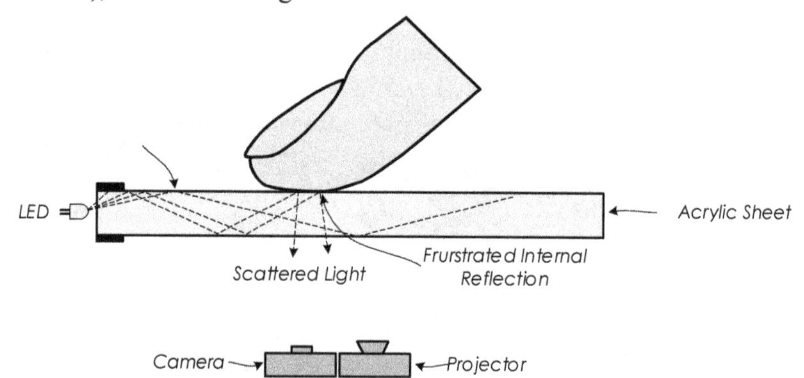

Figure 8.44 – Illustration of the Frustrated Total Internal Reflection principle.

Jefferson Han built a prototype of a multi-touch screen based on this principle using an acrylic sheet[2] (406 mm × 305 mm × 6.4 mm) whose edges were polished. A set of high brightness LEDs (with a power of 460 mW at 880 nm) were placed along the edges of the acrylic sheet to inject light into the material. A video camera with an optical bandpass filter was positioned to film the acrylic sheet. The light injected into the acrylic sheet remains inside the sheet except at points where there is contact with the fingers. In those points, the light is refracted, reaching the camera.

With the use of image processing algorithms, it is possible to detect the position of contact of the fingers with the acrylic sheet. The video image is captured with a resolution of 640 × 480, which allows an accuracy value of 1 mm^2.

An advantage of this technique is that it is not required to push against the surface. A light touch is enough. It is also a low-cost technique with which one can build a large screen that may not even be flat. A

[2]Common glass is not suitable due to attenuation of light transmission of the material.

disadvantage is a need for a large space behind the acrylic sheet for placement of the video camera to be able to shoot the entire surface. Another disadvantage is that it does not work in environments with lots of light, given that this interferes with the refracted light through the fingers.

8.17 Global Positioning System (GPS)

8.17.1 *Introduction*

The global positioning system or GPS is a system that allows the rapid determination of the position of any point on the surface of the planet. It was developed and built by the Department of Defense of the United States of America so that their submarines could be able to easily determine their position when they come to the surface. The initial aim was to allow intercontinental ballistic missiles, launched from submarines, to be able to hit with precision the enemies' missile silos. This accuracy was closely related to the precision with which the initial position of missiles was known.

The GPS system works based on a constellation of 24 satellites placed in space around the Earth. These satellites emit radio frequency signals that serve as a reference for determining the position of any point on the Earth's surface, so analogous to what was done at the time of the Discoveries when stars were used to navigation of vessels.

The power incident on the terrestrial surface (irradiance) of GPS signals from the satellites is very small, on the order of 3×10^{-14} W / m^2 [16]. This requires that sophisticated techniques have been implemented to enable the system to function.

The current technology in GPS receivers allows them to be miniaturized, making them quite practical and economical to use. This enabled GPS systems to be used in many applications from cars, boats, planes, construction equipment, cinematographic equipment, farm equipment, and even mobile phones, computers, and laptops.

There are currently (2009) other systems of the same type being developed by other countries, including:

- The Galileo system was built by the European Union;
- The COMPASS system was built by China;
- The GLONASS system was built by Russia.

8.17.2 *Satellites*

The 24 satellites are spread over six planes with four satellites each. These orbital planes are centered on Earth but are not rotating relative to the distant stars. The planes have an inclination of 55° to the equator and are separated by 60° from each other. In almost all points of the earth's surface, there are always at least six satellites visible. The satellites are at an altitude of 20200 km and have an orbital radius of 26600 km. The orbits are traversed two times per day.

On Earth, there are several stations that monitor the operation and position of the satellites (Figure 8.45).

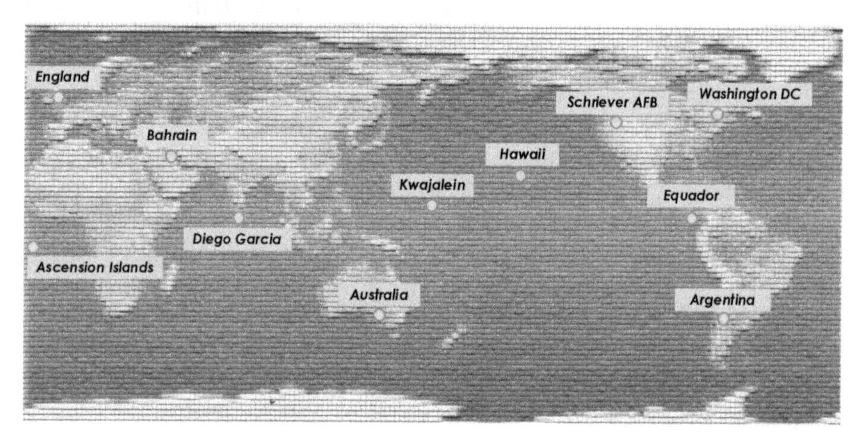

Figure 8.45 – Location of ground stations used in the GPS system.

8.17.3 *Trilateration*

The position of a given point on the Earth's surface is determined from the position of the satellites and the distances of these to the receiver using trilateration. This distance, in turn, is determined from the time of flight of the radio signals emitted by the satellites and received by the GPS knowing the propagation speed (considering the variations due to the atmosphere). The measurement of time is based on atomic clocks.

Trilateration, in graphical terms, can be explained as follows. Imagine that we measure the distance to a given satellite, and it appears that this distance is 11,000 miles. This reduces the possible positions for the receiver within the surface of a sphere, as illustrated in Figure 8.46(a).

Imagine now that you also know the distance from the receiver to a second satellite. The set of possible points for the position of the receiver is reduced to a circle formed by the intersection of two spheres, as illustrated in Figure 8.46(b). Joining the distance information of a third satellite, we decrease the set of possible points to two (Figure 8.46(c)).

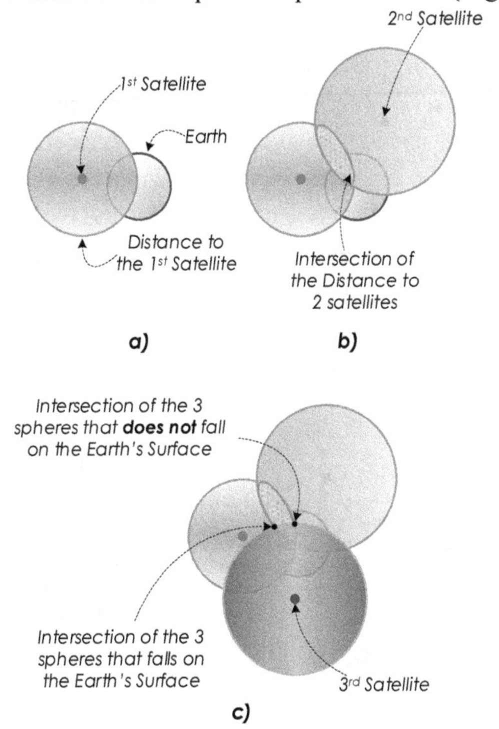

Figure 8.46 – Set of positions that are at the same distance from a satellite.

The implementation of a fourth measurement allows discarding one of these points. However, this is usually not necessary in the case of GPS, for one of two possible points is usually far from the Earth's surface. Anyway, in practice, a fourth measure is used to help correct the measurement of the time of flight of the signals emitted by satellites and received by the GPS receiver, as we will see.

8.17.4 *Measure the Propagation Time of Signals*

Electromagnetic waves are used to measure the distance between the receiver and the various satellites. By knowing the propagation speed and time, it is possible to determine the distance using

$$d = v \cdot t. \tag{8.19}$$

The time of flight in the case of a satellite receiver at 20200 km is approximately 60 ms. To measure this time, both the satellite and the receiver generate the same signal starting at the same time (Figure 8.47, left). At the receiver, the internally generated signal and the signal received from the satellite have an offset because the latter had to travel the distance between the satellite and receiver at a finite speed (Figure 8.47, center). The receiver delays the signal generated internally to be in phase with the received signal (Figure 8.47, right). This delay is the measurement of the time of flight of the signal coming from the satellite.

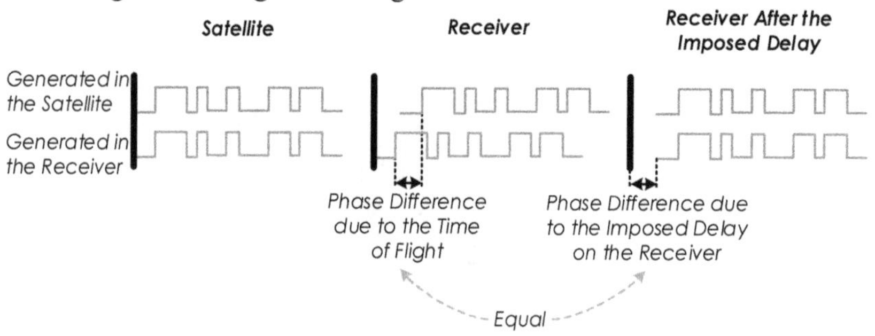

Figure 8.47 – Illustration of how the time of flight of the signal between the satellite and the GPS receiver is measured.

The signals used are created from pseudorandom sequences of "0" and "1". Each satellite uses a different sequence so that the receiver can distinguish the origin of the received signal. The pseudorandom sequence is generated mathematically so that the same sequence can be created in both the satellite and the receivers. The use of a pseudo-random sequence has the advantage that it seems like noise and therefore is little affected by other signals.

For the measurement of the time of flight to work as described, it is necessary that both the signal generated at the satellite and the signal generated in the receiver are in perfect sync. A difference of 1 ms, for example, causes an error of 300 km in the distance estimation considering the propagation speed of electromagnetic waves. The satellite achieves high accuracy at the time of signal generation using atomic clocks. On the receiver side, on the other hand, it is not practical the use atomic clocks due to their cost and size. It is possible to obviate this inconvenience by using information from a fourth satellite. If the clock of the satellites and the receiver are not synchronized, then the estimated position obtained with each set of three satellites is not the same. In other words, the four imaginary spheres centered at each of the four satellites with the radius given by the estimated distance do not intersect at a single point. As a sync error between the satellites and the receiver affects all measures in the same way, it is possible to determine a single correction factor for the receiver clock so that the four spheres intersect with each other at a single point.

8.17.5 *Carrier-Based Synchronization*

The pseudo-random sequence with a rate of 1 Mbps is combined with a navigation message with a rate of 50 bps and used to modulate in amplitude a carrier (Figure 8.48).

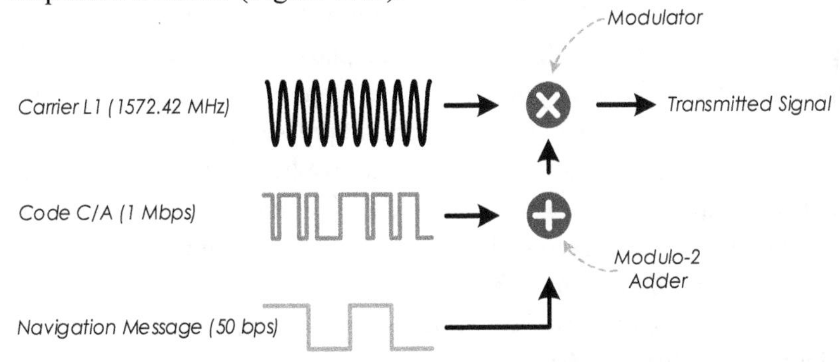

Figure 8.48 – Illustration of the generated signal transmitted by a GPS satellite system by combining the carrier, the pseudo-random sequence, and the navigation message.

The synchronization between the received signal and the internally generated signal in the receiver is typically carried out considering the pseudo-random sequence. Errors less than 2% of the synchronous bit width (1 μs) representing an error of fewer than 6 m in the distance is achieved.

Some receivers use the carrier to improve the timing (Figure 8.49). Since it has a much higher frequency (1.57 GHz), it is possible to get a smaller sync error. An error of 1% in the period of the carrier corresponds to 6.3 ps which leads to a position error of 1.9 mm.

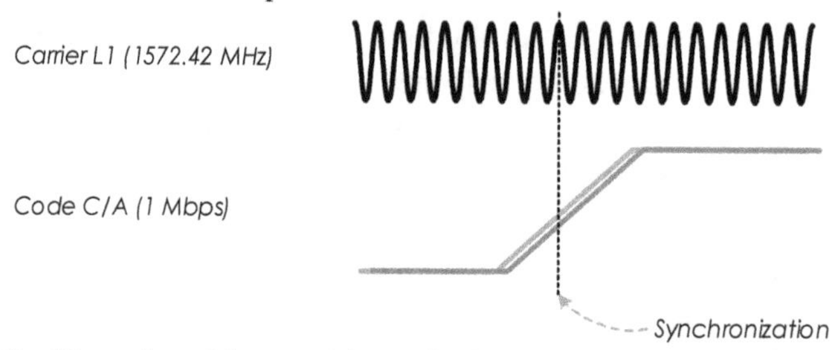

Figure 8.49 – Illustration of the use of the carrier for synchronization between the received signal and the signal generated in the receiver.

8.17.6 *Determination of the Propagation Velocity*

To determine the distance, you need to know, besides the time of flight of the signals, the speed of propagation which in theory would be the speed of propagation of light in vacuum (3×10^8 m/s). This speed is not, however, constant throughout the signal path or the same at all points of the globe. This is due to the atmosphere, particularly due to charged particles that are in the ionosphere and water vapor in the troposphere, in particular their variation with temperature and pressure.

One way to avoid errors caused by the atmosphere is to determine the value that the speed of the waves has in a typical day and use a look-up table that considers the composition of the atmosphere and the incidence angle of the signal in the atmosphere.

Another more sophisticated way (not used in all receivers) is the use of two signals with different frequencies. This is based on the principle that signals with lower frequency get more delayed than the higher frequency ones when going through the atmosphere. All satellites transmit two carriers with timing signals designated L1 and L2 and whose frequencies are 1.57542 GHz and 1.2276 GHz, respectively.

The estimation of such a large distance between a satellite and a point on the earth's surface (more than 20,000 kilometers) with an accuracy of a few centimeters corresponds to a relative error of about ten parts per billion. This is possible using pseudo-random codes, a very accurate clock, and the combination of the measured distance of at least four satellites.

8.17.7 *Satellites Position*

To perform trilateration, however, takes more than the distance between each satellite and the receiver. You also need to know exactly the position of each satellite in space. This is achieved by initially placing the satellites in the desired position and mathematically predicting their trajectory. This is achievable in practice because the satellites are at a distance from Earth where there is no atmosphere and therefore are not affected by it. There are, however, still gaps in the position of the satellites caused by the gravitational attraction of the Sun and Moon and the solar radiation pressure on the satellite. These deviations, called ephemeris errors, are measured by ground radar stations which record in almanacs and send them to each satellite to be transmitted by them along with the signal used to measure the time of flight.

8.17.8 *Theory of Relativity*

Another factor that the system must consider concerns the relative time dilation between two bodies that move with each other at a certain speed (special relativity). In this case, the satellites travel at a speed of about 4 km/s relative to the Earth's surface (an orbit is completed in 11 hours and 58 minutes). The watch satellites are delayed, for this reason, about 7 s per day (compared to receivers terrestrial) [17].

The theory of general relativity has an even greater impact on the GPS system. The curvature of space-time due to the Earth's mass is smaller the farther we are from the center of the planet. Therefore, satellites are in an area with less curvature than the area where the receivers are (land surface). From the point of view of the recipients, satellite clocks appear to be running faster. The accumulated difference in one day reaches 45 s [17].

If these effects were not considered, the position of the receivers would have an accumulated error of 10 km per day. The effect of general relativity is offset by making the clocks in the satellites advance at a slower pace than the receivers. This calibration is performed as soon as the satellites are placed in orbit. The effects of special relativity are compensated by the receivers themselves.

8.17.9 *Differential System*

An improvement of the GPS system is called a differential GPS. This system uses GPS receivers fixed on the earth's surface. Given that its position is well known and remains constant, you can use the information obtained to correct the value of the propagation speed used. This is done for all visible satellites by the fixed receiver. These velocity values are then sent by radio to mobile receivers in proximity.

This makes sense because if a mobile receiver is at a relatively close distance to a fixed receiver, the section of the atmosphere that both signals cross is very similar.

The differential GPS can eliminate all sources of error that are common to the fixed receiver and the mobile receiver. The sources of error that cannot be fixed are due to multi-path routes with the receivers and the errors introduced by the receivers.

This system has, however, the disadvantage of requiring many terrestrial stations to cover a large area. On the other hand, it requires a second communication system between the stations and the receivers, which leads to a higher cost and an increased likelihood of failure.

8.17.10 *Wide Area Augmentation System*

WAAS system (Wide Area Augmentation System) was developed by the United States to improve the GPS system. This system is capable of an accuracy of better than 3 m for 95% of the time.

The system consists of 25 ground stations that monitor the satellites of the GPS system. It also has two major terrestrial stations located on each U.S. coast that receive information from these stations and build the correction messages that include satellite orbits, clocks drift and propagation delays caused by the ionosphere and the atmosphere. The correction messages are transmitted to the receivers using two geostationary satellites instead of using a ground connection, as in the case of differential GPS.

This system has some disadvantages, namely:

- The satellites used to transmit information to receivers are geostationary, implying that they are less than 10° above the horizon for latitudes above 71.4°.
- The reduced number of ground stations limits the corrections of the ionosphere and atmosphere that can be done.
- To be used on airplanes, a certificated receiver is needed, which can cost more than €10,000.
- Their accuracy is not enough to be used to land aircraft autonomously.

Europe also developed a system of the same type designated **European Geostationary Navigation Overlay Service** (EGNOS) is operational since 2005 and is capable of accuracy better than 2 m in 99% of the time.

8.17.11 *Specifications*

GPS modules usually have a positioning accuracy, in normal mode, less than 3 m and less than 1 m if used in differential mode. Note the different operating conditions of the receiver:

- **Power-off start** — Time since power is fed to the module (depends on the time that power takes to reach the nominal value and the location where the module is.
- **Autonomous / cold start** — The receiver is powered on but has no estimate of the time, date, position and does not have a recent almanac.
- ***Warm start*** — The receiver has estimates of time, date, location and has a recent almanac.
- ***Hot start*** — The receiver has estimates of time, date, location and has a recent almanac and ephemeris.
- ***Obscuration recovery*** — The receiver clock did not stop, and so he has a precise estimation of time (to μs).

8.18 Questions

1. Present and compute the relevant values of an electronic circuit containing a 5 V DC source to power a led whose specifications are

presented in such a way that it produces the maximum light intensity continuously.

Table 8.13 – LED specifications.

PARAMETER	TEST CONDITION	SYM.	MIN.	TYP.	MAX.	UNIT
Forward voltage	$I_F = 50$ mA	V_F		2.1	2.7	V
Reverse voltage	$I_R = 10$ μA	V_R	5			V
DC forward current	$T_{amb} \leq 85$°C	I_F		50		mA
Surge forward current	$t_p \leq 10$ μs	I_{FSM}		1		A
Luminous intensity	$I_F = 50$ mA	I_V	4300	11000		mcd
Dominant wavelength	$I_F = 50$ mA	λ_d	611	616	622	nm
Spectral bandwidth at 50% $I_{rel\ max}$	$I_F = 50$ mA	$\Delta\lambda$		18		nm
Angle of half sensitivity	$I_F = 50$ mA	φ		±9		°
Peak wavelength	$I_F = 50$ mA	λ_p		622		nm

2. Present and compute the relevant values of an electronic circuit containing a transconductance amplifier connected to an infrared photodiode (BPW41N from Vishay Semiconductors) operating in photoconductive mode and whose characteristics are shown below. The output voltage is intended to be less than 1 V in the absence of light and greater than 4 V with an irradiance of 0.1 mW/cm^2. Consider that the operational amplifier behaves ideally.

Table 8.14 – Photodiode specifications.

PARAMETER	TEST CONDITION	SYM.	MIN.	TYP.	MAX.	UNIT
Forward voltage	$I_F = 50$ mA	V_F		1.0	1.3	V
Breakdown voltage	$I_R = 100$ μA, E = 0	$V_{(BR)}$	60			V
Reverse dark current	$V_R = 20$ V, E = 0	I_{ro}		2	30	nA
Diode capacitance	$V_R = 0$ V, f = 1 MHz, E = 0	C_D		70		pF
	$V_R = 3$ V, f = 1 MHz, E = 0	C_D		25		pF
Open circuit voltage	$E_e = 1$ mW/cm^2, $\lambda = 950$ nm	V_o		350		mV
Short circuit current	$E_e = 1$ mW/cm^2, $\lambda = 950$ nm	I_K		38		μA
Reverse light current	$E_e = 1$ mW/cm^2, $\lambda = 950$ nm, $V_R = 5$ V	I_{ra}	43	45		μA

Absolute spectral sensitivity	$V_R = 5$ V, $\lambda = 950$ nm	$s(\lambda)$		0.55		A/W
Angle of half sensitivity		η		±65		°
Peak wavelength		λ_p		950		nm
Spectral bandwidth		$\lambda_{0,1}$		870 to 1050		nm

3. In computed tomography, how are the different types of material within the human body distinguished?

4. In computed tomography, radiation passes through the entire body in its path from the emitter to the receiver. The intensity of the received signal, therefore, depends on the material as a whole, where the radiation has passed through the body. How, then, can information be obtained regarding a single point within the body?

5. How does the GPS system work?

6. What factors limit the accuracy of the GPS system?

7. How does the reverse saturation current of a photodiode vary with illuminance?

   a. Linearly.
   b. Sinusoidally.
   c. Quadratically.
   d. The reverse saturation current does not depend on the illuminance.
   e. None of the above.

8. An inversely polarized photodiode is said to work in which mode?

   a. Photovoltaic.
   b. Photoconductive.
   c. Photoelectric.
   d. Electromagnetic.
   e. None of the above.

9. What is the main advantage of an optical pyrometer to measure the temperature?

   a. Very good accuracy.
   b. It allows measuring very high temperatures.
   c. Allows you to make quick measurements.

d. It is very cheap.
e. None of the above.

10. How does the optical mouse of personal computers work?

Chapter 9

Devices Based on Chemical Phenomena

9.1 Introduction

Chemical sensors are used to detect the presence of certain chemical compounds or elements and to measure their concentration. They have applications in various areas such as environmental pollution monitoring, detection of explosives, medicine, and industry.

In industry, for example, chemical sensors are used to monitor the process of manufacturing plastics and metals, for monitoring the air to warn workers to the presence of hazardous substances or in terms of health and explosion risk as electronic noses to determine the degradation of foods, controlling the distribution of pesticides and characterizing different types of beverages. In the field of medicine, chemical sensors are used to measure the amount of oxygen and other gases in the blood and the lungs to diagnose, for example, digestive problems or high levels of alcohol.

A major difficulty in the development of chemical sensors has to do with the fact that the chemical reactions change the sensor, often irreversibly. This type of sensor is typically exposed to a variety of chemicals that can damage it, as is the case with acids, or change it because these compounds are absorbed by the sensor, ultimately leading to a change of its operating conditions.

Chemical sensors can be classified as direct or indirect. In the first case, the sensors use a chemical reaction that directly affects the given electrical characteristic such as resistance (conductometric sensors), voltage (potentiometric sensors), or current (amperometric sensors). In the second case, the sensors use chemical phenomena that do not directly affect an electric characteristic but produce another change as the mass, temperature, or shape, which is then transformed into an electric parameter.

The simplest sensors involve a chemical reaction of the analyte with the sensing element to produce a measurable effect. This reaction is not reversible in most cases, causing a permanent change in the sensor, which leads to an unstable operation and limits its useful life. It also happens that sometimes the chemical reaction consumes the material used by the sensor to cause the chemical reaction. There may also be undesirable reactions with other elements or substances that interfere with the desired reaction.

9.2 Reduction and Oxidation Reactions

The molecules, atoms, and ions have electrons. These electrons can be released or captured. When electrons are released, it is said that the substance has undergone **oxidation** and when electrons are captured, it is said that the substance has been **reduced**.

As an example, see the reaction between one molecule of hydrogen (H_2) and one of fluorine (F_2). The product of this reaction is two molecules of hydrogen fluoride (HF).

$$H_2 + F_2 \rightarrow 2HF. \tag{9.1}$$

This reaction can be written as the result of two half-reactions, namely the oxidation of hydrogen

$$H_2 \rightarrow 2H^+ + 2e^-, \tag{9.2}$$

wherein each hydrogen atom loses an electron, resulting in a positive hydrogen ion, and the reduction of fluorine,

$$F_2 + 2e^- \rightarrow 2F^-, \tag{9.3}$$

in which each fluorine atom captures one electron giving rise to two negative fluorine ions. The two electrons released in the oxidation of hydrogen are therefore captured by fluorine atoms: ions and hydrogen fluoride from these reactions combined to form two molecules of hydrogen fluoride.

The reaction described (9.1) is, therefore, a reduction/oxidation reaction, also known as a **redox reaction**.

The ease of release and capture of electrons differs from substance to substance. To quantify this capacity, we use a quantity called the

Reduction Potential — the greater the reduction potential, the greater the tendency to capture electrons.

In an aqueous solution, for example, the reduction potential is the tendency to gain or lose electrons when it is subjected to change by the introduction of a new chemical species. A solution with a higher (more positive) reduction potential than that of the introduced substance will tend to capture electrons of this substance, i.e., to be reduced by the oxidation of the new substance. A solution with a lower reduction potential (more negative) will tend to lose electrons to the new substance, i.e., it will tend to be oxidized by the reduction of the new substance.

The reduction potential is measured relative to a standard hydrogen electrode because of the difficulty in measuring absolute reduction potential. In this electrode, the following reduction reaction occurs:

$$2H^{+}_{(aq)} + 2e^{-} \rightarrow 2H_{2(g)}, \tag{9.4}$$

where (aq) is the liquid state, and (g) the gaseous state. The absolute value of the reduction potential of this electrode is 4.44 ± 0.02 V at 25°C.

In general, however, the reduction potential of the hydrogen electrode is regarded as 0, and the reduction potential of other substances is referred to as this one. Table 9.1 gives some values of the standard reduction potential that is obtained at equilibrium when there is no current flow and at a concentration of 1 mol/dm^3.

Table 9.1 – Standard reduction potential of some half-reactions.

HALF-REACTION	STANDARD REDUCTION POTENTIAL (V)
$\frac{3}{2}N_{2(g)} + H^+ + e^- \rightarrow NH_{3(aq)}$	−3.09
$Zn^{2+} + 2e^- \rightarrow Zn_{(s)}$	−0.76
$2H^+ + 2e^- \rightarrow H_{2(g)}$	0
$Cu^{2+} + 2e^- \rightarrow Cu_{(s)}$	0.34
$F_{2(g)} + 2H^+ + 2e^- \rightarrow 2HF_{(aq)}$	3.05

The value of the reduction potential depends ultimately on the number of electrons that the atoms involved have and how their orbitals are filled.

9.3 Galvanic Cell

The reactions of reduction and oxidation are used to make batteries and sensors. Either one or the other is based on the galvanic cell, which is formed by two half-cells using different materials and is connected to form an electric circuit, as shown in Figure 9.1.

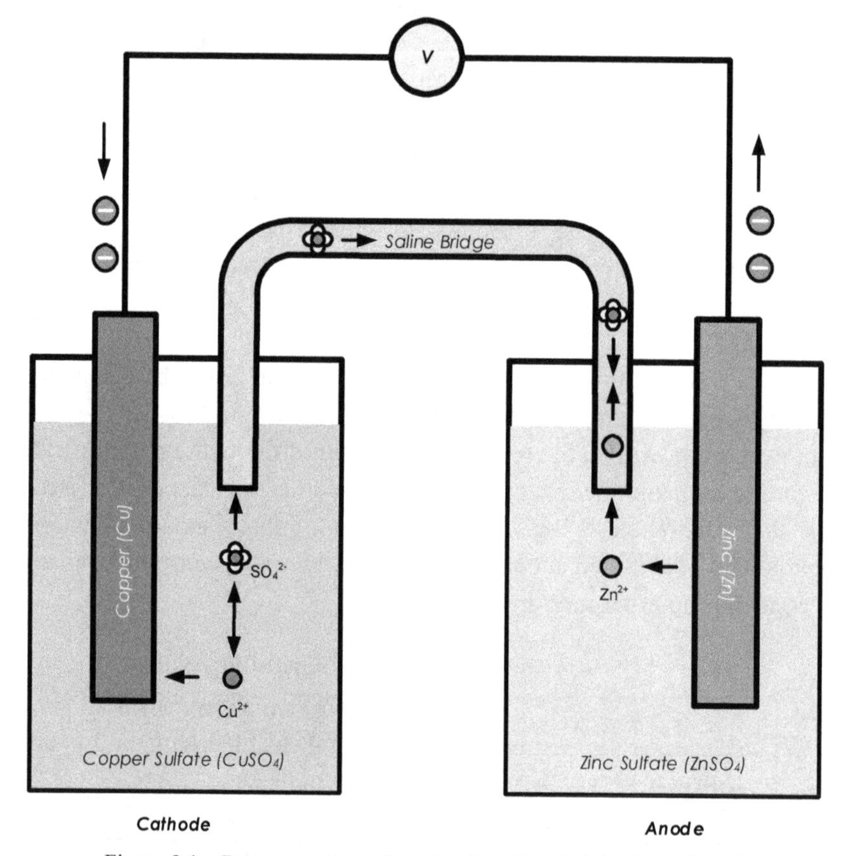

Figure 9.1 – Representation of a galvanic cell called the Daniell Cell.

The use of different materials with different values of reduction potential leads to the emergence of a voltage at the cell terminals.

Figure 9.1 shows a specific example of a Galvanic cell called Daniell cell since it uses copper (Cu) and zinc (Zn) electrodes immersed in aqueous solutions of copper sulfate ($CuSO_4$) and zinc sulfate ($ZnSO_4$), respectively. In each half-cell, a chemical reaction happens. The zinc atoms have a higher tendency to go into the solution than the copper atoms. Thus, the zinc atoms in the anode oxidize, releasing two electrons and yielding positive zinc ions that move to the solution of zinc sulfate.

$$Zn_{(s)} \rightarrow Zn^{2+}_{(aq)} + 2e^{-}. \tag{9.5}$$

On the other hand, the two electrons released travel through the conductor connecting the anode and cathode, where they find positive copper ions, which are reduced to copper by uptake of two electrons.

$$Cu^{2+}_{(aq)} + 2e^{-} \rightarrow Cu_{(s)}. \tag{9.6}$$

This copper eventually accumulates in the copper electrode saturating it. The molecules of copper sulfate in the cathode divide into positive copper ions (which will be reduced in the solid copper) and negative sulfate ions (SO_4^{2-}). Those sulfate ions go through the salt bridge, recombining with positive ions of zinc released by the solid zinc (which is eroded), forming zinc sulfate.

The two half-reactions that take place in the Daniell cell are

$$Cu^{2+}_{(aq)} + Zn_{(s)} \rightarrow Cu_{(s)} + Zn^{2+}_{(aq)}\,. \tag{9.7}$$

The electromotive force of the Daniell cell can be calculated from the reduction potential of the two half-reactions. The reduction of copper, given by (9.6) has a reduction potential of 0.34 V (Table 9.1). The oxidation of zinc, given by (9.5), has an oxidation potential of 0.76 V[1] (Table 9.1). Their sum gives, therefore, 1.10 V, which is the value obtained when there is no current flowing.

This type of construction, called a galvanic cell, is used for the construction of batteries. The use of different materials gives rise to the different open voltages like 1.5 V in the case of alkaline batteries and 1.2 V in the case of nickel/cadmium batteries, for example. The connection of several batteries in series allows for higher values of voltage (4.5 V and 9 V, for example), and the parallel connection allows for higher capacity (size AA and AAA, for example).

A galvanic cell is also the basis for the construction of potentiometric sensors, as discussed below.

---

[1]The oxidation potential value is the negative of the reduction potential.

9.4 Potentiometric Sensor

The standard reduction potentials, such as those presented in Table 9.1, are set to a default situation in which all substances involved have the same concentration of 1 mol/dm^3. When the concentration is different, the reduction potential is also different and can be calculated from the Nernst equation, which specifies the galvanic potential:

$$E = E^{\theta} + \frac{R \cdot T}{n \cdot F} \ln\left(\frac{c_{ox}}{c_{red}}\right), \tag{9.8}$$

where E^{θ} is the reduction potential under standard conditions, R is the gas constant, T is the temperature, F is Faraday's constant, n is the number of transferred electrons, and c_{ox} is c_{red} are the concentrations of the oxidizing and reducing substances, respectively.

Nernst equation is used in chemical sensors to relate the concentration of a substance with the voltage one gets. It is important, however, that the current flows are as small as possible. One should therefore use a voltage measurement circuit that has a very high impedance of some GΩ at least.

Considering that the concentration of a substance is known and kept constant, Eq. (9.8) can be written to room temperature, as

$$E = E^{\theta'} \pm \frac{0.059}{n} \ln(c_x), \tag{9.9}$$

where c_x is the ion concentration of the substance to be measured. The algebraic sign depends on whether these are positive or negative ions. The voltage $E^{\theta'}$ depends on the standard reduction potential and the reference substance and, in practice, is determined experimentally.

Equation (9.9) shows that a 10-fold variation in concentration leads to a variation of 60 mV in voltage if there is an exchange of one electron ($n = 1$).

In each analyte, ions of different substances exist and contribute to the galvanic potential. In order to create a sensor for measuring the concentration of a given substance, it is necessary to use electrodes that are selective to ions. The type of selective ion electrode best known is the glass electrode.

In practice, the electrode lets some undesirable ions also pass through it, which will naturally interfere with the measurement. The galvanic potential can in these cases be expressed as

$$E = E^{\theta'} \pm \frac{0.059}{n_i} \ln\left(c_i + K_{ij} c_j^{\frac{n_i}{n_j}} \right), \tag{9.10}$$

where c_i is the concentration of the substance to be measured and c_j the concentration of the substance that interferes with the measurement. n_i and n_j represent the number of charges of the respective ions. Constant K_{ij} is called the selectivity coefficient. A value of 0.01, for example, means that the interfering substance must have a concentration 100 times higher than the substance to be measured to cause the same effect on the sensor.

9.5 Lambda Probe

The lambda probe is a widely used sensor. It has application in motor vehicles where it is used to measure the oxygen concentration of the air leaving the engine. This is important to regulate the combustion that occurs in the engine to reduce the quantity of hazardous gases emitted.

This oxygen sensor uses a solid electrolyte and operates at a temperature of about 500°C. At these temperatures, a redox reaction occurs, which involves molecular oxygen (O_2). A platinum electrode is used to measure potentiometrically the concentration of oxygen ions. The electrolyte typically used is zirconium dioxide (ZrO_2) and another platinum electrode as the reference, which is in contact with atmospheric oxygen and, therefore, with a known concentration.

A galvanic cell is formed that is constituted by four interfaces of oxygen, platinum, and zirconia (zirconium dioxide), as shown in Figure 9.2.

The platinum electrodes are constructed to be permeable to oxygen. The oxygen from the exhaust from the engine enters the platinum electrode, which is the cathode. At the interface with the electrolyte of zirconia, oxygen is reduced:

$$O_2 + 4e^- \rightarrow 2O^{2-}. \tag{9.11}$$

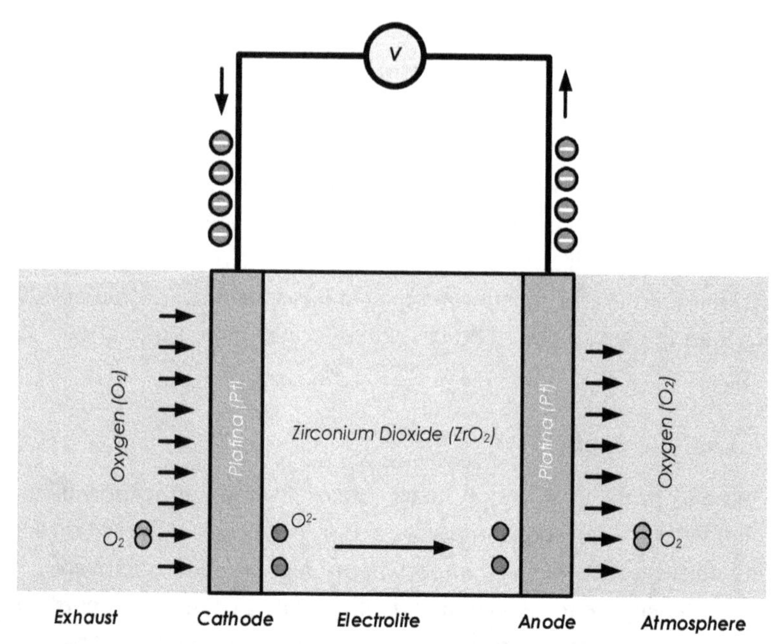

Figure 9.2 – Constitution of a galvanic cell formed in a lambda probe.

As a result, two negative oxygen ions are formed (for each molecule of oxygen) that pass through the electrolyte and reach the anode. Here the reverse reaction takes place, namely the oxidation of oxygen by the release of two electrons (for each ion),

$$2O^{2-} \rightarrow O_2 + 4e^-, \tag{9.12}$$

leading back to molecular oxygen (O_2) that is released into the atmosphere. If the platinum electrodes are connected to each other, the electrons released at the anode travel to the cathode. Otherwise, there is an accumulation of electrons at the anode and a deficiency of electrons at the cathode that gives rise to an electric field in the electrolyte, which precludes the movement of oxygen ions leading to equilibrium.

The potential difference between the electrodes is related to the concentration or partial pressure of oxygen in the exhaust (p_m) and the atmosphere (p_r) given by the Nernst equation:

$$E = E^{\theta'} + \frac{0.059}{4}\log\left(\frac{p_m(O_2)}{p_r(O_2)}\right), \tag{9.13}$$

where the factor 4 is due to four electrons that are released and caught in redox reactions.

Since the atmospheric oxygen pressure is constant, one can write

$$E = E^{\theta'} + \frac{0.059}{4}\log(p_m(O_2)), \tag{9.14}$$

that shows the logarithmic relationship between the voltage and the partial pressure of oxygen in the exhaust air.

Figure 9.3 shows the typical shape of the Lambda probe.

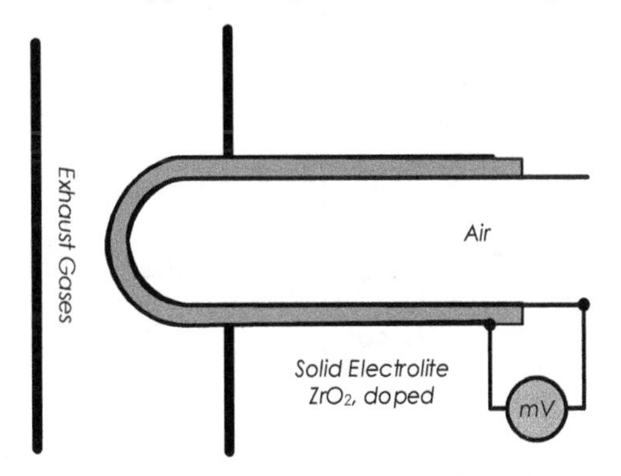

Figure 9.3 – Illustration of a lambda probe.

For lambda probes used in cars, it is common to specify a **Lambda number**. This number is defined as

$$\lambda = \frac{\left(\frac{mass_{air}}{mass_{fuel}}\right)_{actual}}{\left(\frac{mass_{air}}{mass_{fuel}}\right)_{stoichiometric}}. \tag{9.15}$$

It is, therefore, the relationship between the current value and the stoichiometric air mass and fuel ratios. The stoichiometric value is the value that leads to a complete burn of the fuel in the engine cylinders without leaving any residue. This value is about 14.7 and is related to the number of oxygen molecules in the air and the number of hydrocarbon[2] molecules in gasoline.

The value of 14.7 is approximate because gasoline contains different hydrocarbons whose presence in gasoline varies depending on the oil

[2]Hydrocarbons are molecules having hydrogen and carbon atoms.

companies. The value of the stoichiometric ratio may vary between around 13 and 15 depending on the hydrocarbon [11].

In some circumstances, it may be necessary to operate with different values of air/fuel mass ratio (Figure 9.4). The value that leads to greater motor power is 12.6 ($\lambda = 0.86$) while the value that leads to greater fuel economy is 15.4 ($\lambda = 1.05$).

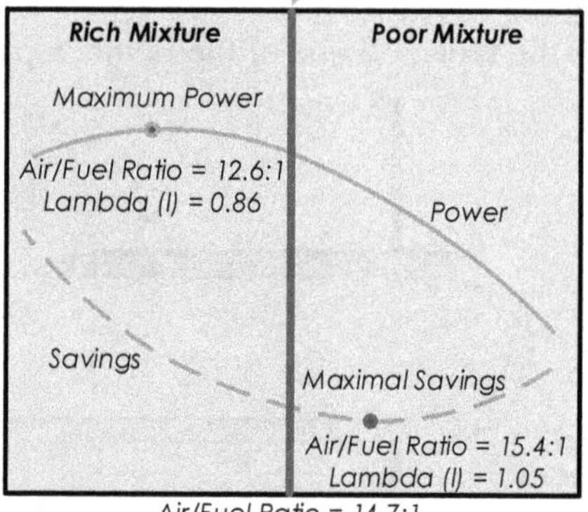

Figure 9.4 – Variation of power and fuel consumption as a function of the air/fuel mass ratio.

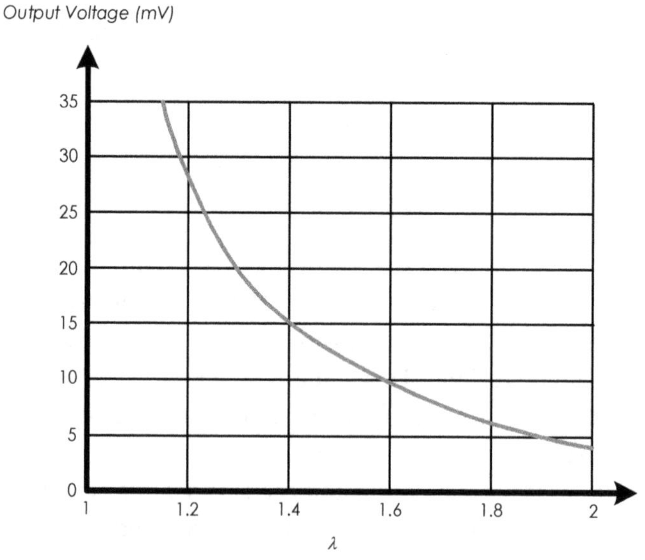

Figure 9.5 – Transfer function of the lambda probe LMS11 manufactured by Bosch.

Figure 9.5 shows the transfer function of a lambda probe as a function of oxygen concentration and the lambda number.

The sensor signal is used by the engine control unit to keep the lambda number close to 1.

9.6 Amperometric Sensor

An amperometric sensor produces a current when a potential difference is applied between two electrodes. The most common example of this type of sensor is the Clark oxygen electrode. It consists of a platinum electrode (Pt) and silver/silver chloride (Ag/AgCl) electrode immersed in an electrolyte of potassium chloride (KCl), as illustrated in Figure 9.6.

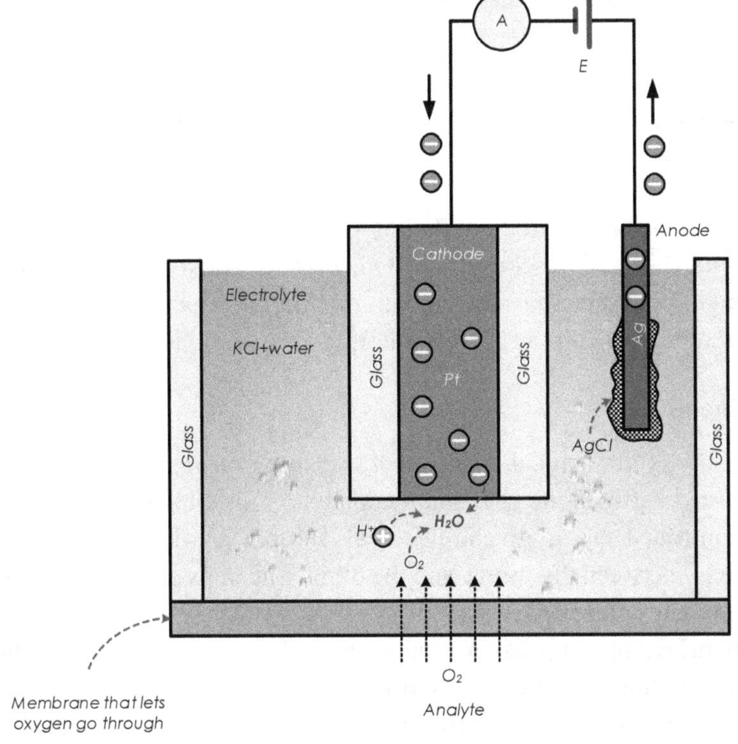

Figure 9.6 – Clark oxygen electrode.

The platinum electrode is surrounded by glass so that only a small surface is exposed to the electrolyte. The oxygen contained in the analyte flows through the permeable membrane (often Teflon) and diffuses to the electrode, where its reduction happens (capture of four electrons). These electrons are then made available by the voltage source, which has an approximate value between 600 mV and 700 mV, which makes them move from the silver/silver chloride electrode (anode) to the platinum electrode (cathode). The reduced oxygen joins with hydrogen ions from the aqueous solution to form water molecules.

In the anode, oxidation happens with the corresponding release of electrons and the creation of silver chloride that deposits on the anode and needs to be periodically removed. The current flowing in the circuit is proportional to the oxygen concentration.

The sensitivity of this sensor is defined as the relationship between current and the oxygen partial pressure.

$$S = \frac{I}{p_0}. \quad (9.16)$$

If one uses, for example, a Teflon membrane with a thickness of 35 μm and a cathode with an area of 2 × 10^6 cm^2, the sensitivity is about 10 pA/mmHg[3].

Clark's oxygen electrode makes it possible to measure the oxygen concentration in the gaseous state. Other amperometric sensors allow measuring blood glucose or the amount of chlorine ions that is important for environmental and industrial monitoring.

9.7 ChemFET

A chemFET is a chemical field-effect transistor made of selective gas layers placed between the gate and the analyte. This allows the transistor to be controlled by a given chemical substance which affects the conductivity between the source and the drain. The transistor behaves thus as a chemically controlled resistance.

A ChemFET operates like a conductance whose value depends on the gas concentration. Figure 9.7 shows the signal conditioning for a ChemFET. The bottom amplifier measures the voltage in the ChemFET, and the top amplifier measures the current that flows through the

[3]The unit of millimeters of mercury (mmHg) is defined as the pressure at the bottom of a liquid column with a height of 1 mm when the density of that liquid is 13.5951 g/cm^3 (density of mercury at 0°C) in a place where the acceleration of gravity is 9.80665 m/s^2.

ChemFET. By dividing one by the other, one can determine the resistance of the ChemFET.

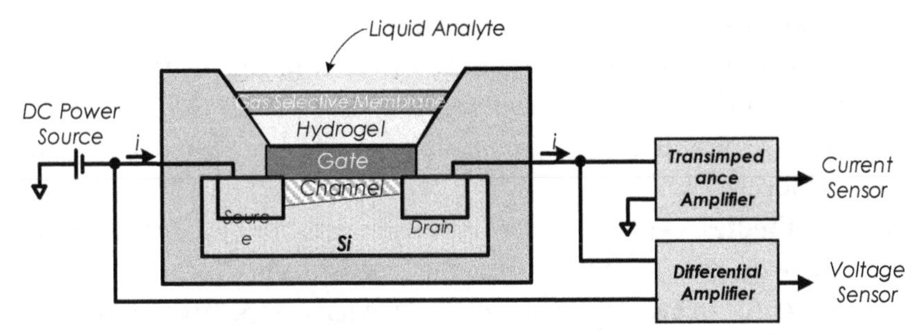

Figure 9.7 – Example of a chemical field-effect transistor (ChemFET).

The use of different materials applied to the gate allows one to make sensors for different gases such as hydrogen (H_2) in air, oxygen (O_2) in blood, some military nerve gases, ammonia (NH_3), carbon dioxide (CO_2), and explosive gases.

9.8 Biosensors

A biosensor is a bioreceptor and a transducer integrated into a single device (Figure 9.8).

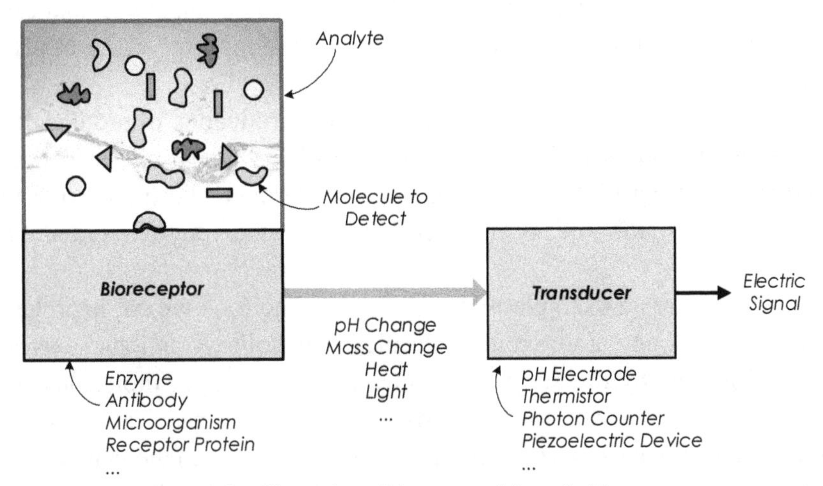

Figure 9.8 – Illustration of the composition of a biosensor.

A bioreceptor is a molecule that recognizes a certain biological material which may be, among others:

- **Enzymes** — proteins are synthesized from amino acids according to the instructions contained in DNA. Enzymes act as catalysts for biochemical reactions occurring in cells. In order to maintain its activity, it is necessary to maintain appropriate temperature and pH levels.
- **Antibodies** — proteins attach themselves to foreign substances in the body to remove them from the bloodstream.
- **Receptor proteins** — are proteins that have a specific affinity for certain compounds. Examples are hormones, taste receptors on the tongue, and receptors of smell on the nose. The receptor protein membranes open and close the channel for the transport of specific metabolites such as insulin.

The transducer converts this recognition into a measurable signal.

The main advantage of the use of biosensors is that there is no need to use reagents which makes the measurement much faster. Simply insert the biosensor in the analysis sample (blood, for example) and read the result. The fact that they also do not need a technician leads to a much lower cost.

Biosensors have numerous applications. In health, for example, they can be used to monitor the biochemical composition of inpatients and can be used even at home or in health centers for monitoring diseases like diabetes and tuberculosis.

In the industry, they are used to monitor the production of proteins for therapeutic purposes such as insulin. In the military arena, biosensors are used for detecting biological agents and in environmental monitoring to monitor the pollution of air or water due to, for example, fertilizers and herbicides.

Enzymes are an example of a bioreceptor. The food we eat is broken down in our body into small molecules (catabolism) and the energy released is used by the body in its functions. These small molecules are used to form the constituent parts of our body, such as proteins (anabolism). Each of these reactions of catabolism and anabolism, which together form metabolism, is catalyzed (to make a more rapid reaction) by a specific enzyme. This is possible because the enzyme can recognize a specific molecule which is advantageous for the realization of biosensors.

In order to make a biosensor, it is necessary that the bioreceptor is positioned in the vicinity of the transducer. This can be done by physical or chemical means. The chemical immobilization involves the creation of a covalent bond with the transducer using appropriate reagents. Note that very small amounts of bioreceptor molecules are needed.

A transducer for use in a biosensor must be able to convert the recognition made by the bioreceptor into a measurable signal. This is typically done by measuring the change that occurs because of the reaction of the bioreceptor. The enzyme glucose oxidase, for example, is used as a bioreceptor in a glucose biosensor because its function is to catalyze the reaction

$$Glicose + O_2 \rightarrow Gluonic\ Acid + H_2O_2. \quad (9.17)$$

For the measurement of glucose in an aqueous solution, three are types of transducers can be used:

- An oxygen sensor that measures the concentration of oxygen (O_2), converting it into an electrical current;
- A pH sensor that measures the gluconic acid (gluconate) that produces a voltage change;
- A sensor that measures the concentration of hydrogen peroxide (H_2O_2) and which produced an electric current.

Figure 9.9 shows the different types of biosensors that can be constructed to measure glucose.

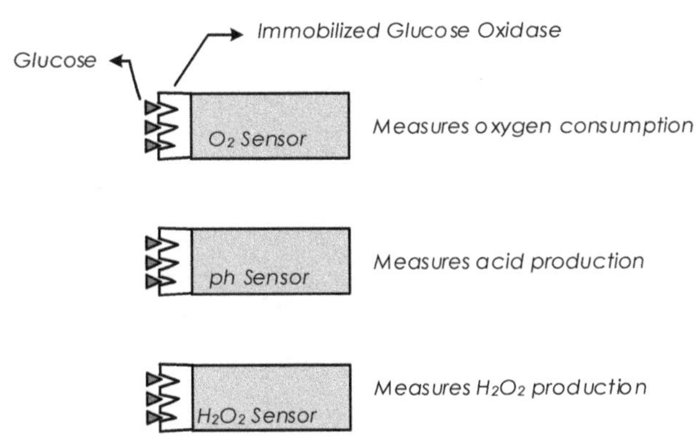

Figure 9.9 – Illustration of the different types of biosensors that can be constructed to measure glucose.

In terms of the transducers used today, three different types can be highlighted:

- Based on the amperometric measurement of hydrogen peroxide or oxygen;
- Based on the potentiometric measurement of pH or ion concentration;
- Photometric using optical fibers.

9.9 Questions

1. What are oxidation and reduction reactions, and why are they useful for making sensors?
2. Explain how a lambda probe works.
3. Explain how an amperometric sensor works.
4. Explain how a Chem-FET works.
5. What is a bioreceptor, and how is it used in a biosensor?
6. Give two examples of bioreceptors.
7. What role does the glucose oxidase enzyme play in a glucose sensor?

Chapter 10
Sensor and Actuator Networks

10.1 Introduction

Until now, we have looked at the sensor and actuator as separate units. In many applications, however, a set of sensors and actuators of the same or different types are used together, forming a network comprising several fixed and/or mobile nodes.

In wireless sensors and actuators networks, there are several aspects that should be considered both in nodes (sensors, actuators, etc.), such as:

- Size,
- Battery power,
- Environmental conditions,
- Reliability of nodes,
- Mobility of nodes

but also within the network itself, such as:

- Dynamic network topology,
- Faulty communication between nodes,
- Heterogeneity of nodes,
- Implementing in large-scale,
- Unsupervised operation.

Ultimately it is intended that the networks of sensors and actuators last as long as possible are inexpensive, robust, fault-tolerant, are reconfigurable (even automatically), and are safe in terms of confidentiality of acquired and transmitted data.

10.2 Applications

In terms of applications where sensors are used, we distinguish two cases:

- event detection;
- spatial and temporal estimation of a given quantity.

In the first case, the purpose of the network is the detection of an event such as the outbreak of a fire in a forest, an earthquake, etc. In the second case, it is intended to measure a certain area over time and a given variable, such as the concentration of a given gas in the atmosphere. With a finite number of steps located at specific points in space, we wish to map the concentration of this gas to study, for example, arises from where or how they dispersed.

Some examples of applications that benefit from the implementation of a network of sensors and actuators are [5]:

- Environmental monitoring;
- Health care;
- Location and tracking of animals;
- Entertainment;
- Logistics;
- Transportation;
- Home and office;
- Industries.

10.2.1 *Environmental Monitoring*

Environmental monitoring is important not only for the scientific community but to society in general. Monitoring can extend for years and cover interior or exterior areas, covering thousands of kilometers. The use of miniature sensors allows detailed measurement in open spaces, which would not be possible otherwise. Often the goal is security and surveillance. With sensor networks, it is possible to detect the imminent occurrence of natural disasters such as fires, floods, earthquakes, tsunamis, volcanic eruptions, and even the imminent collapse of civil engineering structures (buildings, bridges, etc.).

Due to the inherent implications, it is necessary that the networks used operate in real-time and in a particularly reliable manner. Also, considering that, in general, these networks have many sensors and must operate for long periods of time, it is vital to have the lowest possible energy consumption and as autonomously as possible operation.

The early detection of forest fires is essential for a rapid extinction of the same minimizing losses in terms of area burned, human lives lost, and homes destroyed. The networks of sensors are used to measure things like temperature, humidity, atmospheric pressure, and position. The acquired data is sent to a central station so that once the fire is detected, for example, by a sudden rise in temperature, operators can validate the alarm situation by comparing measurements made by other sensors or sending lookout teams.

Sensors may be combined with actuators, for example, in the case of domestic prevention of flooding associated with open faucets or pipe rupture. A set of sensors around the house floor can detect the presence of water, triggering an alarm and commanding the closing of the main water valve of the house.

The registration of data obtained from sensors may be important in diagnosing problems. A set of strain and vibration sensors along with a building, a bridge, or aircraft can determine the structural behavior in the presence of external phenomena such as earthquakes or strong winds. You can not only know the conditions in which they find the buildings after an earthquake or explosion as well as diagnose what went wrong in the event of the loss of a building or bridge.

Environmental monitoring may also have scientific goals, such as the Glacsweb, which is a network of sensors placed in glaciers and that measure temperature, pressure, and subglacial motion to study what happens underneath the glaciers and how they are affected by climate, in particular the phenomenon of global warming.

10.2.2 *Healthcare*

These applications include the remote monitoring of physiological data, locating doctors and patients in a hospital, and administering medications. One difficulty is the heterogeneity due to the different parameters that need to be measured. Another important aspect is that of location since it may be necessary to know exactly where a given person is when, for example, a heart attack is detected by the device that is used to regulate the heartbeat.

The idea of embedding biosensors in the human body is promising but brings important challenges in terms of safety, reliability, maintenance, and energy supply.

One of the needs in terms of health care is the continuous monitoring of patients. Your vital signs must be constantly watched, which can be done with a set of sensors that transmit information to a central computer where a nurse can follow the evolution of the health status of each patient and help them in the event that something abnormal is detected.

This type of monitoring can be extended to the patient's home, where he can stay more comfortably and still be monitored. The sensors can communicate with a mobile phone, for example, and transmit alarming conditions to a physician.

10.2.3 *Logistics*

The field of logistics has benefited especially from the development of micro and nanotechnologies. By using wireless sensor networks, it is possible to follow a product from production through to delivery to the end-user. Due to the diversity of the geographic route traveled by a product, special care must be taken regarding reliability and ease of maintenance. Moreover, to make an economically viable system for tracking products, it is necessary that the sensors and the very infrastructure of the network are low-cost.

One particular application in the area of logistics is the management of stock in a warehouse. Tens or hundreds of thousands of products can be stored awaiting distribution. The possibility of knowing in each moment and automatically, with the accuracy of a few centimeters, where a given product is, allows not only an economy in terms of human resources and enables much faster storage and retrieval of products.

The same type of advantage lies in managing a retail store such as a supermarket. A network of sensors can monitor a huge amount of information about the products such as type, variety, storage status, expiration date, quantity in stock location, and on the consumers like the time spent on each area of the store or in front of a given product or type of products you are interested. All this information can be used to examine the flow of products within the store, the efficiency of your marketing strategy, and the degree of customer satisfaction. By observing in real-time, the impact of products on consumer behavior, one can adapt the strategy used in the presentation of the products available.

10.2.4 *Transportation*

Applications in transportation aim to increase safety and comfort. One hypothesis is the in-car communication to manage traffic. The installation of sensors along the roads also allows the sending of information to drivers and vehicles about the road conditions. Ultimately the goal would be a system of fully autonomous transport.

One of the fundamental requirements of applications in this area is location accuracy and real-time operation. This makes the timing between the various sensors and components of the system are very important.

10.3 Network Organization

The information gathered by the sensors is sent, directly or indirectly, to a central node responsible for collecting such information and controlling the system via commands sent to the actuators that are part of the network. The central node, in turn, may be connected to other networks that together form a heterogeneous wide area network and are not necessarily made up exclusively of sensors and actuators. It may exist, in this extended network, things like computers, mobile phones, camcorders, and other devices with communication capabilities.

This organization, as illustrated in Figure 10.1, is the one typically used since it is simpler to implement. It suffers, however, some limitations, particularly in terms of scalability. As the number of nodes increases, the amount of information flowing in the network and that needs to be processed and/or stored by the central node becomes too high, leading to a depletion of available resources.

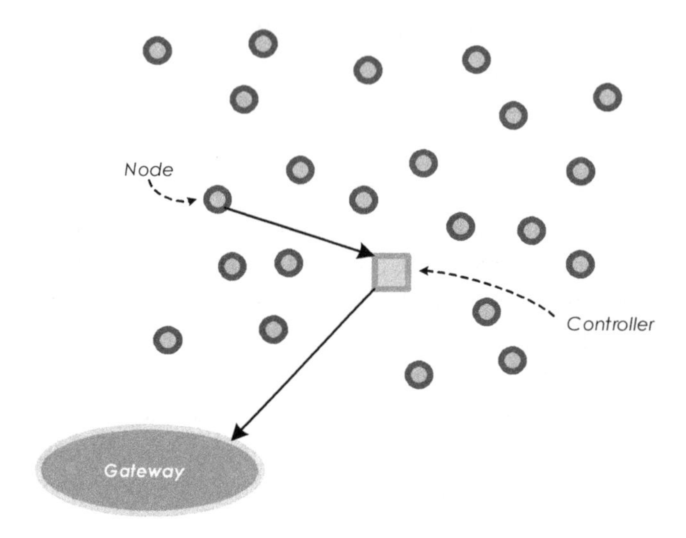

Figure 10.1 – Example of a traditional network that has a single central node (sink/controller).

Network protocols consist of various layers responsible for different aspects of communication (physical layer, medium access layer, etc.). These layers introduce an increase in the information to be transmitted between nodes because, for example, the need for addressing, transmission confirmation, and encryption. These needs lead to a decrease in the transmission rate available for the data itself. To quantify this reduction, we use a factor, α_A, so that the data rate (S_A) is related to the transmission rate as follows

$$S_A = R_b \cdot \alpha_A, \tag{10.1}$$

where α_A has a value between 0 and 1. This value is typically between 0.1 and 0.5.

Imagine a network that consists of *N* nodes and each node transmits or receives *D* bytes every T_R seconds at a rate of R_b bits per second. Assume for the moment that each node communicates directly with the central node. The transmission rate required in this case is

$$S_R = \frac{8 \cdot D \cdot N}{T_R}, \tag{10.2}$$

and must be less than the available data rate given by (10.1),

$$S_R \leq S_A. \tag{10.3}$$

This means that the number of network nodes must be less than or equal to

$$N \leq \frac{\alpha_A \cdot R_b \cdot T_R}{8 \cdot D}. \tag{10.4}$$

The higher the transmission rate between nodes, the more widely spaced the transmissions are, and the smaller the amount of data to transmit, the more nodes the network can have. For example, in the case of a transmission rate of 250 kbit/s and a transmission interval of 10 ms, a factor α_A of 0.1 and the transmission of 3 bytes at a time lead to a maximum number of nodes of 10.

Equation (10.4) assumes that each node communicates directly with the central node. If the distance between nodes and the central node is too large, you can make a communication indirectly using various jumps. If the average number of hops is h_m is then the network capacity is reduced by this value, that is, the maximum number of nodes becomes

$$N \leq \frac{\alpha_A \cdot R_b \cdot T_R}{8 \cdot D \cdot h_m}. \tag{10.5}$$

One way to increase network capacity is to use multiple receiving nodes, as illustrated in Figure 10.2.

Each node can communicate with one of the available receivers. The network is thus easily extended by adding more sensors and more receiving nodes. However, the network complexity increases with the existence of multiple receiving nodes, which can be completely isolated from each other or may form an independent communication network.

In terms of network capacity, if each of the N_s nodes can be connected to a number of sensors given by (10.5) then, the total number of nodes in the network is

$$N \leq \frac{\alpha_A \cdot R_b \cdot T_R \cdot N_s}{8 \cdot D \cdot h_m}. \tag{10.6}$$

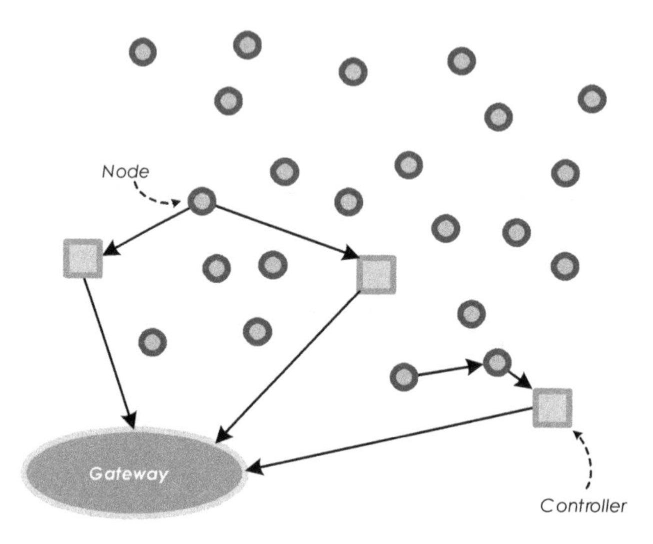

Figure 10.2 – Example of a network of sensors with multiple receiving nodes (sink/controller).

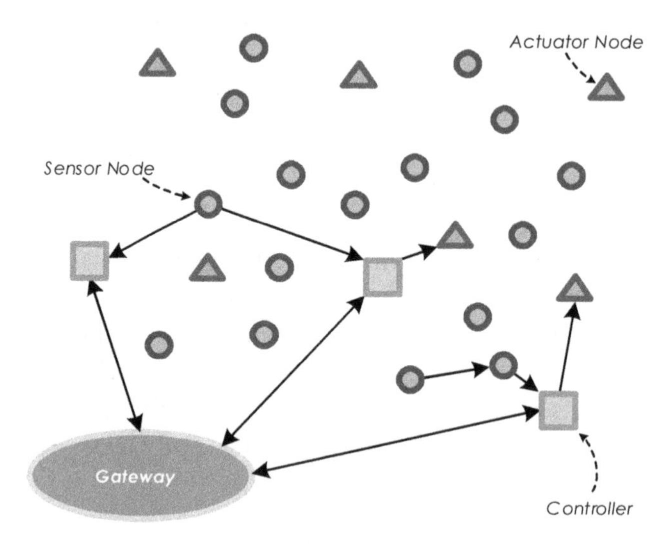

Figure 10.3 – Example of a network of sensors and actuators with multiple receiving nodes/controllers (sink/controller).

A network can, in addition to sensors, be composed also of actuators (Figure 10.3) which is not a trivial extension of a sensor network. The flow of information must be two-way, and there must be many-to-one communication capability for when the sensors provide data and one for when it is necessary to address many actuators or even one-to-one when it is necessary to control a specific actuator.

Networks of sensors and actuators can contain hundreds or even thousands of nodes which brings difficulties in terms of the medium access protocol (MAC layer of the OSI model) and in accordance with requirements that the network must satisfy regarding the ease of adding new nodes and the ability to self-organize and self-repair when some nodes fail.

10.4 Energy

In terms of the complexity of the nodes, the sensors are the simplest and often the most numerous. For this reason, it is particularly important that their cost be as low as possible. Another important requirement for the nodes of a network is to be powered by batteries to facilitate the installation of the network, eliminating the need to use power cords. This requirement implies that power consumption is minimized to increase the operating time of the network. In certain applications, this time may be extended to a few years due to difficult access, such as in hazardous locations (inside volcanoes) or difficult accessibility to the ocean floor or space.

The architecture of a typical sensor node is shown in Figure 10.4. The operation of these nodes revolves around a microcontroller. The sensor element itself can use different principles of operation and can be passive or active. The latter is preferable whenever possible in battery-powered nodes as energy is removed from the measured phenomenon. A battery may, however, be necessary to power the signal conditioning circuit.

A sensor node also includes digital memory for storing the data digitized by the microcontroller and a transceiver (transmitter and receiver) with the respective antenna to send data to the outside or to receive commands from outside for configuration purposes, for example.

Of these components, the one that consumes more energy is the transceiver when it is active, especially when transmitting data, because of the need to create an electromagnetic wave with sufficient energy to propagate the distance required. Note, however, that there are transceivers where the power consumption in reception mode is equal to or higher than the consumption in the transmission mode.

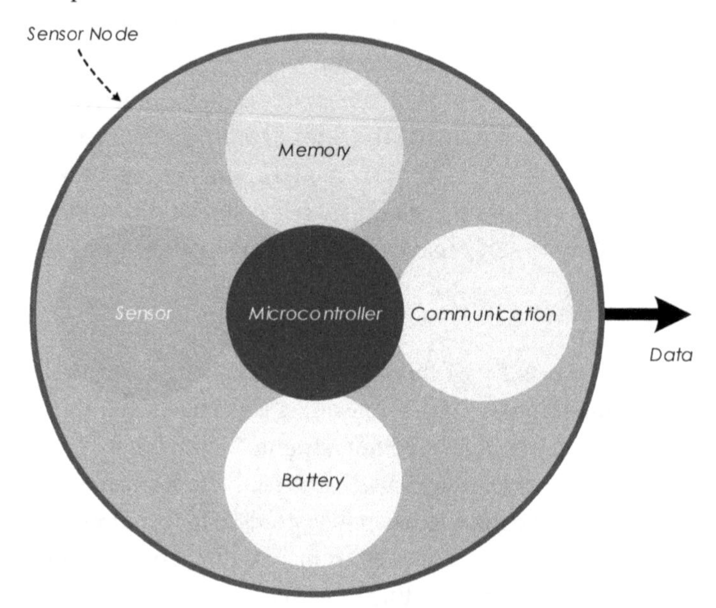

Figure 10.4 – Typical architecture of a sensor node.

As a rule, it is typically accepted that the energy required to transmit one bit of information is more than 100 times greater than the energy required for processing it.

10.5 Communication

There are different ways to communicate information between the nodes in the network and from the nodes to the central station. This communication can be unidirectional or bidirectional. Usually, when one has a network made of nodes that just contain sensors, one can have just a unidirectional communication from the nodes to the central station which purpose is to gather the information from the nodes, store it, process it, or present it to the user.

Successful communication depends on using the right protocols for the environment where the network is installed, having enough data transmission rate to handle the flow of information from the nodes, and implementing the right procedures in terms of data encryption, information validation, and robustness to failures and environment conditions.

The communication protocols used can be divided into wired ones and wireless ones. Depending on the application, one type or the other may be more appropriate. In the category of wired protocols, one can name, for example:

- I^2C
- SPI
- IS2
- SMBus
- I
- Maxim / Dallas 1-Wire and 3-Wire buses
- SSI
- Fieldbuses
- CAN bus
- MODBus (protocol)
- FOUNDATION FieldBus
- IEEE 1451

The I^2C (inter-integrated circuit) protocol is a low-speed serial communication protocol (from 10 kbit/s to 3.4 Mbit/s) developed by Philips, which is used for communication between integrated circuits. The bus uses only two wires — one for the clock, generated by the master, and the other, bidirectional, used for the data. Each of these lines of the open collector type and thus a pull-up resistance must be used. Up to 112 devices may be connected to the bus if the total capacity does not exceed 400 pF. Addressing is done by the software having each device a fixed factory address (7 bits).

The System Management Bus (SMB) protocol is a subset of the I^2C protocol that defines stricter electrical and protocol requirements.

The Serial Peripheral Interface Bus (SPI) protocol is like I^2C but uses four lines: one for the clock, two for data (one for each direction of communication), and one for addressing. The addressing is done by hardware with a different physical connection between the master device and each of the slave devices. Each device, therefore, has no address. This protocol consumes less power than I^2C and is faster. It has, however, the disadvantage of needing more connections.

The CAN-Bus (controller-area network) protocol is a protocol created by Bosch and used especially in vehicles and allows communication between microcontrollers and other devices such as sensors or actuators. In a car, there are several electronic control units for various subsystems such as

- the engine,
- streaming,
- airbags,
- anti-lock brake,
- speed control,
- audio system,
- windows,
- doors,
- mirror adjustment,
- etc.

Some of these systems operate independently, but others need to communicate with each other.

The CAN is a serial bus with multiple masters. Each node can send and receive messages but not simultaneously. It is possible to achieve transmission rates of 1 Mbit/s and a distance of 40 m.

Regarding wireless protocols, one can highlight the following:

- IEEE 802.11
- ZigBee
- Wibree
- Bluetooth
- 6lowpan
- Wi-Fi

Zigbee is low power and low data rate protocol that was designed specifically for sensor networks. Some of those networks operate in a restricted area and must transmit low amounts of information. This protocol is simpler than others like Bluetooth or Wi-Fi that which design is more oriented for the communication between computers and mobile phones. The Zigbee protocol can be used for a distance up to 100 m but is generally used for distances not greater than 10 m.

10.6 Questions

1. Give examples of applications where the use of a sensor network is advantageous.
2. What are the implementation differences between a network that contains only sensor nodes and another network that has sensor nodes and actuating nodes?
3. How many sensors are possible in a network where the transmission rate of the nodes is 250 kbit/s, the useful percentage of transmission is 20%, the transmission interval is 60 ms, and the number of bytes to be transmitted is ten at a time? Consider that all sensors communicate directly with a central node.
4. What challenges are there when implementing a sensor network?
5. Give an example of a sensor network used for environmental monitoring.
6. What are the drawbacks of using a wireless communication protocol?
7. What does a sensor node that is part of a sensor network usually contain?
8. What is the main difference between the I^2C communication protocol and the SPI protocol?

Chapter 11
Summary

11.1 Introduction

This chapter covers the sensors and actuators from a different perspective. The different sensors and actuators that were studied were organized according to the functioning principle and underlying physical phenomena because that is the main aspect when it comes to training future electrical and electronic engineers who may have to design and build their own sensors or use those that were created by others. With this purpose, different signal conditioning circuits, different techniques for building sensors and actuators, different ways to use them, and the meaning of the different features usually found in specification sheets were discussed.

Another task that a future engineer can find is to choose which sensor and actuator are best suited to a particular application. In particular, to measure a given physical or chemical quantity, it must choose which of the sensors available on the market must be used. This choice is not conditioned only by cost but also, and mainly by its metrological characteristics, suitability for the application in question, and ease of installation. An engineer must be able to understand all the specifications provided by the manufacturer and the sensor or the actuator operating mode to be able to draw up a list of advantages and disadvantages for each device on the market. From there, he should weigh the pros and cons of each to reach a conclusion about which specific sensor or actuator should be chosen.

It is in this perspective that this chapter fits. For each type of physical or chemical quantity, selected as belonging to a lot of the most important and representative of the choices to be made, a discussion is carried out on the advantages, and disadvantages of each type of sensor and actuator studied, referring, in certain cases, other devices not covered so far. The following quantities belong to this lot:

- Displacement;
- Temperature;
- Force.

This chapter also works, in a way, as a summary of the different sensors and actuators covered so far. Table 11.1 lists these devices, not exhaustively.

Table 11.1 – List of sensors and actuators covered for each physical/chemical quantity with reference to chapter numbers.

QUANTITY	SENSORS	ACTUATORS
Displacement	Capacitive Displacement Sensor (3.4) Potentiometric Displacement Sensor (4.3) Hall Effect Displacement Sensor (5.6) Inductive Displacement Sensor (5.10) Linear Variable Differential Transformer (5.11)	MEMS Electrostatic Actuator (3.9) Fluid Actuator (6.7) Guckel Thermal Actuator (7.8)
Position	Measuring Distance Using Ultrasounds (6.6)	
Acceleration	Capacitive Accelerometer (3.5) Piezoelectric Accelerometer (6.2)	
Angular Velocity	Gyroscope (3.6)	
Fluid Velocity	Hot-wire Anemometer (7.9)	
Fingerprint	Capacitive Fingerprint Sensor (3.7)	
Acoustic Pressure		Electrostatic Speaker (3.8)
Temperature	Resistive Temperature Detector (4.5) Thermistor (4.6) Integrated Temperature Sensor (4.7) Piezoelectric Temperature Sensor (6.3) Thermocouple (7.5)	Peltier Module (7.5)
Deformation	Strain Gauge (4.9)	
Angular Position	Angular Displacement Sensor – Microsyn (5.12)	Stepper Motor (5.4)
Torque	Magnetostrictive Torque Sensor (5.9)	
Image	Magnetic Resonance Imaging (5.13)	

11.2 Displacement

The displacement is a measurement, made in a straight line, of the distance between two points. It should not be confused with the position, which represents the coordinates of a point relative to a given reference. After knowing the start position and the end position of a given movement, it is possible to obtain the displacement (Figure 11.1). From the starting position and displacement's position, it is possible to know the final position if the direction in which this movement occurred is also known.

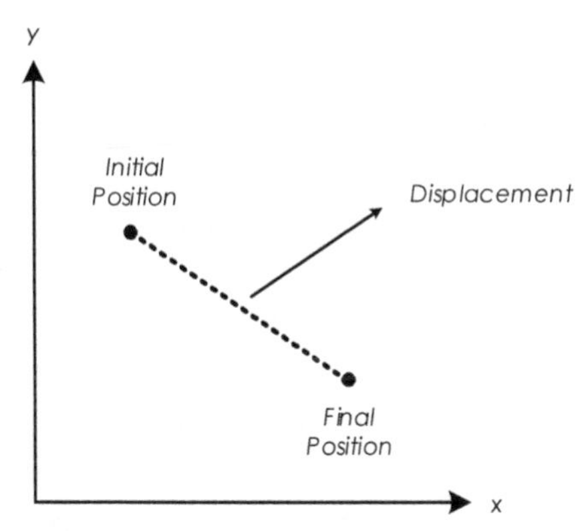

Figure 11.1 – Illustration of the difference between position and displacement.

There is a huge variety of displacement sensors using, for the most part, the electric field (resistive and capacitive), the magnetic field (inductive), mechanical phenomena such as piezoelectricity or acoustic waves, and electromagnetic waves (generally light). Those that are covered here are:

1) Electric Field
   a) Resistive Displacement Sensor — It works like a potentiometer. A cursor moves over a wire coiled on a shaft (winding sensor). An alternative is the use of a shaft covered with a film of conductive material forming a sensor film.
   b) Displacement Sensor by Deformation — one measures the displacement of a shaft that is leaning perpendicularly to the edge of a blade. The other edge is held stationary, which makes the displacement of the shaft cause a deformation of the blade which is measured by a strain gauge. This type of displacement sensor has a high mechanical load.
   c) Capacitive Displacement Sensor — it works as a capacitor. The distance between the plates can be varied (used for small displacement), the dielectric permittivity, or the area of the plates. These last two cases are used for bigger displacement.

2) Magnetic Field
   a) Inductive Linear Displacement Sensor with Dynamic Air Gap — It is composed of a rectangular-shaped magnetic circuit in which one side (beam) is free to move away from the rest of the circuit, creating an air gap. The magnetic circuit forms the core of a coil. The beam is attached to the body whose displacement is to be measured and which causes a change in the circuit's self-induction coefficient.
   b) Linear Variation Differential Transformer (LVDT) — it is a transformer with one primary and two secondary coaxial windings whose core is made of a magnetic material that is free to move as a function of displacement to be measured.
   c) Hall Effect Displacement Sensor — A permanent magnet creates a magnetic field which is measured by a Hall effect sensor. The displacement that is measured varies the distance between the magnet and the sensor, which varies the value of the measured magnetic field.
3) Mechanics
   a) Piezoelectric Displacement Sensor — The body whose displacement is to be measured pushes a mass against a piezoelectric crystal that will deform, creating a voltage. This type of sensor is not used to measure static displacements.
   b) Acoustic Displacement Sensor — An emitter and receiver of acoustic waves, usually ultrasound, is used. The time of flight of the acoustic wave is proportional to the traveled distance. The body whose displacement is to be measured can be associated with the sender or the receiver while the other transducer is in a fixed position. Alternatively, both the sender and the receiver may be stationary, and the moving body may be associated with a surface that reflects the acoustic waves. In this last case, the displacement is half the distance traveled by the waves. The transducers are usually made of piezoelectric crystals.
4) Electromagnetic Radiation (Light)
   a) Optical Displacement Sensor — a light source is used, typically a LED or a laser, and a light sensor (photodiode, for

example). The measured light intensity varies with the distance between the source and the sensor.

Table 11.2 shows a comparison of certain characteristics of displacement sensors.

Table 11.2 – Comparison of some characteristics of displacement sensors.

SPECIFICATION	RESISTIVE DISPLACEMENT SENSOR	CAPACITIVE DISPLACEMENT SENSOR	LINEARLY VARIABLE DIFFERENTIAL TRANSFORMER
Needs to be in physical contact with the object	Yes	Not if the object is metallic	Yes
Maximum Resolution	Good (0.1 μm - film)	Excellent (10 pm)	Excellent
Linearity	Very Good	Very Good	Very Good
Speed	Low	High	High
Environmental Stability	Bad	Good	Good
Robustness	Good	Good	Good
Cost	Low	Average	High

11.3 Temperature

Temperature is the most measured quantity of all physical and chemical quantities due to the impact that it has on the environment, in our body, on the machines, on materials, and on processes of industrial manufacture.

The temperature of a body reflects the kinetic (thermal) energy of the atoms or molecules (in translation, rotation, or vibration). The more the particles are vibrating, the higher the temperature. Heat is the process of transferring thermal energy between two bodies. Heat transfer is synonymous with the transfer of thermal energy. There are three forms of heat transfer:

- **Conduction** — consists of the interaction of atoms and molecules of a body with other atoms and neighboring molecules that leads to the transfer of part of its kinetic energy to these neighbors. This is the main way of heat transfer in solids.
- **Convection** — it is the combined effect of the conduction and the flow of matter. This type of heat transfer only occurs in liquids and gases. An increase in temperature leads to a reduction in density since the same atoms and molecules (same mass) occupy a bigger space (volume increase) due to their increased kinetic

energy. Said in another way — the same volume of the matter corresponds to a smaller mass. Less mass means lower weight. This hot volume of matter (gas or liquid) will rise, and its previously occupied space will be filled by an equal volume of matter with a lower temperature. Convection can be caused artificially by using, for example, a fan.

- **Radiation** — is a form of heat transfer made through electromagnetic radiation. The radiation is produced by the accelerated movement of charged particles (protons and electrons) that form atoms and molecules of the body. This radiation is produced in the same way as it is produced for an antenna using the accelerated motion (oscillations) of electrons in a conductor.

There are two types of temperature sensors:

- **With contact**, the sensor must be in physical contact with the body whose temperature is to be measured. Two bodies in contact with different temperatures tend to alter their temperature by heat exchange so that in a situation of equilibrium, the temperatures are the same. The value of temperature variation of each body until they reach the balance depends on the thermal mass of each. This heat flux is by conduction or, in the case of liquids and gases, also by convection.
- **Without contact**, wherein the temperature measurement is performed indirectly by measuring the infrared radiation emitted by a given body. This type of sensor is only effective in solids and liquids.

The operation of the most common temperature sensors, grouped by used physical principle domain, can be summarized as follows:

1) **Electric Field**
   a) Resistive Temperature Detector (RTD) — Uses an element, usually platinum, whose electrical resistance varies with temperature.
   b) Thermistor — It is similar to RTD but typically made of a metal oxide semiconductor, which gives higher sensitivity to it. There are two types PTC and NTC, as the electrical

resistance increases or decreases with temperature. The CNTs are commonly used as temperature sensors.

c) Integrated Temperature Sensor — it is a diode (or a transistor connected as a diode) whose characteristic is temperature-dependent.

d) Capacitive Temperature Sensor — The temperature changes the isolator's permittivity of a capacitor, thereby altering its capacity.

2) Mechanical Phenomena

a) Piezoelectric Temperature Sensor — uses a piezoelectric crystal in an oscillator. The temperature varies the capacity of the crystal and hence the frequency of oscillation.

b) Acoustic Temperature Sensor — uses the fact that the propagation speed of the acoustic wave depends on the medium temperature. By using ultrasound transmitters and receivers, for example, it is possible to measure the time of flight of an acoustic wave through a known distance and infer the velocity of propagation and the resulting temperature of the environment.

3) Thermal Phenomenon

a) Thermocouple — consists of a junction of two different metals that gives rise to an electromotive force due to the Seebeck effect.

4) Electromagnetic Radiation

a) Pyrometers — measures the radiation from a body using, for example, a photodiode. By knowing the emissivity of the body, it is possible to infer the temperature.

The different sensors have different strengths and weaknesses. The thermocouples can measure a large range of temperatures (−200°C to 2000°C) and are quite robust and can operate in environments with large vibrations (used in the space shuttle, for example). RTDs are quite linear, accurate, and stable but tend to be large and expensive. They may, however, be used to measure temperatures near 0 K. Unlike RTDs, integrated sensors have a low cost, but the range of temperature measurement is limited. In terms of resolution, the best of them all is the thermistors. They have a low cost, but they are nonlinear and have a limited measuring range of temperature. The pyrometers are ideal for applications that do not permit contact of the sensor with the measuring subject's body as in cases where the body is moving (blades of a turbine, for example) or when the corrosive environment is (a measure of the temperature of sulfuric acid in a tank, for example).

A particular feature of the temperature sensors is the response time (Figure 11.2). That time is usually slow because of the need for heat transfer from the measuring body object to the sensor or vice-versa so that they can both have the same temperature. This transfer is much faster if the mass of a sensor is smaller. On the other hand, it is necessary that the temperature sensor changes to the same value as the temperature of the object and not the other way around. Suppose, for example, an application that was intended to measure the temperature of a petal of a flower. The placement of an RTD in contact with the petal would probably make the petal change its temperature until it reaches the temperature of the sensor. The measurement would not be what was intended — the temperature of the petal before the sensor touches it. The thermal mass of the sensor is thus an important factor to consider when the measuring temperature varies by more than 1°C per minute.

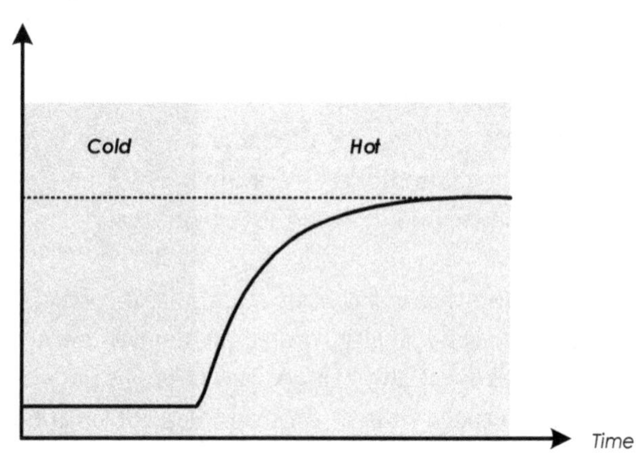

Figure 11.2 – Response time of a temperature sensor.

Table 11.3 presents the main advantages and disadvantages of some types of temperature sensors.

Table 11.3 – Advantages and disadvantages of different types of temperature sensors.

SENSOR TYPE	ADVANTAGES	DISADVANTAGES
Thermocouple	Self-powered Simple Robust Inexpensive Wide Temperature Range	Non-linear Low Voltage Reference Required Least Stable Least Sensitive
RTD	Most Stable Most Accurate More Linear Than Thermocouple	Expensive Slow Requires Current Source Small Resistance Change 4-wire Measurement
Thermistor	High Output Fast 2-wire Measurement	Non-linear Limited Range Fragile Required Current Source Self-heating
Integrated	Most Linear Highest Output Inexpensive	Temperatures Lower than 250°C Requires Power Source Slow Self-heating

11.4 Force

Force is a concept used to explain the acceleration experienced by a body. The mathematical relationship between these three quantities is given by Newton's second law

$$F = m \cdot a, \tag{11.1}$$

where F is the force, m the mass, and a is the acceleration.

There are different ways to measure an unknown force:

- Balancing it with a known force;
- Measuring the acceleration caused a test mass;
- Measuring the deformation caused in a known body;
- Measuring the pressure caused by force on a fluid.

The balance of the force to be measured with a known force can be made with gravity, with force produced in an electromagnetic way or even with a spring. In this last case, the force to be measured causes a compression of the spring, which is then measured by a displacement sensor. Whatever way is used, a Force sensor is a complex sensor since it uses another sensor for measuring a quantity that is related to the force (displacement, deformation, pressure, or acceleration). Of the four ways listed, the most common is the measurement of deformation of a known body. For this reason, it is important to understand how a given force applied at a given point of a body affects its geometry.

A force applied to a body creates mechanical stresses. These stresses are not directly measurable — only the displacements and deformations which are associated with them.

Closely connected with the deformation of a material (ϵ) is the mechanical stress (σ) to which it is exposed. Mechanical stress is defined as the relationship between the applied force and the area over which it is applied,

$$\sigma = \frac{F}{A}. \tag{11.2}$$

In the case of elastic materials, the relationship between the deformation and the stress is linear,

$$\sigma = E \cdot \epsilon, \tag{11.3}$$

where E is the constant of proportionality characteristic of the material and generally designated by Young's modulus or modulus of elasticity. Equation (11.3) is known as Hooke's Law, and it represents an approximation to the general case of a material that may or may not have an elastic behavior, as illustrated in Figure 11.3.

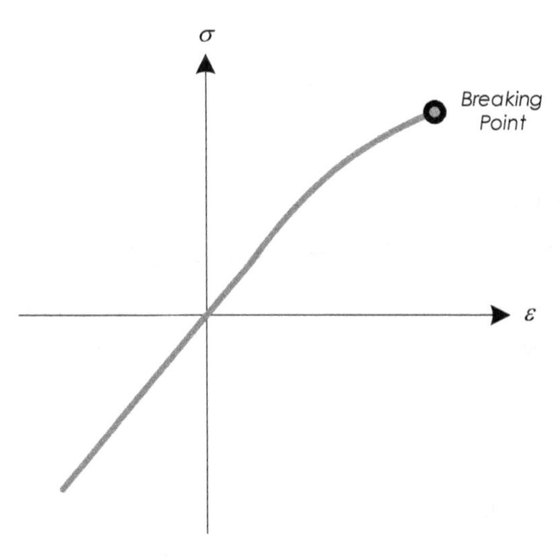

Figure 11.3 – Example of the relationship between mechanical stress and deformation of a material.

Table 11.4 presents the values of the modulus of elasticity of some materials.

Table 11.4 – Value of the modulus of elasticity of some materials.

MATERIAL	ELASTICITY (GPa)
Steel	200
Aluminum	70
Glass	70
Brick	20
Bone	16 to 9
Wood	10
Tendon	0.02

As an example, consider the case of human leg bone with a value of Young's modulus of 9 GPa, a length of 50 cm, and a cross-section of 3 cm^2. Imagining that the weight above the waist is 54 kg, the mechanical stress to which it is exposed, using (226),

$$\sigma = \frac{54\times 9,8}{3\times 10^{-4}} = 1.8 \text{ MPa}. \quad (11.4)$$

Using Hooke's Law, (11.3) one can determine the change in the length of the bone

$$\epsilon = \frac{\sigma}{E} = \frac{1.8\times10^{6}}{9\times10^{9}} = 2 \times 10^{-4}. \tag{11.5}$$

Finally, using (4.33), $\epsilon = dl/l$, a change of the length of the leg bone of 0.1 mm is obtained:

$$\Delta l = \epsilon \times l = 2 \times 10^{-4} \times 0.5 = 0.1 \text{ mm}. \tag{11.6}$$

The sensor which must operate in the region where the body has an elastic behavior depends essentially on the application. The following parameters are important:

- Size;
- Shape;
- Modulus of elasticity of the material;
- Sensitivity to local deformation and large deflections;
- Dynamic response;
- Load effect.

For the same body type, the change in its geometry and the material it is made of allow one to create force transducers with different measurement ranges.

There are three different types of objects that are used:

- Beam — in which the rod length is much greater than the width and the thickness, and the force is applied perpendicularly to the axial axis. Used for small forces;
- Ring — Used for forces between 1 kN and 1 MN;
- Column — Used to forces from 10 kN to 50 MN.

The beam can be fixed in one point, two points, or simply supported on two points (Figure 11.4).

In the case of fixed configurations, the highest deflection occurs in the fixed ends. In the supported configuration, the highest deflection occurs in the middle.

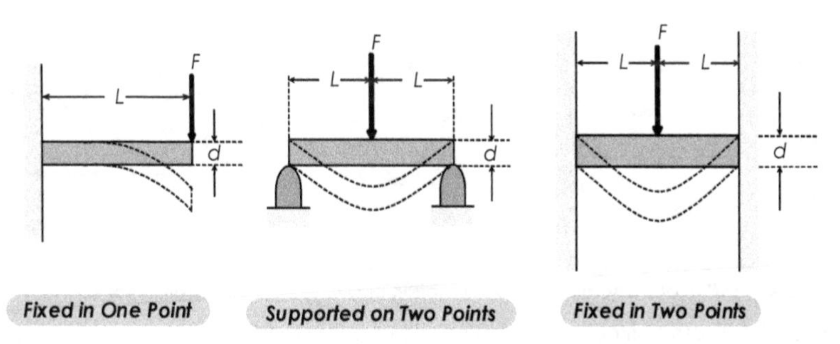

Figure 11.4 – Illustration of three types of beams used to measure a force.

Another possible configuration is the ring, as shown in Figure 11.5.

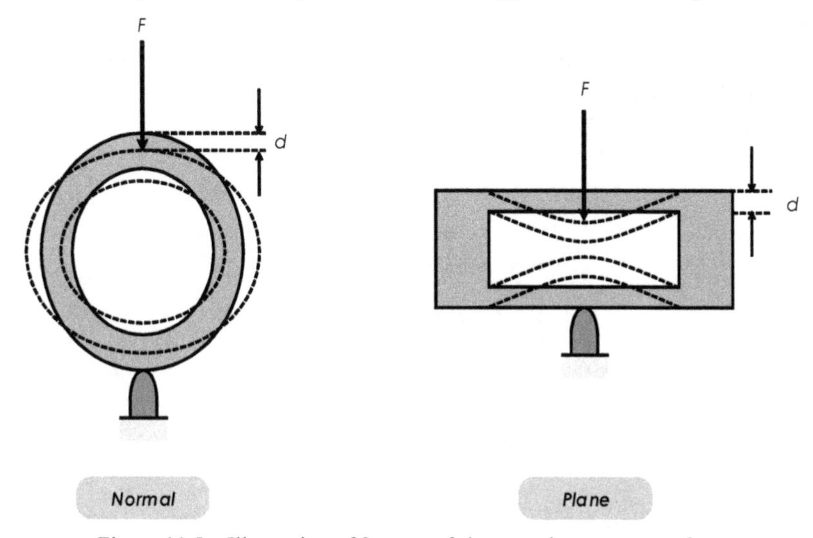

Figure 11.5 – Illustration of 2 types of rings used to measure a force.

Figure 11.6 illustrates the position where four strain gauges can be placed in a ring for measurement of force.

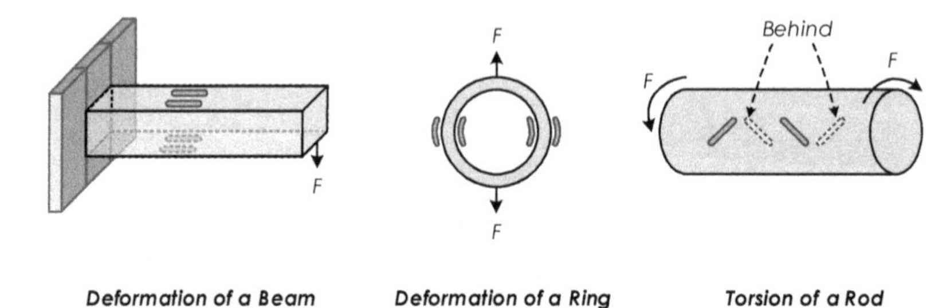

Figure 11.6 – Illustration of the position of 4 strain gauges placed on a beam, on a ring, and on a rod for measuring force and torsion.

Yet another configuration used to measure very large forces is the column (Figure 11.7). The strain gauges are placed at the midpoint of the column, two in compression (vertically oriented) and two in traction (horizontally oriented).

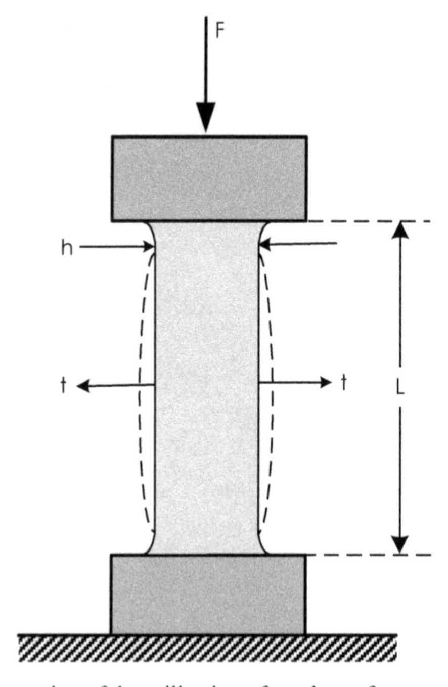

Figure 11.7 – Illustration of the utilization of a column for measurement of force.

11.5 Signal Conditioning

As seen in Chapter 1, smart sensor or actuators requires not only the transducer itself but an electronic circuit also able to create electronic signals needed to stimulate a sensor and actuator and able to transform the signals produced by the sensor into a form suitable for its analog-to-digital conversion for further digital processing. In the case of sensors that generate voltages themselves or currents, such as piezoelectric crystals and photodiodes, respectively, the signal produced can be very weak, noisy, or containing undesirable components. In these cases, the signal conditioning includes amplification and filtering of signals from the sensor. There are, however, many other functions that are typically performed by a signal conditioning circuit:

- Demodulation.
- Current-to-voltage conversion.
- Charge-to-voltage conversion.
- Amplification.
- Level and impedance adaptation.
- Galvanic isolation.
- Linearization.
- Filtering.
- Analog-to-digital conversion.

11.5.1 *Demodulation*

Demodulate is to retrieve the parameter value of a sinusoidal signal that depends on the quantity to be measured. Depending on the sensor, it may be necessary to perform demodulation of amplitude (AM), frequency (FM), or phase (PM) depending on the affected parameter. Modulations of this type are widely used in radio, and thus some of the circuits used in this area are also employed for the signal conditioning of sensors. Examples of these are the heterodyning (for AM and PM) and phase-locked loop (for FM). There are, however, demodulation techniques that are particularly used in the field of sensors. They are RMS/DC conversion (AM), peak detection (AM), synchronous demodulation (AM and PM), and frequency-to-voltage conversion (FM).

AM demodulation is used when the quantity to be measured determines the amplitude of a sinewave. An example of this type of sensor is a capacitive displacement sensor based on the variation of distance between the plates. As seen previously, the impedance of this type of sensor is directly proportional to the displacement. Possible conditioning consists of applying a sinusoidal alternating current and measuring the voltage amplitude at the sensor terminals. The way to produce a direct voltage that is proportional to the displacement is to use an RMS/DC converter or a peak detector.

Recall that there are several ways to perform an RMS / DC conversion:

- **Using Form Factor** — The effective value of a signal is equal to the average value of its absolute value multiplied by the form factor of the signal. This factor depends only on the shape of the signal, which is 1.11 in the case of a sine wave. A converter of this type is, therefore, a rectifier to obtain the absolute value, a low-pass filter to perform the average, and an amplifier to multiply by 1.11.
- **Using the mathematical definition of Effective Value** — The effective value of a signal is, by definition, the square root of the mean squared value. Therefore, an electronic circuit for performing this is a multiplier to compute the square of the signal, a low pass filter to calculate the average value, and a circuit for calculating the square root. The square root can be obtained with an operational amplifier in voltage follower assembly with a multiplier performing the square in the feedback loop.
- **Using the physical definition of Effective Value** — The effective value of a signal is the DC voltage value which produces the same thermal effects that the signal in question. An electronic circuit for this purpose comprises two resistors to which the input voltage is applied (signal to be converted) and a DC voltage produced internally (output signal of the RMS/DC converter). This voltage is produced by an operational amplifier whose inputs are connected to two temperature transducers measuring the temperature of each resistor. The amplifier will thus produce a DC voltage which produces in the electric resistor the same thermal

effects (the Joule effect) that the input signal produces in a resistor with the same value.

Another technique for amplitude demodulation is the use of heterodyning, that is, in the multiplication of the sensor signal by a sinusoid with the same frequency and initial phase. Imagine a sensor whose output is a sine wave with an amplitude is proportional to the quantity to be measured ($x(t)$):

$$v_s(t) = K \cdot x(t) \cdot \cos(\omega \cdot t). \tag{11.7}$$

The heterodyning is about multiplying this signal by a sine wave with the same frequency (ω) and with a known magnitude (B):

$$v_p(t) = B \cdot \cos(\omega \cdot t). \tag{11.8}$$

The output voltage is the product of (11.7) by (11.8):

$$v_o(t) = v_s(t) \cdot v_p(t) = x(t) \cdot K \cdot B \cdot \cos(\omega \cdot t) \cdot \cos(\omega \cdot t). \tag{11.9}$$

This is equivalent to

$$v_o(t) = x(t) \cdot \frac{K \cdot B}{2} + x(t) \cdot \frac{K \cdot B}{2} \cdot \cos(2 \cdot \omega \cdot t). \tag{11.10}$$

Performing low-pass filtering, with a cut-off frequency of less than 2ω, one has:

$$v_o(t) = x(t) \cdot \frac{K \cdot B}{2}, \tag{11.11}$$

that is, a voltage v_o which is proportional to the quantity to be measured x.

In the case of displacement sensors like LVDTs, the output voltage is a sine wave with a magnitude that is proportional to the displacement. The direction of the offset, positive or negative, is given, however by the relative phase shift between the signal of the primary and secondary — "in phase" means a shift in one direction, and in "phase opposition" means a shift in the opposite direction. The creation of a DC voltage that is proportional to the displacement (if it does not vary over time) cannot be made with the aid of an RMS/DC converter or a peak detector because the value of the algebraic sign of the displacement is not determined.

The solution normally applied in this case is the use of synchronous demodulation, where the signal is rectified by a switch that selects it, or it's symmetric depending on the signal applied to primary of the LVDT is in the positive or negative alternation. Then, a low-pass filter is used to take the average value of this rectified signal.

Although heterodyning could, in this case, be used, the use of a synchronous demodulator makes the use of a multiplier unnecessary which is typically a more expensive component.

Often the quantity to be measured directly affects the frequency of a sinusoidal signal, as in the case of capacitive or piezoelectric sensors, in which the sensor element is placed in an oscillator in order to make the oscillation frequency depend on the quantity to be measured. An example is the piezoelectric temperature sensor (Section 6.1). In this case, the frequency demodulation may be carried out using a phase-locked loop.

11.5.2 *Amplification*

Amplification is one of the most common typical functions of signal conditioning. The goal is to increase the signal level so that it can be used in the downstream circuit. Nowadays, with the advent of smart sensors and the processing, storage, and transmission of information in digital form, it is common to use an analog-to-digital converter (ADC) connected to the output of the sensor. These converters are typically voltage converters, and they have a predetermined range which is typically between 1 and 5 V. Therefore, it is necessary to convert the signal present in other forms, like charge or current into a voltage, and amplify it to take values that use as much as possible the entire range of the ADC.

The use of a current-to-voltage converter is typically required when photodiodes are used since the effect that radiation has upon impinging on them is to create an electrical current that flows through them. That current is usually very small of the order of nA or μA. A converter of this type, known as a trans-impedance amplifier, is often based on an operational amplifier with a high-value feedback resistor where the photodiode current flows, resulting in a voltage drop.

The charge-to-voltage converters are used in capacitive sensors or those that use piezoelectric materials. An example seen was the case of accelerometers using piezoelectric materials. This type of amplifier, also known as charge amplifiers, uses an operational amplifier and a capacitor in the feedback loop with a reduced value to produce, with little load, a significant voltage.

Besides having a pre-established range, ADCs have a finite resolution; that is, they perform quantification of the signal with a given number of bits. A converter of 10 bits, for example, divides its range in 1024 (2^{10}) distinct sub-ranges. The value of the input signal at each instant is attributed to one of these ranges (quantization), which corresponds to a digital word (in this case, a number between 0 and 1023). If the range is ±2.5 V, for example, the quantization step, that is, the resolution is approximately 4.88 mV (5/1024). An input signal that varies, for example, between 0 and 1 mV, is converted, under these circumstances, always into the same digital word. All the information about the quantity to be measured is therefore lost in this example. This loss of information in the ADC conversion is inevitable but can be minimized if an ADC with the highest number of possible bits is chosen and the signal to be converted is previously amplified so that its range fits as best as possible to the range of the ADC.

Even when there is no analog-to-digital conversion, it may also be required to amplify the signal to be used to stimulate an indicator or trigger an alarm. In the latter case, amplifiers/comparators that may or may not have hysteresis are used to minimize the effect of noise.

Amplification by itself does not enhance the signal/noise ratio since the amplification affects both. In fact, if the noise is broadband so that the occupied bandwidth is bigger than the bandwidth of the amplifier, there will be a noise reduction at the output compared to the signal (improved signal/noise ratio). On the other hand, the amplifier itself generates some noise which will be added to the amplified signal with the consequent degradation of the signal/noise ratio.

The decrease of the amount of noise that is present in the signal can be achieved using filters. If the quantity to be measured has a limited bandwidth, a low-pass filter can be used to eliminate all the noise that is at higher frequencies. Imagine a thermistor temperature sensor. This type of sensor has a response time longer than several seconds. The bandwidth of the signal at its output will never exceed 1 Hz. It makes sense in this case, the use of a low-pass filter with a cut-off frequency of 1 Hz to eliminate the noise of higher frequencies.

Depending on the application and the bandwidth occupied by the signal at the output of the sensor, bandpass and high-pass filters can also be used. In the case of an accelerometer used to detect a vehicle collision, the signal of interest has a high frequency. A high-pass filter aids in eliminating low-frequency noise which could accidentally trigger a system that uses that information, such as an airbag.

Finally, regarding the signal's amplification performed by the sensor, the instrumentation amplifier must be taken into account. This amplifier formed by three operational amplifiers has two features that are particularly important when working with sensors. The first is the low input current (high input impedance), allowing the amplifier to be connected to a signal conditioning circuit such as a Wheatstone bridge without altering the operation of the upstream circuit or the circuit where the sensor is located.

The other outstanding feature of an instrumentation amplifier is its high common-mode rejection ratio (CMRR). Any differential amplifier has an output voltage that, besides depending on the difference between the voltages at the two inputs (differential gain) it also depends, undesirably, on the common value of these two voltages, that is, the common-mode voltage (common-mode gain). An instrumentation amplifier has a particularly low common-mode gain (A_c) when compared to the differential mode gain (A_d). The common-mode rejection ratio, given by,

$$CMRR = 20 \cdot \log\left(\frac{A_d}{A_c}\right), \tag{11.12}$$

typically assumes values bigger than 100 dB in the instrumentation amplifier.

The use of instrumentation amplifiers when you want to amplify a Wheatstone bridge's voltage unbalance is virtually mandatory since the common-mode voltage is typically one or more orders of magnitude higher than the differential voltage.

11.5.3 *Linearization*

One of the concerns of an engineer when building a sensor is that the relationship between the output voltage and the quantity to be measured is as linear as possible. On the one hand, this makes it easy to determine the quantity to be measured from the voltage obtained — using just two constants. These constants are, in general, the slope of the transfer function and the amount of voltage that is obtained when the quantity to be measured has a null value (offset voltage). Another advantage of having a linear relationship is that the sensor sensitivity is the same for all operating points.

When the sensor itself does not have a linear characteristic, it is necessary to make it linear. This can be done in the sensor or in the signal conditioning circuit (linearization at source) or, alternatively, in downstream circuits.

Linearization in the source may be done in several ways.

- It can be achieved by changing the operating point of the sensor by proper polarization, as illustrated in the case of the magnetic field sensor using a giant magneto-resistive effect.
- Another solution is to change the sensor circuit as illustrated in the case of thermistors by placing a suitably sized resistor in parallel with the sensor.
- Another solution also includes the association of two or more sensors, as illustrated in the case of linear inductive displacement sensors with a variable air gap. The same type of linearization was used in the case of multiple gauges in a Wheatstone bridge measurement of deformation.
- Finally, using the feedback in the signal conditioning circuit is another way of linearization of the characteristic of the sensor, as illustrated in the case of the thermistor included in a Wheatstone bridge.

Also, downstream from the source, there are different linearization techniques that can use, and which are classified as analog or digital.

Imagine a sensor whose relationship between the output voltage (v_m) and the value of the quantity to be measured (m) can be expressed by a polynomial

$$m = a_0 + a_1 \cdot v_m + a_2 \cdot v_m^2 + a_3 \cdot v_m^3 + \cdots \qquad (11.13)$$

where the coefficients a_i can be determined by adjusting a polynomial of a given order to a set of points (pairs output voltage/measured quantity) obtained experimentally. One wants to create an electronic circuit whose output, v_s, is proportional to the quantity to be measured, that is,

$$v_s = b \cdot m, \tag{11.14}$$

where m is the proportionality constant (sensor's sensitivity). Inserting (11.13) into (11.14) leads to

$$v_s = a_0 \cdot b + a_1 \cdot b \cdot v_m + a_2 \cdot b \cdot v_m^2 + a_3 \cdot b \cdot v_m^3 + \cdots \tag{11.15}$$

This mathematical function can be implemented electronically using multipliers and an adder with multiple inputs weighted by the coefficients $a_i b$.

In special cases where the function relating the output voltage of the sensor, v_m, and the quantity to be measured m, is of a known type, such as an exponential function, an electronic circuit that implements the inverse function can be built, in this case, a logarithm.

In the case of a Wheatstone bridge with a single sensor element, it is possible to linearize the characteristic of the sensor using a single multiplier and one adder with two inputs, as shown in the case of a resistive temperature detector.

The downstream linearization can also be carried out digitally using two methods. The first is to store on a non-volatile memory the math function needed to transform the voltage value obtained at the output of the sensor to a value proportional to the measured quantity. The value of the output signal of the sensor in each instant is converted to a digital value that is used to address a memory containing the corrected digital value, which will have a linear relationship with the quantity to be measured (Figure 11.8).

The larger the memory size, the more points can be used to approximate the mathematical function, and the smallest will be the error incurred. In the example given in Figure 11.8, the function m (v_s) is discretized into 8 intervals. Each interval is attributed to a decimal number between 0 and 7. This number is used to address the memory. In this case, the memory will need eight words. The quantity value of each of these intervals is quantized and, in the example shown, used 16 quantization levels (0 to 15). The value of quantized quantity m is stored in the memory at the position corresponding to the interval to compute. In interval number

6, for example, the quantity to be measured is attributed to the digital value of 9.

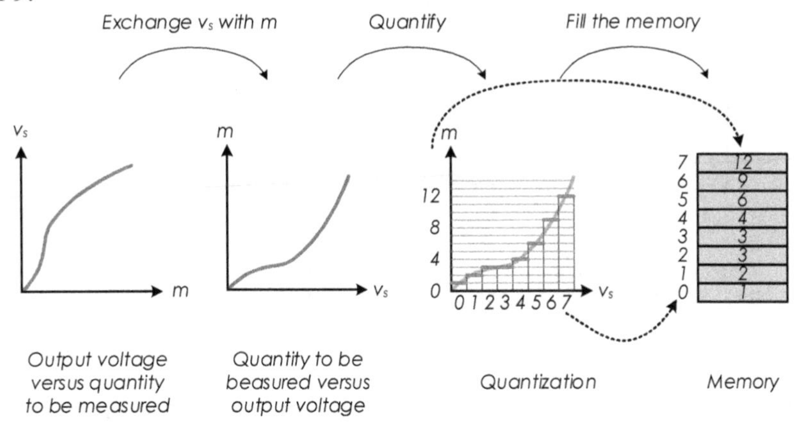

Figure 11.8 – Illustration of memory filling process to be used in sensor linearization.

Figure 11.9 shows the simplified block diagram of the circuitry used in this case. The ADC has 3 bits corresponding to 8 quantization intervals. These bits are used to address memory which has 8 words (2^3). The word stored in memory has 4 bits and is the output of the circuit, that is, the digital value of the measured quantity (m).

In this example, the memory has 32 bits in total. In real situations, the memory will have a much higher capacity, such as 1024 words of 2 bytes (16 bits) each. The ADC would have10 bits (2^{10} = 1024) and a total memory capacity of 2 Kbytes (1024 bytes × 2).

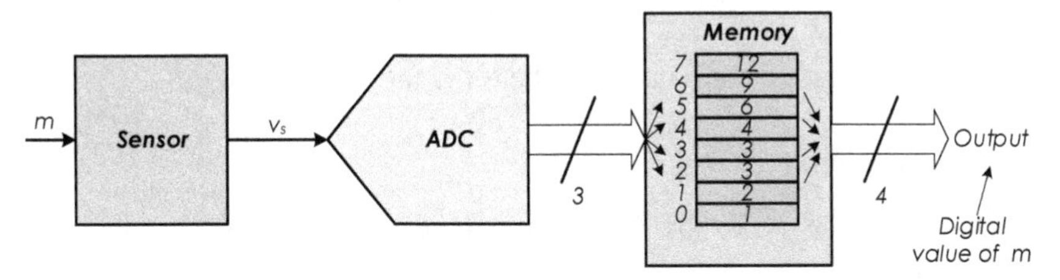

Figure 11.9 – Block diagram of the digital linearization circuit of a sensor using a memory.

The second method of digital linearization is to use a processor to perform a linear interpolation from a set of predefined points, which describe, by segments, the mathematical function required to convert the obtained voltage output into a value of the measured quantity (Figure 11.10).

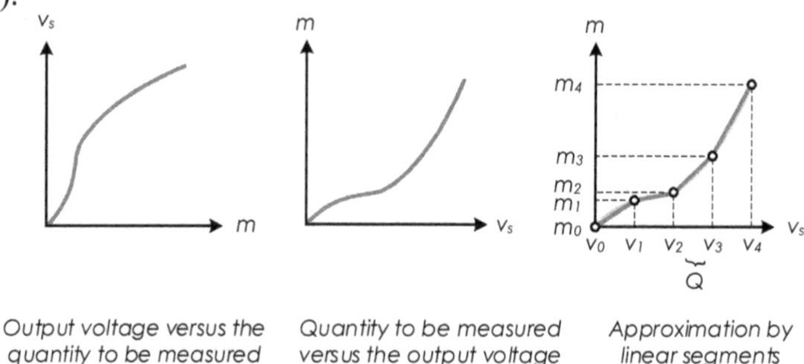

Figure 11.10 – Illustration of the process of approximation of a function by segment.

In the example of Figure 11.10, four segments defined by 5 points (v_i, m_i) are used. The microprocessor uses these four segments to determine the value of the quantity to be measured (m) from the output voltage of the sensor element (v_s). The points defining the segments are equally spaced horizontally (Q). The value of a segment for a given voltage v_s is determined by

$$i = int\left[\frac{v_s}{Q}\right], \tag{11.16}$$

where "int" represents the integer part of a number. The leftmost segment is number 0. The value of the quantity to be measured (m) is determined by linear interpolation using

$$m = \frac{v_s - v_i}{v_{i+1} - v_i}(m_{i+1} - m_i). \tag{11.17}$$

The more segments are used, the better is the approximation.

11.6 Questions

1. Describe two types of displacement sensors that use mechanical phenomena.
2. What are the three forms of heat transfer?
3. Describe an application in which the use of a non-contact temperature sensor is imperative.

4. Why is the response time of a temperature sensor, in general (but not always), long?

5. Where should a strain gauge be placed to measure the deformation caused in a ring by a vertical force from top to bottom?

   a. At the base of the ring.
   b. At the top of the ring.
   c. On the right and left sides of the ring.
   d. It is indifferent.
   e. None of the above.

6. Where should an LVDT be placed to measure the vertical force exerted on a ring from top to bottom?

   a. At the base of the ring.
   b. At the top of the ring.
   c. On the right side of the ring.
   d. On the left side of the ring.
   e. None of the above.

7. Where should an extensometer be placed to measure the deformation caused in a column by a vertical force from top to bottom?

   a. At the base of the column.
   b. At the top of the column.
   c. On the column side.
   d. It is indifferent.
   e. None of the above.

8. Which of the following bodies used to measure a force allows to measure greater forces?

   a. Ring.
   b. Column.
   c. Beam.
   d. Everyone can measure the same force.
   e. None of the above.

9. What are the different forms of signal conditioning?

10. What does the common-mode rejection ratio (CMRR) of an amplifier mean?

11. What are the advantages of having a linear relationship between the output voltage of a sensor and the value of the measured quantity?

12. In what ways can the relationship between the output voltage of a sensor and the value of the measured quantity be linear?

Chapter 12
Laboratory Guides

12.1 Introduction

This chapter presents six different project ideas for the students to carry out in the laboratory. These use one sensor and one actuator each. The projects are carried out in groups and should be completed in under 2.5 hours.

At the beginning of class, the students are questioned about the project they are about to carry out to see if they know how the sensors and actuators they are going to use work internally and how they should be used. They are also questioned about other sensors and actuators that measure and actuate over the same quantities. The goal is for them to point out their advantages and disadvantages and appropriateness for the application at hand. They also must prepare the operational part of the project by thinking about the circuits that they are going to need to assemble during class and choose the right parameters for the electronic components used.

During class, the students are asked to assemble the circuit, show them working, take some measurements and draw some conclusions about their operation. A pre-made form is supplied to the students to document that work.

The proposed projects deal with:

- Measuring distance with an ultrasound emitter and one ultrasound receiver.
- Changing the temperature of a Peltier module and measuring it using two smart sensors.
- Operating a stepper motor.

- Using a led and a photodiode to detect when someone crosses a door.
- Measuring the rotational velocity of a DC motor using Hall effect sensors.
- Using an accelerometer to determine how much a servo motor rotor has moved.

12.2 Ultrasound

12.2.1 *Introduction*

An acoustic wave is alternating compression and expansion of a medium that can be solid, liquid, or gaseous. These waves are said to be sound if the frequency is between 20 Hz and 20 kHz. Below 20 Hz, they are called infrasound, and above 20 kHz are called ultrasound.

Detection of infrasound is used in structural analysis of buildings, forecast earthquakes, and other large sources. When infrasound has a significant magnitude, it can be felt by humans, even causing psychological effects (panic, fear, etc.).

Audible waves can be created by vibrating strings (stringed musical instruments), vibrating columns of air (wind instruments), and vibratory plates (some vibrating tools, vocal cords, and loudspeakers).

Ultrasounds are typically used to measure distance or proximity. A wave is emitted at a certain point, and the time it takes for it to return to that point, after being reflected in each object, is measured to determine the distance traveled by the wave.

The ultrasonic transducers, in general, can act as transmitters or receivers. These transducers can be constructed using piezoelectric materials.

The emitters convert a voltage into a deformation which in turn displaces the air (or another medium), causing a wave. The receiver does the opposite. The pressure change causes the medium to deform the piezoelectric material, which in turn produces a voltage.

There are several methods used to estimate the distance using ultrasound. This work uses the time of flight of a sinusoidal burst determined using the correlation peak of the signal sent and received.

A target structure consisting of a wooden square with a width of 20 cm is available, as well as a printed circuit board containing the ultrasound transmitter and receiver. You are free to move them on a horizontal track. It is possible to manually adjust the distance between the ultrasonic transducer and the target.

12.2.2 *Connecting the Ultrasound Module to a Microcontroller*

Connect the module that contains the ultrasound emitter and receiver (HC-SR04 from Cytron Technologies) to the Arduino microcontroller. Present a drawing of the connections with the pin names/numbers used. Describe the function of each of the four pins of the ultrasonic module.

12.2.3 *Programming the Microcontroller*

Create a program for the microcontroller using the Arduino IDE that can obtain from the ultrasound module the information about the distance to a target and display it in the PC console as a distance in centimeters. Present only the main parts of that program. What is the signal read from the ultrasound module? How is it converted into a distance-to-the target?

12.2.4 *Characterizing the Performance of the Ultrasound Module*

Using a ruler and the application created, determine the relationship between real and measured distance-to-the target (use 5 to 10 different values of distance). Plot a chart with that relationship. Determine an upper bound for the linearity.

12.2.5 *Upsetting the Normal Operation of the Distance Measurement System*

Make a hypothesis about what could affect the operation of the distance measurement system. Experimentally evaluate as many of those hypotheses as possible. Write down some conclusions.

12.2.6 *Material*

The material required for this project is:

- Ultrasonic module HC-SR04 (Cytron Technologies).
- Aluminum track.
- Wooden target.
- Arduino Uno microcontroller and personal computer.

12.3 Temperature

12.3.1 *Introduction*

A smart sensor is an integrated circuit without external components that includes a sensing element, an interface with the outside world, signal processing, and some "intelligence," including self-test, self-identification, and self-validation.

This project is intended to serve as an introduction to the use of smart sensors, with two smart temperature sensors from Maxim, model DS18B20. One Peltier module is used, which will have a changing temperature dependent on the DC voltage used to power it. A personal computer fan is used to remove heat from the hot side. The cooler the hot side is kept, the lower is the temperature that the cold side of the Peltier module can reach.

12.3.2 *Circuit Assembly*

Assemble the temperature sensors in a breadboard and connect them to the Arduino microcontroller. Identify which side of the Peltier module is the hot one. Attached the sensors on the opposite sides of the Peltier Module and power the module with an adjustable DC voltage. Attach a computer fan, powered with 5 V, to the hot side of the Peltier Module. Draw the connections made.

12.3.3 *Communication with the Sensor*

Create software that reads the temperature from the sensors using the microcontroller and shows it on the computer screen. Present the flowchart of the application created.

12.3.4 *Activate the Peltier Module*

Power the Peltier module (with the fan turned off) using the adjustable power supply. Start registering the temperatures measured by both sensors while increasing the DC voltage applied. Comment what you observe.

Repeat the procedure with the fan turned on.

12.3.5 *Final Remarks*

Comment on the operation of the system, including the accuracy, response time of the measurement, and additional features implemented.

12.3.6 *Material*

The material required for this project is:

- 1 Peltier Module from CUI Devices model CP20351.
- 2 temperature sensors from Maxim, model DS18B20.
- 1 personal computer fan (5V).
- 1 dual DC power supply with fixed 5 V and adjustable 0 to 15 V outputs.
- 1 Arduino Uno microcontroller and personal computer.

12.4 Stepper Motor

12.4.1 *Introduction*

A stepper motor is used to make things move. This movement can aim to change the position of the load or give it a given speed or acceleration. There are several forces that naturally oppose the change of position or speed of a body, including inertia, friction, gravity, the force of a spring, etc. It is, therefore, necessary that a motor can exert sufficient torque to overcome these other forces that may be acting on the load. In general, the motors used are rotary motors, i.e., the bodies are rotated around a fixed axis. When one wishes to have a linear movement, one often uses a rotary motor and a mechanical device that turns this movement into a linear movement. An example is an automobile.

A stepper motor is distinguished from other motors to the extent that the movement of rotation is not uniform. In DC or AC electric motors, for example, the applied voltage is proportional to the rotation speed. If the voltage is constant, the rotational speed will remain constant as well. If you wish to have a given acceleration to a body, you just linearly increase the voltage applied. If the goal, however, is the position of a body in a certain position using these types of motors, then a more complex control circuit is required together with a sensor that indicates the position of the body. The control circuit uses this information to increase or decrease the voltage applied to the motor until the desired position is reached. A closed-loop control system is required.

It is exactly in applications where easy positioning of a body is desired that stepper motors are used. The control of these motors is simpler in that they can operate in an open-loop.

Stepper motors are constructed in such a way that the motion of the axis (rotor) occurs in well-defined steps, i.e., each time it "sends a command" to the motor, it rotates by a fixed amount. In a stepper motor with a step of 7.5° where one wants to turn the rotor by 30°, one just sends this "command" 4 times to produce four steps.

12.4.2 *Type of Motor*

Perform the tests that you find necessary to determine if the type of stepper motor that you have in the laboratory. Describe the test and indicate the type of motor you think it is.

12.4.3 *Connector Identification*

Perform tests that you find necessary to identify the terminals of the stepper motor and the internal constitution of the windings. Describe the test performed and the values measured and show a diagram containing the connection of the terminals to the motor windings.

12.4.4 *Engine Control System*

Devise a system of control of the stepper motor using the ULN2003A integrated circuit that contains a set of transistors in Darlington assembly and the Arduino Uno microcontroller. Build it and present your wiring diagram.

12.4.5 *Control Application Development*

Develop an application using the Arduino IDE that allows you to control the position of the motor rotor (number of steps in an absolute reference frame). Create a simplified block diagram of the code of this application.

12.4.6 *Number of Motor Steps*

Use the application developed to determine the number of motor steps necessary to complete a full revolution. Indicate the value obtained and the corresponding step angle.

12.4.7 *Material*

The material required for this project is:

- 1 stepper motor.
- 1 integrated circuit ULN2003A, which has a set of transistors in a Darlington assembly.
- 1 Arduino Uno microcontroller and personal computer.

12.5 LED and Photodiode

12.5.1 *Introduction*

A LED is a semiconductive P-N junction that emits light when energized (light-emitting diode). That light is not monochromatic like in a laser. Instead, its spectrum occupies a narrow bandwidth. The light emitted by a LED is produced by the energetic interaction of the electrons on the material. In any directly polarized P-N junction, holes and electrons recombine. That recombination leads to a release of energy as light or heat by the free electron when it binds with the material. The wavelength of the emitted light depends on the bandgap of the semiconductor material; by directly polarizing the junction, one "pushes" the electrons and holes together, leading to their recombination.

A photodiode is a semiconductive device used to convert light into an electrical signal. There are different types of photodiodes, all based on a junction between doped semiconductor materials. Their operation is based on the photoelectric effect; that is, the incident radiation makes some electrons in the valence band jump into the conduction band and be part of an electric current. Electromagnetic radiation (including light) is thus converted into an electric current.

If a P-N junction is directly polarized (positive battery terminal at the "p" side), one has a current flowing from anode (p side) to cathode (n side). If electromagnetic radiation impinges in that junction with the appropriate wavelength, there is going to be an additional current in the same direction. This current will, however, be much smaller than the current one had when there was no electromagnetic radiation (dark current).

If the junction is inversely polarized, the current that flows through it in the absence of radiation is almost null. When radiation impinges in the junction, electron-hole pairs are created on both sides of the junction, and the electrons that are in the conduction band flow toward the cathode (n side) and the holes toward the anode (p side). One has, therefore, a current flowing from cathode to anode.

A wooden structure is available in the shape of an inverted T that supports a LED placed horizontally. There is an identical structure with a photodiode on it. Both structures are meant to slide along a straight aluminum rail. It is, therefore, possible to manually adjust the distance between LED and photodiode. The photodiode support also allows one to rotate it in the horizontal plane to change the angle at which light from the LED reaches it.

You should use the LED and photodiode to detect when someone crosses the threshold of a door. You should design the electronic circuits so that the distance between LED and photodiode be as large as possible.

12.5.2 *Energizing the LED*

Design the electronic circuit for the LED (Fairchild QED423) so that it can emit as much light as possible in a continuous fashion. Draw the schematic. Present the nominal LED current and the radiant intensity in those conditions.

12.5.3 *Photodiode Signal Conditioning*

Design a transconductance amplifier based on the AD8541 integrated circuit to convert the photodiode current (OSRAM SFH 213FA) into a voltage. You should use it in photovoltaic mode. Connect the output of the amplifier to an Arduino microcontroller that displays the output voltage in the PC console. Present the schematic and the computations for the output voltage of the amplifier when i) no Light reaches the photodiode and ii) when light from the LED placed 70 cm away reaches the photodiode.

12.5.4 *Experimental Setup*

Assemble the designed circuit and measure the voltage at the output of the transconductance amplifier in the two conditions mentioned earlier. Compare those values with the ones computed before.

12.5.5 *Maximum Operating Distance*

Determine the maximum distance between LED and photodiode that you can have and still be able to detect when someone crosses the door threshold. Point out any changes that you might have made to the electronic circuits used.

12.5.6 *Photodiode Directivity*

By measuring the amplifier's output voltage and rotating the photodiode in the horizontal plane, determine the half-angle of the photodiode. Compare it with the value read from the datasheet.

12.5.7 *Material*

The material required for this project is:

- 1 infrared LED from Fairchild, model QED423.
- 1 infrared photodiode from OSRAM, model SFH 213FA.
- 1 differential amplifier from Analog Devices, model AD8541.
- Several electrical resistances (1/4 W) with different values.
- 2 wooden structures in the shape of an inverted T.
- 1 aluminum rail.

12.6 Hall Effect and DC Motor

12.6.1 *Introduction*

The Hall Effect is a rise in voltage between two opposite sides of a conductor traversed by the current when immersed in a magnetic field. The electrons moving inside a conductor are subject to a force due to the magnetic field that pushes them towards one side of the conductor.

A DC motor also uses the interaction between a magnetic field and an electric current to produce, in this case, a movement of rotation. The stator consists of a permanent magnet. The rotor includes one coil driven by a DC current. The current direction switches every half revolution of the rotor so that torque is always in the same direction, which leads to a continuous rotation of the motor.

This work it one wants intended to measure the rotational speed of the DC motor using the Hall Effect sensor to detect the change in the magnetic field of a permanent magnet attached to the motor shaft. A microcontroller is used, which receives information from the Hall Effect sensor and generates a PWM (pulse width modulation) signal to control the speed and direction of rotation of the motor.

12.6.2 *Driving the DC Motor*

Use an Arduino microcontroller with an Ardumoto Motor Driver Shield (H-Bridge) to make the DC motor rotate in both directions. Present the electrical circuit assembled.

12.6.3 *Measuring rotation speed with the Hall Effect sensor*

Attach the permanent magnets available to the shaft of the DC motor. Place the Hall Effect sensor in proximity to detect the presence or absence of the magnetic field created by these magnets in various stages of the rotation. Connect the sensor to the microcontroller and develop the software to determine the rotation speed, and display it to the user. Draw the structure built. Discuss the operation of the measurement system.

12.6.4 *Angular Velocity versus Voltage Transfer Function*

For five different values of the voltage applied to the motor (determined by the duty cycle of the PWM signal), measure the rotational speed. Plot the graph of the rotational speed versus voltage. Comment on the result.

12.6.5 *Material*

The material required for this project is:

- 1 DC motor.
- 1 Hall Effect sensor HAMLIN 55110.
- 1 Ardumoto Motor Driver Shield (H-Bridge).
- A set of permanent magnets.
- 1 Arduino Uno microcontroller and personal computer.

12.7 Accelerometer and Servo Motor

12.7.1 *Introduction*

An accelerometer is a sensor that allows one to measure the acceleration of a body. It is used, for example, to trigger the airbag in a car when a collision occurs or to adjust the orientation of the screen image in a smartphone.

There are several physical phenomena that can be used to measure acceleration. One of the most common in low-cost accelerometers is the use of a test mass coupled to a spring which causes the plate of a capacitor to move when accelerated. The change in distance between capacitor plates leads to a change in its capacitance which is then used to measure acceleration. The manufacture of this type of sensor using MEMS technology allows one to have very small and low-cost sensors that can measure the acceleration in 3 perpendicular axes simultaneously.

A servo motor is a motor (usually a DC one) and a position sensor connected in a close loop. It allows one to rotate the shaft by the desired angle. The control is usually carried out using a PWM signal whose duty cycle controls the rotation angle.

In this project, we want to make the accelerometer rotate in the XY plane using the servo motor. A microcontroller is going to be used to obtain the acceleration that the accelerometer is subject to during one-half turn of the servo motor shaft.

12.7.2 *Connecting the accelerometer*

Connect the supplied accelerometer to the Arduino microcontroller. Draw a schematic with the connections made.

12.7.3 *Connecting the servo motor*

Connect the supplied servo motor to the Arduino microcontroller. Draw a schematic with the connections made.

12.7.4 *Software Application*

Create, using the Arduino IDE, an application that acquires the electrical signals from the accelerometer and shows them in the Arduino IDE Serial Plotter (only in Arduino IDE 1.6.6 and above). This application should also make the servo motor complete one-half turn.

12.7.5 *Maximum acceleration measurement*

Plot the three components of acceleration and determine the maximum acceleration it registers during the one-half turn of the servo motor shaft. Comment the values obtained.

12.7.6 *Material*

The material required for this project is:

- 1 accelerometer from Analog Devices, model ADXL330, mounted on a PCB.
- 1 PWM-controlled servo motor from SunFounder.
- 1 Arduino Uno microcontroller and personal computer.

References

[1] R. H. Bishop (editor), *Mechatronics System, Sensors and Actuators*, CRC Press, 2008.

[2] S. Kasap, "Thermoelectric Effects in Metals: Thermocouples," *Web-Materials*, November 6th, 2001.

[3] R.F. Queirós, *High Resolution Ultrasonic Measurements*, PhD Thesis, Instituto Superior Técnico, 2008.

[4] W.G. Jung (editor), *Op Amp Applications Handbook*, Newnes, 2004.

[5] R. Verdone, D. Dardari, G. Mazzini, A. Cont, *Wireless Sensor and Actuator Networks – Technologies, Analysis, and Design*, Academic Press, 2007.

[6] S.Y. Yurish, M.T.S. R. Gomes (editors), "Smart Sensors and MEMS," Kluwer Academic Publishers, 2004.

[7] J. Fraden, *Handbook of Modern Sensors – Physics, Designs and Applications*, 3rd edition, Springer, 2004.

[8] J. Wilson (editor), "Sensor Technology Handbook," Newnes/Elsevier, 2005.

[9] N.V. Kirianaki, S.Y. Yurish, N.O. Shpak, V.P. Deynega, "Data Acquisition and Signal Processing for Smart Sensors," Wiley, 2002.

[10] S. Fox (editor), "Measurement, Instrumentation and Sensors Handbook," CRC Press, 1999.

[11] A. Forbes, "Bosch Fuel Injection Systems," HPBooks, 2001.

[12] R.W. Johnstone, M. Parameswaran, *An introduction to surface-micromachining*, Springer, 2004.

[13] H. Ziegler, "A low-cost digital sensor system". *Sensors Actuators*, 5, 169–178, 1984.

[14] . Asch, *Les capteurs en instrumentation industrielle*, 6th edition, Dunod, 2006.

[15] J. Matlack, "Modern LVDTs in New Applications, in the Air, Ground, and Sea," Fierce Electronics, September 1st, 2010.

[16] N. Ashby, "Relativity and the Global Positioning System," *Physics Today*, May 2002, pp. 41-47.

[17] R. Pogge, "Real-World Relativity: The GPS Navigation System," March 2017.

Index

CPSIA information can be obtained
at www.ICGtesting.com
Printed in the USA
LVHW051232220723
753115LV00006B/116